Water resources and agricultural development in the tropics

Longman Development Studies
Edited by Professor D. J. Dwyer, University of Keele

Chris Barrow: Water resources and agricultural development in the tropics

Piers Blaikie: The political economy of soil erosion in developing countries

David Drakakis-Smith and Chris Rogerson: Urban employment in the Third World

Bjorn Hettne: Development theory and the three worlds

David Phillips: Health and health care in the Third World

Water resources and agricultural development in the tropics

Chris Barrow

Longman Scientific & Technical

Copublished in the United States with
John Wiley & Sons, Inc., New York

Longman Scientific & Technical,
Longman Group UK Limited,
Longman House, Burnt Mill, Harlow,
Essex CM20 2JE, England
and Associated Companies throughout the world.

Copublished in the United States with
John Wiley & Sons, Inc., 605 Third Avenue, New York, NY 10158

First published 1987
Reprinted 1992

British Library Cataloguing in Publication Data
Barrow, Christopher J.
Water resources and agricultural development in the tropics. — (Longman development studies)
1. Water supply, Agricultural — Tropics
I. Title
631.7'0913 S994.5.W3

ISBN 0-582-30137-8

Library of Congress Cataloging-in-Publication Data
Barrow, Christopher J.
Water resources and agricultural development in the tropics.

(Longman development studies)
Bibliography: p.
Includes index.
1. Water in agriculture — Tropics. 2. Irrigation — Tropics. 3. Agriculture — Tropics. 4. Water resources development — Tropics. I. Title. II. Series.
S494.5.W3B36 1987 631.5'87'0913 86–28724
ISBN 0-470-20795-7 (USA only).

Set in Linotron 202 10/11 Plantin
Printed in Malaysia

To Anna

Contents

Foreword

There are many excellent texts on water supply and irrigation engineering, irrigation economics, agricultural development and the problems which often plague such efforts. Few syntheses of such writings have been made, despite a clear need for them from people interested in water resources and agricultural development: students of geography, economics, development studies and agricultural management, administrators, planners and aid agency staff.

This book attempts to provide a broad interdisciplinary introduction for such people. The first five chapters consider background and principles, the character of tropical agriculture and water resource systems and broadly how they might be managed. Chapters 6 to 10 review the technology and practice of water resources and agricultural development in the tropics, concentrating on where water may be obtained, where savings might be made or moisture better used. Irrigation methods are briefly examined, especially those which may be adopted by farmers in developing countries, and the problems associated with water supply, conveyance, application and disposal are considered.

In writing this book I have tried to make it clear that a considerable, if not overwhelming, proportion of the difficulties associated with developing agriculture in the tropics are neither environmental nor technological: they are obstructive bureaucratic systems and social attitudes, economic problems and human institutions. Ensuring that crops get sufficient water is important, but alone it cannot assure successful agriculture.

Chris Barrow

Acknowledgements

The idea of producing this book grew, in part, out of discussions on the environmental consequences of water resources development in the tropics during the Commonwealth Geographical Bureau Workshop on Natural Resources in the Third World, held at the University College of Swansea in 1981. I would like to express my thanks of the then Director of the Commonwealth Geographical Bureau, Professor R.W. Steel, and to all the contributors to the 1981 Workshop.

I am greatly indebted to Professor D.J. Dwyer for his encouragement and support throughout the production of this book, especially in editing the draft. To my wife Anne, my thanks for reading the manuscript and for her patience during the many evenings and weekends when I was working on it.

I am also grateful to the staff of the Library, University College of Swansea, particularly the Inter-Library Loans Department.

Chris Barrow, Centre for Development Studies,
University College of Swansea, University of Wales, April 1986.

We are grateful to the following for permission to reproduce copyright material:

Academic Press Inc and the author, B A Krantz for fig 6.5 (c)(d), (Manassah & Briskey 1981); Academic Press Inc (London) Ltd for fig 1.3 (Harris 1980) Copyright Academic Press Inc (London) Ltd; George Allen & Unwin (Publishers) Ltd for the map 'The Sahelian-Sudanic Zone' (Van Apeldoorn 1981); the editor for fig 1.5 (Falkenmark 1977); the American Society of Civil Engineers for fig 2.7 (Gunnerson & Kalbermatten 1978); Angus & Robertson (UK) Ltd for figs 3.2, 7.6 & tables 7.2, 7.8 (Wiesner 1970); the editor for fig 2.1 (Keller 1984); the editor for fig 6.10 (Pacey *et al* 1977); Edward Arnold (Publishers) Ltd for fig 1.1 (Price 1983) & fig. 7.3 (Oliver 1972); Cambridge University Press for fig 2.3 (modified), (Porter 1978); Center for International Studies, Massachusetts Institute of Technology for fig 2.6 (Millikan & Hapgood 1967)

Copyright Center for International Studies at MIT; Marcel Dekker Inc for figs 4.1, 4.3 (Yaron 1981); the editor for table 2.1 (Hamming 1978); Food and Agriculture Organization of the United Nations for figs 6.2(b)(c), 6.3, 6.4 (Sheng 1977) & table 1.4 (Finkel 1977); the author, Gary W Frasier for table 6.3 (Frasier 1975); W H Freeman and Company for fig 6.8(b), (Cox & Atkins 1979) Copyright (c) 1979 by Scientific American Inc. All rights reserved; German Foundation for International Development for table 8.2 (German Foundation for International Development 1977); Interim Committee for Coordination of Investigations of the Lower Mekong Basin for fig 2.4 (Interim Mekong Committee 1982); Intermediate Technology Publications for figs 6.8(c), 7.8, & tables 3.1, 7.3 (Stern 1979); International Union for the Conservation of Nature and Natural Resources for figs 1.6 & 2.8 (IUCN, UNEP, WWF 1980); The Johns Hopkins University Press for fig 4.5 (Howe & Easter 1971) published for Resources for the Future, Inc by The Johns Hopkins University Press; Longman Group UK Ltd for table 7.1 (Heathcote 1983); McGraw-Hill Book Company (UK) Ltd for fig 3.3 (Tolman 1937) & fig 3.5(a)(b), (Ward 1967); Methuen & Co Ltd for an extract from p 79 (Redclift 1984); Middle East and North African Studies Press Oxford for fig 6.7(a), (Gischler 1979); Ministry of Planning and National Development for fig 6.7(b)(c), (Lewis 1984); National Research Council for figs 6.6, 6.8(a), (National Academy of Sciences 1974); the editor for fig 9.5 (Charnock 1983); the editor for fig 1.2 & table 1.2 (Harrison 1983); Pergamon Press Ltd for table 1.5 (Heathcote 1983); Schweizerbart'sche Verlagsbuchhandlung for fig 6.1 (Hall *et al* 1979) and figs 9.1, 9.2, 9.3 (Petr 1978); Singapore University Press for fig 2.9 (Barrow 1983); Tycooly International Publishing Ltd for fig 9.6 (Zaman 1983); The Unesco Press and FAO for fig 7.2 & table 7.7 (Kovda *et al* 1973) (c) Unesco/FAO 1973; Westview Press Inc for fig 9.4 (Falkenmark & Lindh 1976); John Wiley & Sons Inc for fig 7.1 (Israelsen & Hansen 1962); John Wiley & Sons Ltd for fig 3.4 (Israelsen & Hansen 1962) & table 1.3 (Dasmann et al 1973); The World Bank for fig 4.4 (Bromley 1982), fig. 4.6 & table 4.1 (Bottrall 1981(a)).

We have been unable to trace the copyright holders in fig 2.2 (Park 1980) and fig 7.5 (Adams *et al* 1978) and would appreciate any information that would enable us to do so.

Part I

WATER RESOURCES AND AGRICULTURAL DEVELOPMENT: BACKGROUND AND PRINCIPLES

Chapter 1

Factors affecting tropical agricultural development

Introduction

Roughly three-quarters of the world's population live in some 160 countries collectively referred to as the South, the Third World, less-developed countries, underdeveloped or developing countries. About two-thirds of the total population of these countries depend directly or indirectly upon agriculture for livelihood, often merely to subsist, sometimes barely to survive (Fig. 1.1) (OECD 1982: 11; World Bank 1982: 39).

Agricultural development, although still concerned with the opening-up of virgin land in a few countries, more often involves the improvement and, increasingly, the rehabilitation of existing traditional agriculture and sometimes of failed development efforts. Traditional agricultural strategies are deteriorating in many regions as a result of a range of 'development pressures', some generated within developing countries, some from outside. These include population growth, the adoption of inappropriate agricultural strategies and/or unsuitable techniques, new tastes or attitudes, or the failure to modify outmoded, harmful local ways (Mabogunje 1980; Todaro 1981: 50–80; Dickenson *et al.* 1983).

The degeneration of agricultural systems, whether traditional or recently introduced, has frequently caused serious environmental and socio-economic impacts. From the early 1970s ecologists and environmentalists studying development-related problems in the tropics have called for 'ecodevelopment', that is, environmentally informed development (Dasmann *et al.* 1973; Eckholm 1976; Ehrlich *et al.* 1977; Glaeser 1984). Realization that difficulties exist comes much swifter than the information, expertise, trained personnel, funding and changes in people's attitudes that are needed to avoid or mitigate such problems (Independent Commission on International Development Issues 1980: 20, 47; IUCN, UNEP, and WWF 1980; Brandt Commission 1983). The environmental degradation in developing countries so clearly linked with agricultural production has become so obvious that many writers refer to the 'South's environmental crisis' (Riddell 1981; Redclift

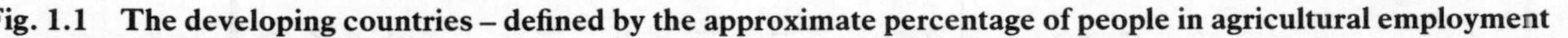

Fig. 1.1 The developing countries – defined by the approximate percentage of people in agricultural employment

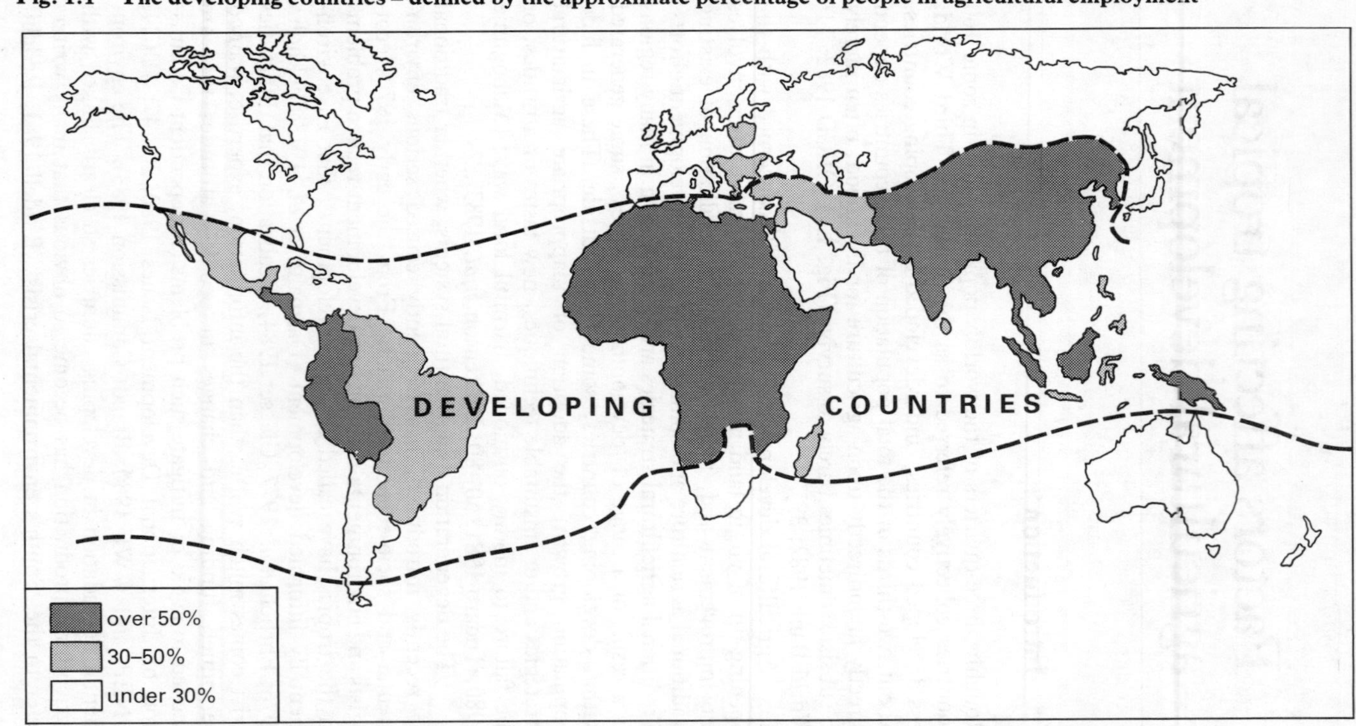

Source: Price, G. (1983) *Patterns of Development: Population and Food Resources in the Developing World*, Edward Arnold, London, p. 9 (Fig. 9).

1984). Combatting environmental deterioration and improving agricultural production are very important closely related issues and often a key factor in achieving either is sound water resources management.

Although there may be broad similarities between developing countries, there are nevertheless tremendous differences in environmental conditions and agricultural practices from locality to locality (Golley & Medina 1975). Traditional agricultural systems in the tropics are outlined in Table 1.1.

Socio-economic and socio-political factors affecting tropical agriculture

Traditional farmers and herders

Although well over half the population of developing countries depends upon small-scale, mainly subsistence agriculture, until relatively recently investors, researchers and planners paid pitifully little attention to this sector. In contrast massive investment, research and development support has been given to large-scale, mainly commercial production by state-run agricultural organizations or multinational corporations of crops for export or to feed urban populations. Also, a disproportionate amount of research has been done to improve commercial crops like: rubber or coffee. Much less has been done to improve subsistence crops and the farming methods of small-scale producers (Richards 1985: 19).

The dramatic gains in agricultural production which *should* be possible when large-scale irrigation schemes are established has led the governments of many developing countries and funding agencies to favour this form of development. Unfortunately the employment opportunities and profits (if any) generated by such developments frequently do little to help those who practise small-scale rainfed cultivation, small-scale irrigation or pastoralism, often in relatively remote locations (the majority of farmers in developing countries) (Redclift 1984: 31). Some countries which invested in large-scale, often irrigated, production of export crops like cotton have seen their returns cut by falling world market prices beyond their control and have found it difficult to pay for agricultural infrastructure, the inputs it requires and other necessary imports.

Although highly variable, available figures show that small farmers work about 40 per cent of the world's total cultivated land, some practising shifting cultivation, and some sedentary farming (Arnon 1981: 1). Few small farmers own their land; most are tenants who sometimes pay exorbitant rents, or are sharecroppers

Table 1.1 Traditional agricultural systems

Broad Subdivision	Environment/extent	Method/technology	Type of crop surplus
Shifting cultivation (land rotation or bush fallow)			
Slash-and-burn – random shift – ordered shift	Humid tropics/ subhumid tropics/ open woodland savanna. Poor soils forest lands/lowland or highlands. Over 200 million people depend on shifting cultivation. Figures suggest shifting cultivation is increasing – Lanly (1985:20) estimated at a rate of *ca*. 125% since 1980. 'Fall-back' strategy for refugees and disrupted farmers. Practised by some sedentary farmers on poor lands near their holdings to grow crops for profit	Small plots cleared by axe and fire. Ash provides nutrients for crops. Hoe or digging stick used to cultivate, no mechanization. Very diverse crop mix. Plot abandoned when weeds, pests and/or soil exhaustion become a problem. Abandoned plots usually continue to provide tree crops while forest/scrub regenerates. Restoration of soil fertility depends on plot being left fallow *long enough* and on adequate regeneration of natural vegetation during the fallow. New plot may be selected in random manner, according to cultural criteria, in an orderly fashion or where certain plants or other indicators show the soil is suitable	Small surplus or none, no particular harvest season due to diversity of crops. May be some rice or maize for sale or barter. Sometimes high-value crops produced, e.g. opium
Many terms used: ladang, swidden, milpa			
Shifting cultivation without fire	Humid/subhumid tropics	Trees felled by axe, but instead of fire vegetation is left for termites to break down	
Dryland bush fallow (chitamene)	Open savanna woodland: Zambia, N. Zimbabwe	Vegetation collected from a greater area than that cropped, often branches lopped from trees which are left alive to recover. The collected vegetation is burnt, concentrating nutrients where the crops are planted. Crops sown in ash immediately after seasonal rains break	
Pastoralism (livestock herding)			
Humid tropical herding	Central and South America, Malaysia, Africa	Cattle, goats, buffalo, pigs. Recently much forest clearance for beef ranching, often multinational companies involved	Export of beef to USA/ Europe ('hamburger connection'). Dairy produce difficult to keep in tropics. Cow/buffalo milk cheese produced locally, e.g. Amazonia, S. India.
Subhumid and tropical dryland cattle herding	Widespread, especially in savannas. In Africa south of Sahara there are at least 50 million herders	Nomadic or semi-nomadic, usually a seasonal pattern of movement. Mobility depends on whether herders are mounted or on foot	Meat/hides, quality generally poor, marketing a problem because sales of cattle tend to be when drought threatens so market is glutted
Tropical dryland goat and camel herding	Goat very widespread, camel confined to Africa/Asia		Meat/hide, cheese
Humid/subhumid tropical pig rearing	Widespread, especially in the Pacific islands and Papua New Guinea	Pigs generally forage around settlements	Local consumption, important role in culture of some peoples, e.g. in Papua New Guinea

Advantages	Disadvantages	Possibilities for improvement
Output to input ratio (energy value of crop: labour input) is as good as, often better than, modern alternatives. Can function where soils are very poor. Crop diversity gives security against pests, diseases and drought. Farmer's family have a varied diet	Not intensive – the fallow to cropped land ratio is often 20:1. If population increases shifting cultivation degenerates leading to environmental damage. Seldom sustainable in humid tropics if population density exceeds 40 to 60 persons/km^2. In dry regions maximum carrying capacity is likely to be about 5 persons/km^2. People scattered so communications usually poor. Shifting cultivation strategies are closely tied to culture, change may therefore be difficult to bring about	Improved rainfed crop varieties could boost yields – ideally non-perishable and high value to weight crops. Plant soil-improving crops to speed up fallow. Encourage cultivators to plant useful tree species on land left fallow (*agri-silviculture*). Make the pattern of shifting more rational. Conserve moisture by terracing, contour bunds, strip cropping, microcatchments. Improve storage/crop drying
Independence, few outside inputs needed	Cattle disease a problem. Grass generally poor quality – carrying rates are low. Seasonal fodder shortage. Climate may debilitate cattle. Humid tropics cattle ranching seldom sustained for long Amazonia/Central America vampire bats a problem. Herders tend to keep large, poor-quality herds, for socio-cultural reasons: number of cattle = status. No commons rights in many regions so no control over grazing, breeding, disease transmission. Overgrazing a major problem	Better breeds of livestock. Improved veterinary care. Stock control (fencing). Pasture improvement by seeding and/or runoff control. Fodder production by irrigation. Fodder from shelter-belts/ hedges. Greater care over provision of wells for stock watering. Use of trucks to transport livestock to pasture/market or to take water to livestock. Aid with marketing livestock or animal products. Development of new livestock from wild game species. Control of trypanosomiasis in Africa would open up about 6.5 million km^2 not presently used. Integrate arable farmers and herders to prevent conflict and enable both groups to benefit: arable farmers provide fodder or fallow grazing, herders provide meat and dung
Can be possible where it is too dry for arable cropping or cattle raising, or where cattle disease is a problem	Blamed for much overgrazing in tropics. Poorest people often are goatherders, difficult to control such groups. Less veterinary work done on goats compared with cattle	
Uses waste food, or forage around settlements	Pig unacceptable to some cultures. Disease transmission from pig to man may be a problem	

Table 1.1 (continued)

Broad Subdivision	Environment/extent	Method/technology	Type of crop surplus
Pastoralism (continued)			
Tropical highland llama/alpaca/yak herding	Latin America, Himalayas (yak)	Seasonal transhumance to find best pastures/avoid bad weather	High-quality wool/ hides, meat
Sedentary cultivation (to varying degree crop rotation) rainfed			
Rainfed cultivation – hand tillage	Widespread, especially in the subhumid tropics. Practice of majority of developing country farmers. 80% of world's rice (81 million ha) is produced by rainfed cultivation	Land tilled and planted just after rains break (where soil is easily workable tillage may be possible before rain softens soil). Digging-stick or hand-hoe. Success depends on rain after sowing plus any moisture stored in soil before sowing. Rotation generally poor. Fallowing may be practised to accumulate moisture in drier regions	Mainly subsistence, may be occasional or seasonal grain or root crop surplus. Dryland rice, wheat, millet, maize, sorghum, cassava
Rainfed cultivation – draught animal tillage	Widespread	Oxen, bullock, buffalo, mule, donkey, camel used to prepare soil for sowing, usually with simple plough. Farm has to be large enough to support the animals	
Plantain cultivation	Tropical highlands East and Central Africa. Less developed but widespread throughout tropical lowlands	Hoe or digging-stick, compost or nightsoil may be applied	Local subsistence, quite sustained
Irrigated			
Wet rice cultivation – natural wetlands	Widespread, dambo (wet depressions of Africa), swamp areas	Wet depressions sown with rice	Often considerable surplus, may be two crops or more a year where wet rice has been improved. Can be considerable. Risk of crop loss typically one in every six harvests
Wet rice cultivation – padi rice (sawah)	Very important in S. and South East Asia. Spreading to Africa and Latin America. 59 million ha worldwide	Usually small-scale/medium-scale farmers. Skilful control of water using bunds, in some cases water-lifting devices. Skilled tillage and timing of cultivation activities. Increasing mechanization. In some regions wet rice is grown by large-scale operators and may be highly mechanized. Traditional small-scale wet rice farming needs little input of fertilizer, but fertilizer boost yields so is increasingly adopted. Human labour is an important input – if increased it can boost crops with no other inputs altered	
Wet rice cultivation – 'flood' rice	Natural floodlands, e.g. Bangladesh, Burma, Thailand, parts of India	Uses rice varieties that can grow fast enough to survive floods or which can withstand brief submergence (some can grow to 6 m tall)	Subsistence, potential for surplus

Advantages	Disadvantages	Possibilities for improvement
World demand for good-quality wools. Little else possible at very high altitudes	Easy to overgraze or trample soil in harsh highland environments	
Freedom from depending on inputs which might fail or increase in cost. Will function in remote areas with poor communications	Vulnerable to rainfall fluctuations, and can have problems with insect and other pests. Soil erosion a risk. May be periodic shortages of food which weakens farmers and hinders cultivation	Tremendous potential for improved high-yielding varieties (HYVs) rainfed crops, especially sorghum, millet and cassava. Soil improvement by growing nitrogen-fixing crops, or seeding soil with nitrifying bacteria or breeding crops to fix nitrogen. Better crop rotations needed. Moisture conservation/partial (supplementary) irrigation. Zero-tillage?
	As above. Livestock may be debilitated if there is occasional or seasonal shortage of fodder	Improve draught animal harnesses to increase efficiency. Improve hand-tool and plough designs. Crop mixes which discourage pests. Improve communications
Independence, sustainability, needs little input, material or labour. Maintains a cover which reduces erosion	Not all environments are suitable, must be adequate rainfall Dambos can supper erosion	Improved plantain varieties. Attempt to get plantain cultivation more generally used
	Dambos can suffer erosion	
Probably most efficient crop production systems presently available. Soil quality not especially important. Stable and sustained. Provides employment. Fish, ducks, crabs, frogs, prawns may be produced in padi-fields and associated channels and tanks – source of protein for rural people. Little risk of erosion once established	Water must be available. Farmers need to have a high degree of skill. Rice uses a lot of water compared with other cereals, may be better to use water for some other grain. May have problems drying and storing rice in some environments. Due to relative complexity of wet rice cultivation it may sometimes be difficult to transfer to new areas	Better rice varieties, Better water management. 'Green manure' use of *Azolla* spp. floating on padi to fix nitrogen for rice
	Planting sometimes delayed	Develop HYVs. Encourage adoption of 'flood rice' where presently not grown

Table 1.1 (continued)

Broad subdivision	Environment/extent	Method/technology	Type of crop surplus
Irrigated (continued)			
Wet rice cultivation flood recession rice	In many areas where land near rivers is flooded or swamps/ lakes rise and fall. Large reservoir drawdown areas	Rice sown on land as flood-waters recede. Guyana, Amazonia, Mali, S. and South East Asia	Subsistence, potential for surplus
Traditional irrigated agriculture – water spreading	Localities in receipt of periodic spates	Bunds used to spread water to increase soil moisture, wet soil then planted	Pasture improvement/ subsistence crops
– tank irrigation	South Asia, 'monsoon lands'	Tanks store excess wet season runoff for dry season crops	Sometimes marketable surplus
– channelling floodwaters	Riverine lowlands	Channels convey floodwaters to cropland	Subsistence mainly
– water lifting	Widespread	Various man-, animal-, wind-powered lifting devices to raise irrigation water	Subsistence
– runoff collection and concentration	N. Africa and Israel but spreading	Collection and concentration of rainwater shed from a suitable catchment	Subsistence
– floating or raised fields		Planting bed floats or is raised above flood level	Subsistence

Sources:
Shifting cultivation: Nye & Greenland 1960; Conklin 1961; Allan 1965, Spencer 1966; Janzen 1973; Grandstaff 1978; 1981; Hunter & Kwakuntiri 1978; Ruthenberg 1980; Rambo 1982; Lanly 1985.
Pastoralism: Johnson 1969; Box 1971; Swift 1977a; McCown *et al.* 1979; Walker 1979; Shaffer 1980; Heathcote 1983; Manassah & Briskey 1981; von Maydell & Spatz 1981; Sandford 1983.
Sedentary Cultivation: Bayliss-Smith, 1982; Grist 1959; Geertz 1963; Grigg 1970; Gourou 1980; Spencer 1974, Fernando 1980a; Wrigley 1981; Byres *et al.* 1983; Swaminathan 1984.

or even serfs bound to feudal masters, and in some countries farmers work land which is communally owned. Such farmers have little incentive to invest effort or money in improving production for fear it may benefit others rather than themselves or their families (George 1977: 34; Redclift 1984: 33).

Most developing country farmers work plots of less than 5 ha and many cultivate less than 2 ha. Given irrigation and good soil,

Advantages	Disadvantages	Possibilities for improvement
	Risk of crop loss if floods come early	Adopt flood-resistant or flood-adaptive rice. Construct bunds
	Use of seasonal flows may entail risk of erosion of cropland or deposition of infertile silt. Flows may be inadequate or come at wrong time	Better water control. New waterproofing materials for tanks, canals, etc. Better pumps, pipes and water application technology. Improved crops
	Tank irrigation depends on the amount of surplus runoff, in some years tanks only partially fill and cropped area is restricted	
	Human disease may be a problem with irrigation	
	Can function in very low rainfall environments. Small farmers could probably adopt techniques without needing a lot of funding	Suitable collecting areas are required. Soil may be unsuitable and fail to shed water

1 ha in the tropics could support a typical smallholder family. However, circumstances are seldom that favourable and a 1 or even a 2 ha plot under rainfed cultivation is likely to provide a livelihood well below the World Bank absolute poverty line (meaning an existence somewhere between a barely adequate diet and starvation). While small farmers frequently have miserable life-styles, there are some who are even less fortunate – the landless agricultural

labourers who comprise as much as one-third of the rural population in some regions. Although not necessarily precluding agricultural improvement, small farm size usually means the farmer is poor and poverty is a major retardant on development (Schultz 1964: 105).

The small farmer's predicament is difficult not only because he/she is poor and lacks secure access to sufficient land, but because the people who control financial aid, pass legislation, administer taxation, decide agricultural research priorities and who control the national market prices for farm produce, indeed those who determine almost everything that can help or hinder small farmers, are likely to be based in cities and seldom have an adequate understanding of the problems faced by rural people; they may regard them as 'backward' and 'inefficient' (Lipton 1977; Harriss 1982: 94). It is not unusual for farm prices to be held artificially low, exacerbating rural poverty in order to ensure cheap food supplies so as to avoid unrest in cities and demands for increased wages in industry. Without equitable domestic terms of trade there will be little reinvestment in small-scale agriculture and through that long-term development (Bunting 1970). Unfortunately, when developing country decision-makers do show concern for rural peoples they often do so because of tribal or kinship ties or political motives, rather than true concern for the well-being of countryfolk (Falkenmark & Lindh 1976: 177; Todaro 1981: 482).

Expatriate or expatriate-trained personnel concerned with agricultural development have commonly associated ordered, neat planting with good husbandry, even where such practices are inappropriate. Small farmers with irregularly spaced crops, who are reluctant to waste effort on non-essential activity get dismissed as 'lazy'. Generations of experience have been accumulated by small farmers and herders, but it is passed on by word of mouth and often missed by those setting out to develop tropical agriculture. This experience is, according to Richards (1985: 40): '. . . the single largest knowledge resource not yet mobilized in the development enterprise.'

It is true that those practising traditional agriculture are sometimes slow to adopt new ways, but this may often be because they have little or nothing with which to pay for inputs. They have no security against failed experimentation, and advice is seldom available to them. Sometimes it may be to the advantage of the traditional farmer to keep a 'low profile' because the image of a 'subsistence backwater' is '. . . as good as any to fend off the tax gatherer and others interested in taking a share of what may be a thriving trade in basic foodstuffs' (Richards 1985: 111).

Given adequate support, new crops and techniques which are seen to work and which are affordable are usually adopted sooner or later. Indeed, in Africa, and very likely elsewhere, traditional farmers have a history of inventive self-reliance. Nevertheless, there are practices and attitudes which hinder development, and there may be no relatively fast 'technological fix' for these. For example some peoples scorn arable cultivation. Taboos and tastes may preclude the raising of otherwise suitable crops or livestock. The importance of women in traditional agriculture has been neglected by those charged with developing agriculture. The Independent Commission on International Development Issues (1980: 60) noted that: 'Africa's food producers – the women – continue largely to be ignored.' Agricultural funding or education directed towards menfolk can therefore be largely wasted.

The attraction of wage labour in mines or cities can deplete rural areas of their strongest, most able males, leaving a less dynamic, often elderly and conservative workforce which hinders any development of agriculture. The replacement of a subsistence economy with a cash economy is fraught with danger, for those with little experience of handling money can easily get into debt and lose their crops or land.

Population increase and the breakdown of traditional agricultural practices

In many developing countries population growth is such that increases in food production have little impact on per capita food consumption. In some regions population growth is doing far more damage by causing the degeneration of traditional agriculture. The problem is particularly marked in tropical drylands where recent estimates suggest population rose by roughly two-thirds between 1960 and 1974 (Harris 1980: vi).

Once-stable crop and livestock production systems break down when a population increases enough to cause a reduction of fallow periods or overstocking. Land rotation farming systems (*shifting cultivation*) and pastoralism are especially vulnerable and their vulnerability increases where soils are poor and/or there are periods with so little rainfall that vegetation is under stress. Once degeneration starts things tend to go from bad to worse; poor yields from the land means that it is put under further pressure and so suffers more. The incidence of poverty, malnutrition and poor health increases; people's morale deteriorates – and in a short time they can do little to rehabilitate the land and boost agricultural output unaided. This scenario may be found in large parts of Bangladesh, Thailand, Vietnam, North-East Brazil, Malagasy, South Yemen and Bolivia; but in Africa the situation seems especially desperate. Clearly there are tremendous difficulties in the

Sahel, the Sudan and Ethiopia, but much greater parts of the continent seem destined to join the suffering. According to Higgins *et al.* (1981), all parts of Africa with less than a 149–day growing season were already carrying over twice the population that can be sustained with the techniques and inputs traditionally used (Table 1.2; Fig. 1.2). More recent studies reinforce these warnings (Higgins *et al.* 1984).

Attempts to remedy the situation in the Sahel, the Sudan and Ethiopia have not been at all successful (Adefolalu 1983; Anon 1983a). Despite massive foreign aid since the late 1970s, cereal production in the Sahelian countries rose by only 1 per cent between 1973 and 1980, and while money and effort were being expended the Sahel's population rose by roughly 2.5 per cent (Grainger 1982: 34). In wetter environments as well as drylands, population increase can cause problems if traditional inheritance laws allow excessive subdivision and dispersal of farmland.

Traditional agriculture in marginal lands

Traditional cultivators and herders frequently have to make a living in difficult environments: were they as incompetent as some believe it would be amazing that they have survived unaided for so long. Their survival is proof that their strategies 'fit' their physical and socio-economic environments; indeed, given minimal inputs, little support and poor communications they perform remarkably well. In the same environments large-scale producers with modern equipment are often content with much lower yields. Those setting out to develop tropical agriculture should first make some effort to study existing land use.

Land near the limits of cultivation or pastoralism (*marginal land*) normally provides a low economic return to those using it – a return which is a function of parameters which include: moisture availability; soil characteristics; quality of seeds; fertilizer availability; pest-control problems; law and order; and market price of produce. If but one critical factor alters, raising of crops and livestock may become more worthwhile and production may well advance into new areas, or it may become less worthwhile and producers retreat, hopefully, to more favourable localities (Biswas 1979: 257). Improving moisture supplies for agriculture is often very important but may not be the key factor needed to boost agriculture, or if it does it may be negated by some other parameter changing.

Those who depend on their own labour to cultivate the soil are likely to be greatly hindered if malnourished or, as is often the case, both malnourished and debilitated by illness. In many parts of the tropics, especially marginal lands, there are seasonal or

Table 1.2 Population supporting ratios. The figures in this table indicate the number of times by which the potential population carrying capacity of the land exceeds the actual or expected population for the year stated

Input level	Africa	Asia	SW Asia	South America	Central America	Total
A. Year 1975 populations, all cultivable land under food crops						
Low	2.7	1.1	**0.8**	5.9	1.6	2.0
Intermediate	10.8	3.0	1.3	23.9	4.2	6.8
High	31.6	5.1	2.0	57.2	11.5	16.3
B. Year 1975 populations, actually cultivated land, one-third under non-food crops						
Low	**0.4**	**0.7**	**0.8**	**0.6**	**0.5**	**0.4**
Intermediate	1.5	1.9	1.2	2.4	1.3	1.5
High	4.5	3.1	1.9	5.8	3.7	3.6
C. Year 2000 populations, all cultivable land under food crops						
Low	1.5	1.1	**0.7**	3.5	1.4	1.5
Intermediate	5.4	2.3	**0.9**	13.3	2.6	4.1
High	15.6	3.3	1.2	31.5	6.0	9.1
D. Projected stable populations in 2150, all cultivable land, one-third under non-food crops						
Low	**0.4**	**0.4**	**0.3**	1.5	**0.6**	**0.5**
Intermediate	1.4	**0.9**	**0.3**	5.5	1.1	1.44
High	4.2	1.3	**0.5**	13.1	2.5	3.2

Note: The carrying capacity was established with reference to soil, climate, length of growing season, optimum crop, slope, and recommended calorie intake for people in each country. Databases are not accurate so this study arrives at general conclusions, not accurate predictions.
Bold type indicates critical figures, where a region cannot support its population. Any figure below 2 indicates a situation which merits concern.
Low input level corresponds to what might be expected in most rural areas of developing countries, i.e. subsistence farming inputs. *Intermediate input level* is a basic 'package' of fertilizers and pesticides, with some improved crop varieties, simple soil and water conservation measures and the most productive crop mixes possible on half the farmed area. *High input level* would equate to that presently used by farmers in the USA.

Source: Harrison 1983: 2.

Fig. 1.2a to 1.2e Developing country areas (by world regions) able to support more than one person per hectare with low inputs to agriculture (cross-hatched); and those areas unable to support their 1975 populations with low levels of inputs even if all their land were used for food crops (solid shading)

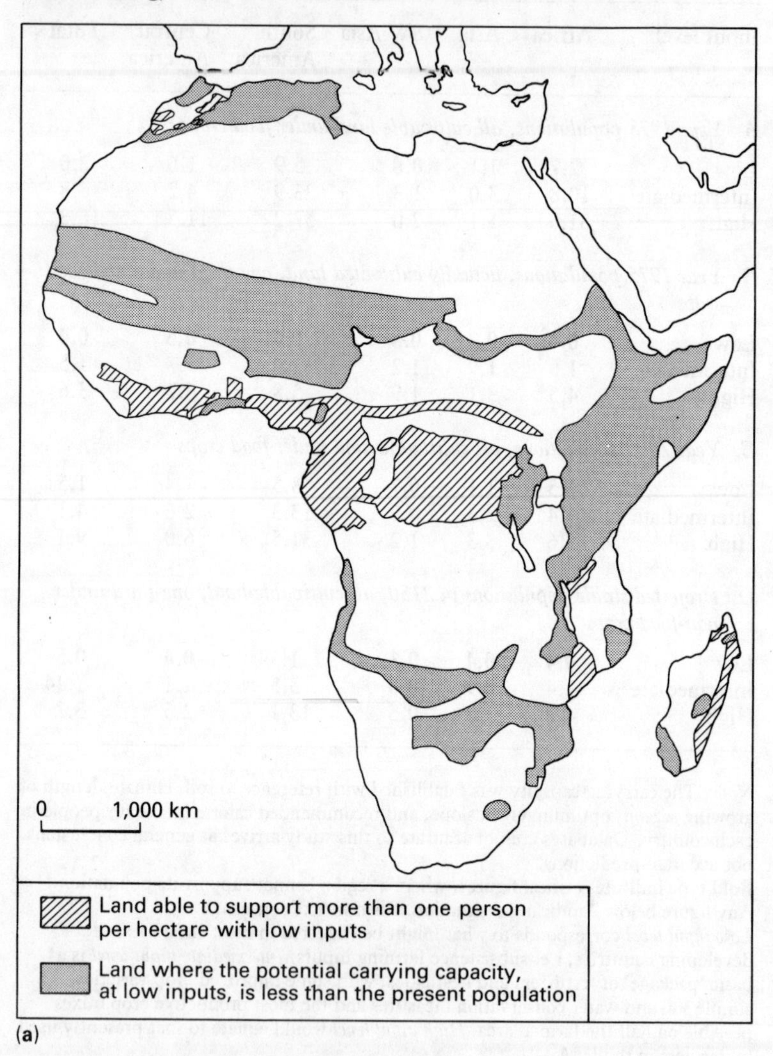

(a)

Note: Low level of inputs equates with the practices of most subsistence farmers in developing countries today.

(a) Africa (b) The 'Middle East' (c) Central America (d) South Asia and South East Asia (e) South America.

Source: Harrison 1983– Africa map from p. 3; Southern Asia + western South East Asia pp. 4/5; Central America p. 6; S. America p. 7; Middle East p. 8 (there were no Figure numbers in Harrison, 1983).

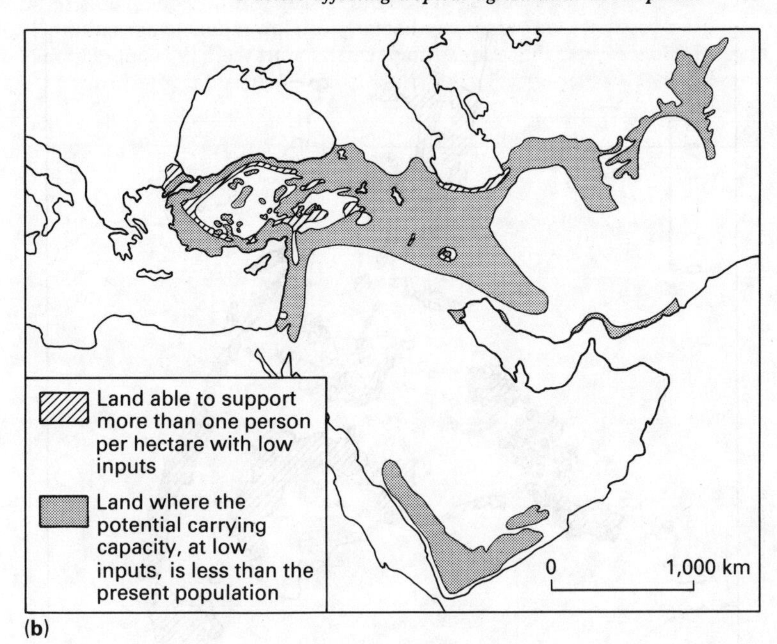
Land able to support more than one person per hectare with low inputs
Land where the potential carrying capacity, at low inputs, is less than the present population
0
1,000 km

(b)

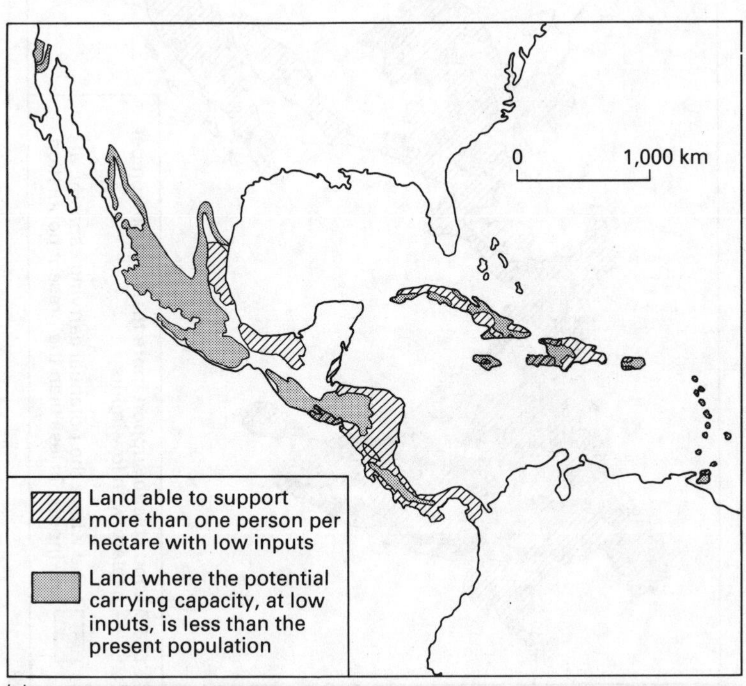
0
1,000 km
Land able to support more than one person per hectare with low inputs
Land where the potential carrying capacity, at low inputs, is less than the present population

(c)

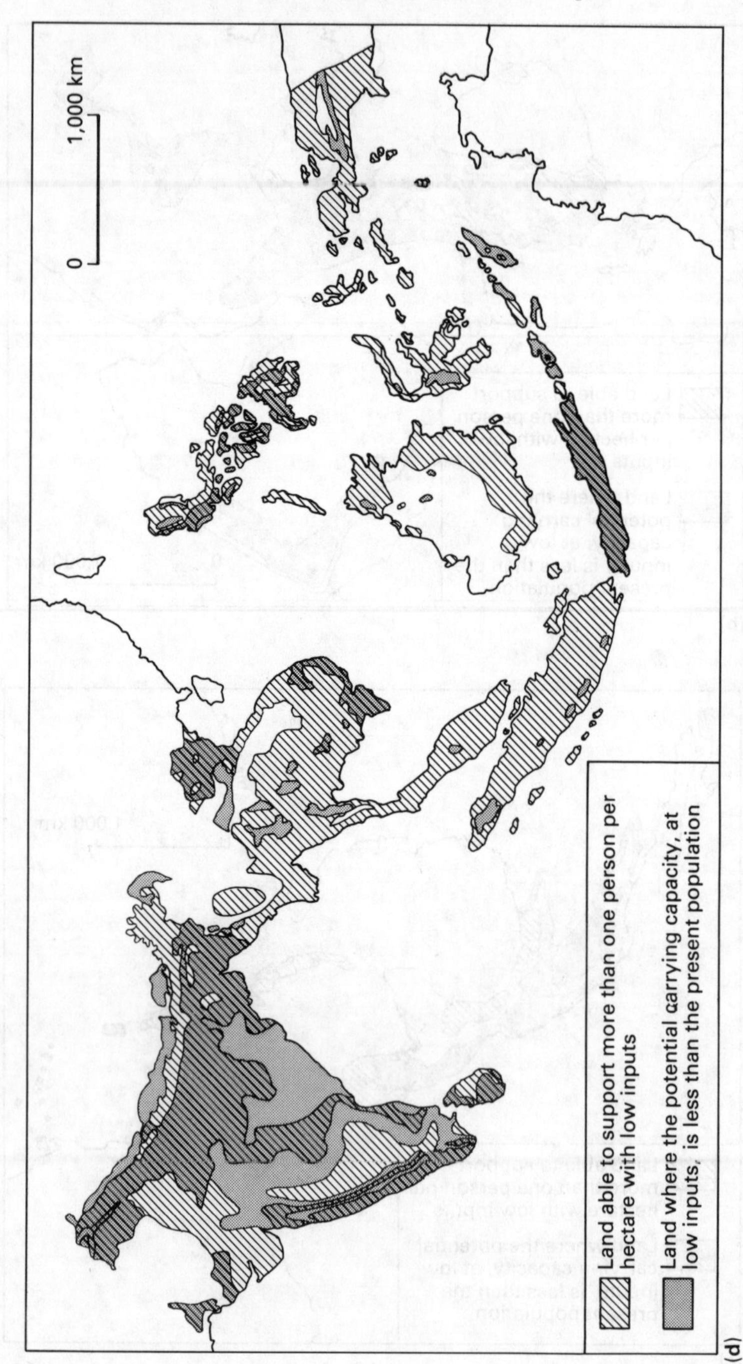
0
1,000 km
Land able to support more than one person per hectare with low inputs
Land where the potential carrying capacity, at low inputs, is less than the present population
(d)

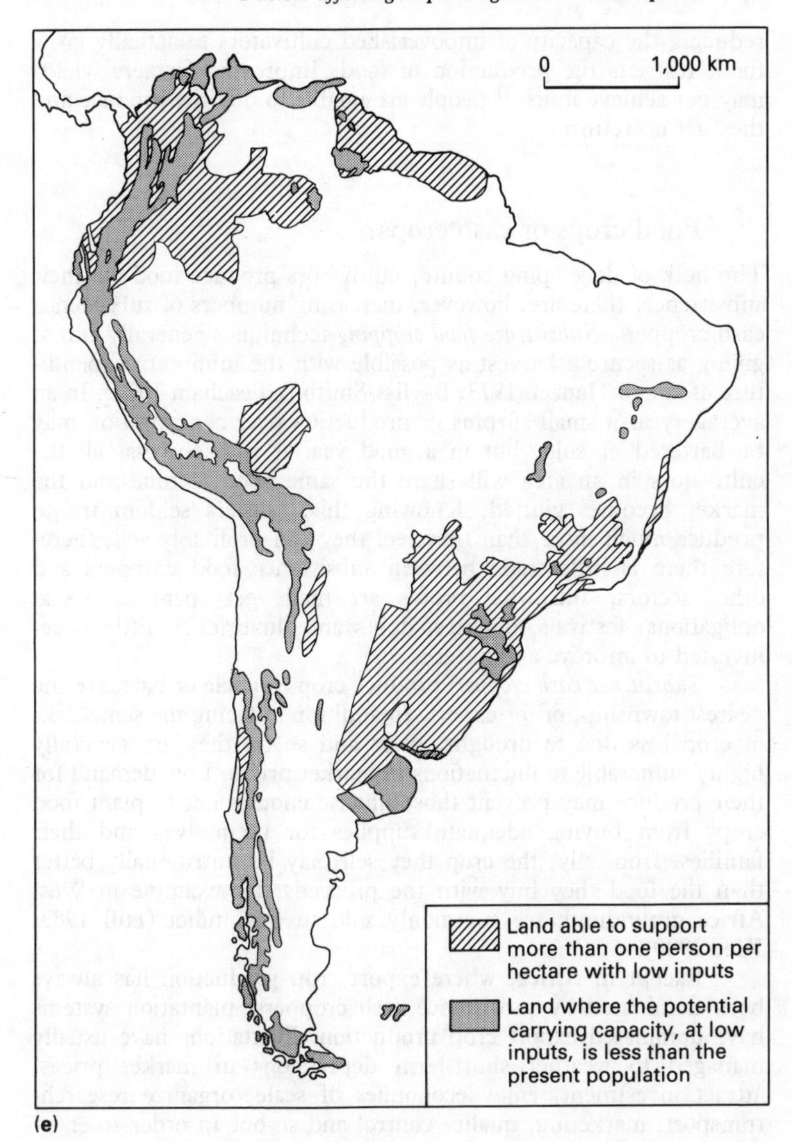

periodic occurrences of malnourishment and/or illness which can trigger of a vicious cycle of debilitation, leading to poor agriculture, to debilitation and so on. Except in times of environmental catastrophe or civil unrest, the foremost cause of hunger is poverty, and the same poverty, by limiting the effective demand for food and by reducing the capacity of impoverished cultivators to actually grow food, restricts the production of food. Improving farmers' yields may not achieve much if people are unable to buy the produce and they get no return.

Food crops or cash crops?

The bulk of developing country cultivators produce food for their subsistence: there are, however, increasing numbers of subsistence cash croppers. *Subsistence food cropping* techniques generally aim at giving as secure a harvest as possible with the minimum expenditure of labour (Janzen 1973; Bayliss-Smith & Feacham 1977). In an average year a small surplus of production over consumption may be bartered or sold, but in a good year it is likely that all the cultivators in an area will share the same good fortune and the market becomes glutted. Knowing this, farmers seldom try to produce much more than they feel they can profitably sell, therefore there is little trade between subsistence food croppers and other sectors, and what profits are made get spent on social obligations: festivals, dowries, debts and 'luxuries' – little is re-invested to improve agriculture.

Subsistence cash croppers produce crops for sale or barter to the nearest township, or for export. In addition to facing the same risks of crop loss due to drought, pests and so on they are generally highly vulnerable to fluctuations in market prices. Low demand for their produce may prevent those unwise enough not to plant food crops from buying adequate supplies for themselves and their families. Ironically, the crop they sell may be nutritionally better than the food they buy with the proceeds, for example in West Africa groundnuts are commonly sold to buy millet (Hill 1963; Klee 1980).

Except in Africa, where export crop production has always been dominated by subsistence cash croppers, plantation systems have dominated export crop production. Plantations have usually managed to weather short-term depressions in market prices, attract investment, enjoy economies of scale, organize research, transport, marketing, quality control and so on. In order to enjoy similar benefits small producers must either work for some sort of marketing agent or combine into a cooperative. In some developing countries this has been encouraged, for example in Malaysia small producers of rubber have been supported. A few developing countries now have what might be described as 'world-market-orientated peasant economies' as in Ecuador, where since about 1947 roughly a quarter of banana production for export has been in the hands of small producers. In both Senegal and the Gambia about 90 per cent of export earnings are from groundnuts produced by small-scale cultivators.

Unfortunately, most of the traditional products of small-scale cultivators and herders in the tropics give poor profits, are bulky and not especially saleable (with the exception of narcotics). High-value-to-weight, non-perishable crops for which there is a stable demand would be ideal for developing country farmers, especially those in remote areas. Such crops would provide valuable supplemental incomes.

Given the limited availability of financial aid, the restricted range of **proven** agricultural improvements available, and the slowness with which they are often adopted, it is unlikely that more than a fraction of those presently practising subsistence food cropping will successfully make the transition to regularly producing and selling moderate crop surpluses (Harwood 1979). It seems more realistic to hope for the transformation of many more small producers from a harsh, precarious subsistence to a better and more secure subsistence, which can be sustained indefinitely without significant environmental degradation, perhaps with occasional crop surpluses for sale, but not regular commerce.

Environmental factors affecting tropical agricultural development

Some 35 to 40 per cent of the Earth's land mass is *tropical* (for a discussion of what constitutes tropical conditions see Jackson 1977: 7 and Almeyra 1979). For the purposes of this book the tropics are considered to be those regions where seasonal water availability is the critical factor controlling biotic activity; the non-tropics are where seasonal low temperatures are critical for biotic activity. It should be noted that at least 25 per cent of the land between the Tropics of Capricorn and Cancer lies above 900 m altitude and is subject to temperatures low enough to affect biotic activity (Manshard 1974: 4–49, 77; Harrison 1979; Lugo & Brown 1981). The problems of delimiting the tropics become more pronounced as one moves away from the 'core' (equatorial) tropics and tries to map the outer subtropics or semitropics. Any boundaries based on moisture regime or vegetation cover are imprecise and prone to fluctuation, especially in drylands where rainfall can vary greatly from year to year. Nevertheless, average annual rainfall or length of dry or wet season(s) (Fig. 1.3) are often used to subdivide the tropics. To facilitate the discussion of water resources and agricultural development in the tropics, it is helpful to have some convenient subdivision. The most appropriate is to divide the

Fig. 1.3 The tropics subdivided according to the number of months during which evapotranspiration exceeds precipitation

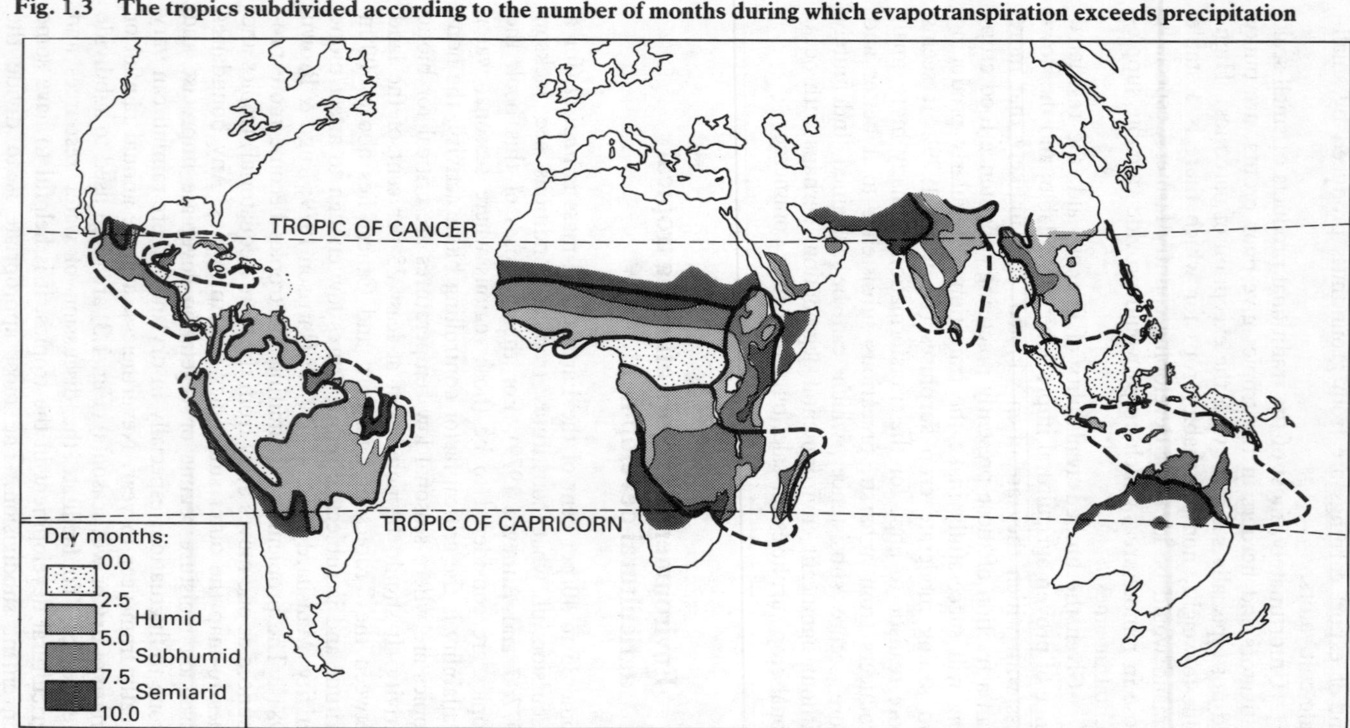

Source: based on Harris 1980 (Fig. 1).

tropics into: (1) *the humid (wet or equatorial) tropics*; (2) *the subhumid (semi-humid or wet/dry) tropics*; and (3) *the tropical drylands (savannas, semi-arid and arid regions)*. These three subdivisions reflect rainfall availability: usually abundant, sometimes problematical and usually difficult respectively. Each of these subdivisions presents the developer with a different set of challenges.

The humid tropics

The humid tropics occupy a belt roughly 10 deg. north and south of the equator. On the whole they are perennially well watered with over 2,000 mm/y (and in some regions 10,000 mm/y) rainfall, and, except above 900 m altitude, have year-round temperatures typically about 27 °C. Precipitation is generally assumed to nearly always equal potential evapotranspiration, i.e. there is no marked dry period during which vegetative growth is inhibited. There are, however, regions described as humid tropical where precipitation is less than evapotranspiration for up to five months of the year, and most of the humid tropics has at least one such month a year (Ruthenberg 1980: 1). In addition, factors such as topography can cause arid conditions within the humid tropics, for example in Dominica, Central Java and even parts of the Amazon Basin.

Rainfall above roughly 2,000 mm/y provided there is no marked dry period or steep slopes and given fertile, not too rapidly draining soils, can support rainforest vegetation below about 1,000 m altitude (Richards 1964; Kershaw 1973). Formerly such forests covered most humid tropical lowlands, but human activity has considerably reduced the extent, and is presently removing forest at an ever-increasing rate (Lanley & Clement 1979).

The subhumid tropics

The subhumid tropics are characterized by a wet season of four and a half to seven months, during which precipitation exceeds evapotranspiration, i.e. they are seasonally well watered (Balek 1977; Ruthenberg 1980). In Central and South America the subhumid tropics lie between roughly 5 deg. south and 25 deg. south, in Asia between about 3 deg. north and 10 deg. north.

Two generalized precipitation regimes may be recognized: a double rainfall maxima separated by a dry season during the low-sun period, but only a slight dip in rainfall during the high-sun period since the equatorial trough would not be far away; and a single rainy season/single dry season regime (Jackson 1977: 34). Parts of the subhumid tropics, notably India, Pakistan, Bangladesh,

Sri Lanka, South East Asia, southern China and Japan (roughly between 7 deg. north and 40 deg. north) receive much of their rainfall as a consequence of seasonal reversals of winds. These 'monsoon lands' are notoriously subject to seasonal and year-to-year fluctuations in time of onset, quantity and distribution of rainfall; not suprisingly there has been much irrigation development in these regions designed to combat rainfall uncertainty, rather than boost crop yields.

Wherever subhumid tropical soils dry out for more than a brief time, or where they are very poor or fast-draining, the climax vegetation changes from tropical rainforest to woodland or scrub. Depending on the criteria used to describe these plant associations they are variously termed subhumid, seasonally dry, deciduous tropical or (probably the least satisfactory term) 'monsoon forests'. Structurally these are less complex than humid tropical rainforests, the canopy is thinner and there are fewer tree species per unit area. Typical examples of subhumid forest are the dry Miombo and Mopane forests of East and Southern Africa – these are to be found where there is a rainfall of about 1,000 mm/y and a five- to six-month dry season. In South America the Zebil forest may be described as subhumid, and in South East Asia trees of the Dipterocarpaceae family dominate under similar conditions (Troll 1963).

A wide range of 'rain-green', i.e. bearing new leaves shortly after the beginning of the wet season, open forest, savanna woodland and scrub form the climax vegetation in regions too dry to support subhumid forest. In Africa, *Acacia* and *Euphorbia* species can form deciduous woodland, provided the soil is suitable, where there is as little as 1,000 mm/y (so long as the rainfall is reasonably distributed throughout the year). In North-East Brazil caatinga scrub could be described as 'rain-green'; however, this and many other 'natural' vegetation associations in the tropics may reflect factors other than precipitation.

The tropical drylands

Lands which may be described as tropical drylands lie mainly between 23 deg. north and 35 deg. south of the equator. Recent estimates suggest roughly 35 per cent of the Earth's total land surface is tropical dryland (Andreae 1981: 57). Whatever subdivision is made, for example into savannas, semi-arid and arid, all share the characteristic of being perennially ill-watered. Broadly speaking as one moves from the equator, subhumid forest grades, sometimes abruptly, into wooded or grassland savanna and ultimately this is replaced by vegetation adapted to semi-arid environments.

Savanna is a generic term which covers a wide variety of ecosystems in which '. . . grasses and woody plants compete . . .' (Balek 1977: 50). The grasses seem to get the upper hand as rainfall decreases. Certainly many savanna regions coincide with short wet/long dry season climates, but poor drainage, low soil fertility, excess salinity, wildlife grazing and bushfires may be important causative factors, at least locally (Golley & Medina 1975: 179; Furtado 1980; Gourou 1980: 79; Goudie 1981: 41). Whatever the causes, savannas are widespread, covering an estimated 18 million km^2 (Grigg 1970).

Savannas are seen by some planners as regions of great potential for agricultural development (Glantz 1977; Kowal & Kassam 1978). In theory, because solar radiation is higher and daylength longer than in the humid tropics, tropical drylands are likely to produce more crops than lands nearer the equator – *if adequate moisture is available*. The clearance of vegetation can present problems in the humid and subhumid tropics, in the tropical drylands it is less demanding. Although termite mounds can be a nuisance, mechanization is generally easier in tropical drylands and, because humidity is lower, crop drying and storage are also easier.

Savanna soils are commonly quite fertile, yet less than 1 per cent of the world's savannas are presently cropped (Garlick & Keay 1970: 60). Possibly around 31 million ha of savanna could be developed for arable cropping, but it would be sensible to expect a one in five risk of crop failure if such land were under rainfed cultivation using conventional strategies of production (Arnon 1981: 24). Irrigation or improved rainfed cultivation is therefore going to be very important for the realization of the agricultural potential of the savannas. Projects like the Sabi–Limpopo Scheme in south-eastern Zimbabwe have already indicated the value of savanna irrigation (Harris 1980: 346).

Poleward from the savannas are the semi-arid and arid tropics. The former have a short wet season when the sun is low, due largely to extra-tropical disturbances; the latter gets little or no rainfall mainly due to the constant presence of subtropical high pressure cells or topographical features (Jackson 1977: 34).

Mapping semi-arid and arid lands is not straightforward because of the confusion over their exact definitions. Some consider regions with rainfall between 400 and 500 mm/y and 100 to 250 mm/y as arid, but reliance on average annual rainfall gives only a rough picture. Length of dry season (Fig. 1.3), rainfall reliability (Fig. 1.4) and botanical criteria such as special adaptations of plants to water shortage and sparsity of vegetation, especially of trees, are more useful measures (Webster & Wilson 1966: 12). It is possible to map semi-arid and arid lands according

Fig. 1.4 Annual rainfall variability. Map indicates relative variability. On the whole, the darker the shading, the more variable the areas' rainfall. The large variability of the semi-arid areas is emphasized by the technique of analysis (see Jackson 1977: 55–6).

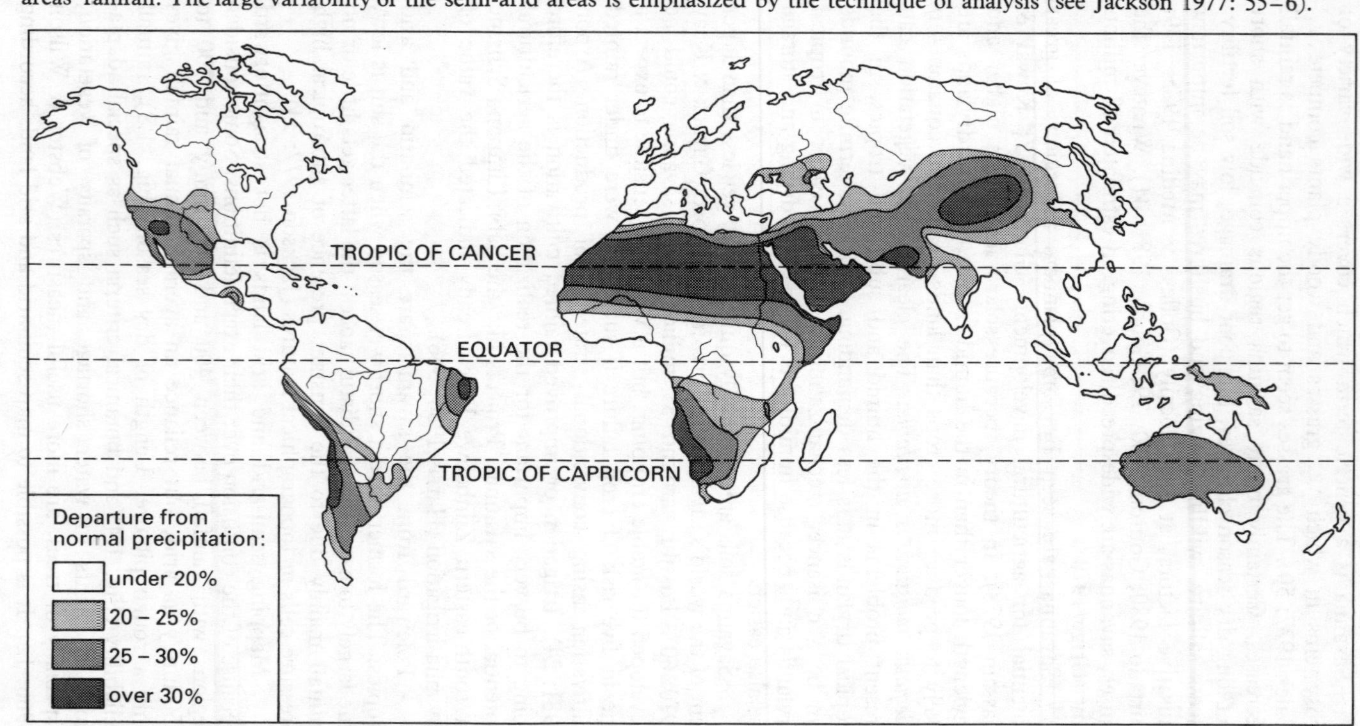

Source: redrawn from several sources.

to geological or geomorphological evidence, for example the occurrence of soils of low organic content (less than 0.5 to 0.8 per cent). The term *desert* is often applied to regions where vegetation is sparse or absent, the implication being that conditions are dry; however, sparse vegetation can reflect other causes as previously mentioned. A useful definition of semi-arid might be: *environments which allow the development of a more-or-less continuous vegetation cover, but which are too dry and variable to permit secure, regular rainfed cultivation of cereal or other crops* (Walker 1979: 7). An alternative, but more unwieldy definition is: *those environments where monthly average precipitation exceeds the potential evapotranspiration on average for three to four months (but no more than seven), regardless of total annual precipitation* (Lal & Russell 1981: 257; Finkel 1982: 1). Roughly 16 per cent of the Earth's land surface has conditions which would satisfy the foregoing definitions of semi-aridity (Nir 1974: xv; Matlock 1981: 4).

Wherever *total potential evapotranspiration exceeds actual precipitation* a region may be said to be arid (Kaduma 1982: 418). Under such conditions vegetation is likely to be very sparse or absent and there is no season in which crops can be raised using natural precipitation alone (Yaron *et al.* 1973: 2). Even if moisture can be provided to supplement natural precipitation, arid regions generally have high levels of solar radiation during the day and night-time temperatures which fall well below those of daytime, conditions which are not suitable for many crops.

The environmental constraints on tropical agriculture

If one marks off a band a few thousand kilometres in width encircling the equator one finds within it virtually no developed countries (Ooi Jin Bee 1983: 2). Some have suggested that this cannot be entirely an accident of history, but must have to do with some special handicaps faced in the tropics, directly or indirectly related to climate (Huntington 1915; Myrdal 1970; Kamark 1976). Such environmental determinism has been dismissed by critics, who point out that far from offering poor opportunities the tropics should have great potential for agricultural production because high year-round temperatures and abundant sunshine support plant growth better than higher-latitude environments – some go so far as to blame poor husbandry and/or inadequate agricultural strategies (Tosi and Voertman 1964; Gourou 1980: 34; Ellen 1982: 1–20). It is also a mistake to assume that poverty and tropicality always go together, some poor countries lie partly – or wholly – outside the tropics, and there can be less difference between conditions in some parts of the tropics and some parts of the extra-tropical world than there is between different parts of the tropics (Dickenson *et al.* 1983: 2).

If the tropical environment has such potential, the reader may ask why does temperate agriculture produce more dry matter with a higher edible proportion than tropical agriculture? Those who point out the favourable temperature regimes and abundance of sunlight frequently seem to forget that moisture is vital for plant growth and, as Fig. 1.5 illustrates, moisture availability **is** a problem over much of the tropics. Although environmental factors undoubtedly hinder agriculture in the tropics socio-economic and socio-cultural factors must not be overlooked. Redclift (1984: 79) adopted the most sensible stance, commenting that: 'Human poverty makes physical environments poorer, just as poor physical environments make for greater human poverty.' Those charged with development, if they are to avoid implementing schemes which fail to yield optimum returns and/or generate adverse environmental and/or socio-economic impacts, must understand the structure and function of tropical environments and the societies and economies they are working with. Until recently, knowledge about tropical environments was rather limited and is still some way from being satisfactory for planning purposes. Rugged terrain, dense vegetation, high temperatures and humidity, poor communications, disease hazards and a whole range of socio-economic and political problems makes research and the collection of data for planners to use difficult. Some tropical environments are so modified by fire, grazing or logging that existing conditions are not a good indication of the real potential of a locality, merely a reflection of past and present management and its shortcomings (Dasmann *et al.* 1973).

Information on environmental and social conditions may be wasted if development planners or administrators make no effort to seek it. Not infrequently warnings from experts, which could have averted environmental or human problems, are not heeded. This may be due to lack of finance, haste, ignorance of the necessity, political or personal expediency, or because planning and decision-making teams are dominated by personnel preoccupied with engineering or economic aspects of development (Sachs 1976).

Environmental hindrances to agricultural development in the humid tropics

In most environments there are seasonally dry or cold conditions which reduce or cure weed, insect pest and disease problems. In

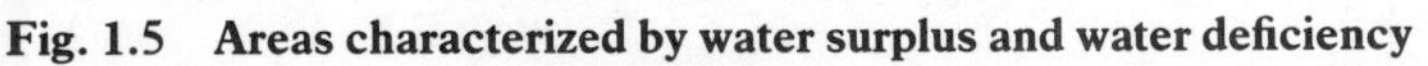

Fig. 1.5 Areas characterized by water surplus and water deficiency

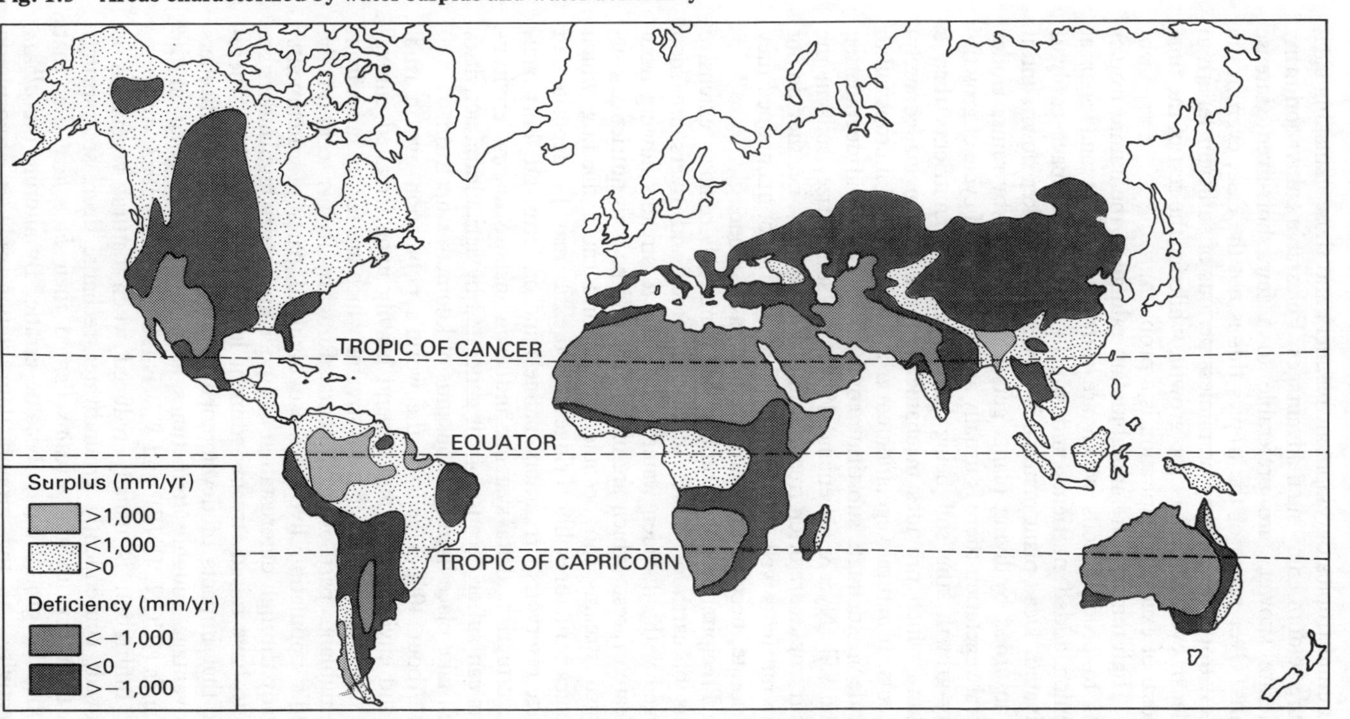

Source: based on Falkenmark, M. (1977) 'Water and mankind – a complex system of mutual interaction, *Ambio*, VI (1), 5–9 (p. 5 unnumbered Fig.).

the humid tropics development planners and those managing agriculture seldom enjoy such advantages. For arable cultivation many moderate showers are preferable to a few short-lived, intense storms. Over much of the tropics this is not the case; even in the wettest tropics there may be rainless periods of sufficient length to hinder or even endanger crop growth, while for the rest of the time disposal of excess precipitation is a problem.

High temperature and abundant rainfall in the humid tropics tends to produce soils which are nutrient-deficient and poor at retaining added nutrients and which tend to compact easily if cultivated. Dead plant material is very rapidly broken down, and if not absorbed by living plants gets leached away by rains. Under natural vegetation there is usually an almost closed cycle of growth–decay–growth, the soil being less a source of nutrients than a medium which provides mechanical support (often inadequately) for roots. If natural vegetation cover is removed, and if crops fail to provide an adequate substitute cover, there is a rapid degradation of the soil (Nye & Greenland 1960). To summarize, soils in the humid tropics are often free-draining and 'hungry', i.e. they do not retain nutrients well, and if vegetation cover is not maintained they deteriorate, become unproductive and easily erode.

Precipitation falling on bare soil commonly removes extensive sheets of earth or carves gullies. Streams and rivers are soon charged with silt-laden water in contrast to streams flowing from uncleared forests which are often clear and poor in nutrients. Silt-charged streams can become choked and cause flooding which damages structures like bridges, or the silt may be deposited in storage reservoirs, irrigation channels and on farmland. Port facilities, estuaries, delta systems and even inshore marine environments can suffer because of silt generated by agricultural activities which take place literally thousands of kilometres upstream.

About 40 per cent of the world's population live in rural lowland environments and depend heavily upon irrigated cultivation supplied with water from streams which are fed by flows from the highland part of catchments. Roughly 10 per cent of the world's population live in those highland parts of catchments. Already through deforestation and poor husbandry highlanders are said to have raised stream sediment loads. One could say, therefore, that impacts of development by tropical highland dwellers disadvantage roughly four times as many people living in the lowlands (IUCN, UNEP and WWF 1980; Redclift 1983: 49).

Infiltration is usually reduced when natural vegetation is removed, consequently groundwater recharge decreases, springs may then dry up and streams denied their *base flow* (i.e. that supplied by groundwater) become erratic. By altering the *albedo* (reflectivity) of the ground deforestation reduces atmospheric moisture over that ground making renewed vegetation growth difficult. Such alteration of the water balance can make pre-clearance forecasts of water demand and supply for irrigation useless and, if a large enough area of forest is removed, even if it is replaced by crops, rainfall may be reduced downwind because of lowered evapotranspiration from the once forested land. Fears have, for example, been expressed that productive farmlands in Brazil's centre-south could suffer from decreased rainfall if too much forest is cleared in Amazonia (Barrow 1985).

Table 1.3 lists the characteristic rainfall, soils, natural vegetation and commonly observed land-use in humid, subhumid and dryland environments and the watershed management necessary to conserve soil and maintain stream and groundwater resources.

Daylength changes little from month to month in the lower latitudes; many crops, including until recently rice, demand some fluctuation in hours of daylight to mature and were therefore unsuitable for the humid tropics (Webster & Wilson 1966: 26; Williams & Joseph 1973). Another brake on agricultural productivity throughout the tropics is that nitrogen fixation by soil organisms is lower than it is at higher latitudes.

In the drier tropics there are often 'hungry seasons' during which human food supplies, fodder, grazing and often water are in short supply. In the humid tropics there may be a similar season of shortage; grass is often difficult to manage and of relatively poor nutritional value and, if pasture is restricted to valley lands (as it often is) flooding may interrupt grazing. These tropical 'hungry seasons' can easily delay cultivation and planting, so reducing agricultural productivity. Better hand-tools, or, when it can be afforded, the introduction of tractor-pulled ploughs can help overcome this problem, although there is a risk that the employment of seasonal labour or women may be reduced (Chambers, *et al.* 1981: 22).

The humid tropical environment favours the transmission of crop, livestock and human diseases. Many development efforts have failed or have had difficulties because these risks were not adequately considered. Natural vegetation in the humid tropics commonly has a wide species diversity; individuals of a plant species are widely scattered and are frequently toxic or distasteful to pests. The *monocultures* – fields of a single crop variety – often grown when agricultural development takes place provide excellent opportunities for pests to multiply. Even in the drier tropics, weeds quickly colonize cleared land and hinder cultivation; in the wetter tropics the rate of weed growth can be a real problem. Soils in the humid tropics are sometimes infested with nematode worms which attack certain crops, and often carry plant disease viruses. Pollination can be a problem with many tropical crops. They may require

Table 1.3 Considerations necessary when developing tropical catchments

Environmental factor	General characteristics	Management consideration
Humid tropical catchments		
Rainfall	Exceeds potential evapotranspiration in most months. May characterize middle and upper watersheds as subject to orographic influences. Short dry season if any	Maintenance of natural vegetative cover of paramount importance, in order to absorb rainfall, release clean runoff and stabilize flow.
Soils	Mature soils may be deeply weathered clays with good structure and internal drainage, but low natural fertility. Podzolic soils in cooler zones	Leaching rapidly depletes exposed soils. Ill-suited for cultivation except on younger fertile soils. Erosion danger variable but slumping and landslides a danger, especially after deforestation
Natural vegetation	Evergreen forest, including 'rain forest' and 'cloud forest'	Best left in natural state, especially in steep catchment areas, unless economic perennial crop providing equivalent protection is possible
Commonly observed land-use	Little permanent farming except on exceptionally fertile soils. Shifting cultivation likely	Soil and forest exploitation technologies not well developed for this environment, except on best soils. Road construction and maintenance costly; fungal diseases serious problem in agriculture
Subhumid tropical catchments		
Rainfall	Equal to or slightly less than potential evapotranspiration on annual average, but higher than potential evapotranspiration during rainy season, including intense downpours	Runoff control and regulation measures to maximize both clean runoff to reservoir and availability of soil water for crops and pasturage

Table 1.3 (continued)

Environmental factor	General characteristics	Management consideration
Soils	Variable in structure, depth and fertility, but sometimes with good potential for agriculture. Erosion a potentially serious problem. Rapid runoff may also cause flooding	Great care required on cultivated land to prevent erosion and rapid runoff, especially on sloping terrain with clay soils
Natural vegetation	Semideciduous forests, savanna 'parklands' (fire climax)	Can include valuable tropical woods. Natural forest should be left on steep catchment areas
Commonly observed land-use	Short cycle and perennial crops, grazing, forestry. Conflicts can be expected between agricultural and forest uses of lands. High population densities and intensive land-use is likely	Productive agricultural and animal-raising uses possible under resource-conservative management. Vigorous soil and water conservation programmes needed in high-intensity-use, dense population zones. In steep areas natural forests should be conserved, but plantation crops may provide adequate cover
Tropical drylands catchments		
Rainfall	On annual average much less than potential evapotranspiration. Heavy, intense showers possible	Control phreatophytes and other vegetation so as to maximize clean runoff, but at same time provide protective vegetative cover. Also to stabilize flow

Table 1.3 (continued)

Environmental factor	General characteristics	Management consideration
Tropical drylands catchments (continued)		
Soils	Characteristically shallow with high pH, poorly developed structure. Low field capacity for moisture storage. Highly susceptible to erosion	Terracing and other earthworks, as well as retention of vegetative cover to stabilize soil, especially on steeper lands
Natural vegetation	Xerophytic, sparse, slow growth, low	Prevent excessive burning and cutting of woody species for charcoal and firewood
Commonly observed land use	Grazing and browsing by cattle, goats, sheep. Charcoal production. Ephemeral cultivation of short-cycle, drought-resistant crops. These environments are characteristically deteriorating	Maintain animal populations at carrying capacity or less. Reduce use intensity where leading to erosion and degradation of vegetation

Source (with modifications): Dasmann *et al.* 1973: 209–10 (Table 4).

an insect, bird or bat which feeds or roosts in very vulnerable sites like caves or mangrove swamps. For example, in Malaysia a prized fruit the durian (*Durio zebethinus* and related species) is pollinated by bats which roost in only a few caves which are increasingly subject to human disturbance. Successful agricultural development will require some forest land to be conserved in order to provide habitats for animals which pollinate or which prey on pests and to protect steep slopes from erosion.

Environmental hindrances to agricultural development in the subhumid tropics

Wherever vegetation is under moisture stress, which is the case in the subhumid tropics and semi-arid tropics, if there is any overgrazing, fire, trampling or unwise cultivation the re-establishment of plant cover may be very difficult. Rainfall is often of high intensity; for example, in northern Nigeria's savannas 90 per cent of the precipitation falls in storms of 25 mm/h or greater intensity. Soil unprotected by vegetation quickly sheds water, groundwater recharge is reduced and erosion occurs, as in the humid tropics silt reaches streams and rivers and causes considerable problems.

Bushfires and, in some regions, intensely drying winds can quickly remove plant cover and erode the land. In some drylands dry winds may debilitate livestock and man. However, the major problems faced by agriculture are to conserve and use wisely whatever soil moisture rainfall provides, and/or to find water for irrigation or stock watering. Salinization is much more of a problem than in the humid tropics, although it can occur there as well. As in the humid tropics insect pests can be a tremendous nuisance. Although crop damage by ants is less of a problem outside equatorial environments, pests like the desert locust, which can wreak havoc with crops, and the tsetse-fly, vector of livestock and human trypanosomiasis (sleeping sickness), make cultivation and herding difficult or impossible over great tracts of the subhumid lands.

A major advantage enjoyed by the subhumid tropics and dryland tropics over the humid tropics is that there is some part of the year when conditions become dry enough to make road transport easier and mechanized cultivation possible. Just as important the dry season(s) slow or halt biotic activity enough to give managers of development projects or farmers a chance to overcome pest and disease problems.

Farming and herding in the tropical drylands

People farm or graze livestock in drylands for various reasons: some because they were born into a tradition of living in such environments; some are refugees, dislodged from better lands by civil unrest or population pressure; some are attracted by possibilities for growing cash crops. Cash cropping in dryland may be due to other reasons than hope for profit. Sometimes subsistence cultivators start to grow saleable crops to pay taxes imposed upon them by government. Government or aid donors may promote cash crop production. If the market price of a widely grown dryland cash crop falls farmers intensify their cropping and environmental damage can easily result. This has been the case in Senegal, Nigeria, Niger and the Sudan, where considerable environmental damage has been caused in regions in receipt of about 500 mm/y rainfall (Grainger 1982: 16).

When farmers turn their attention to cultivating what are

often inedible cash crops, or if they sell surplus food crops they may hoard insufficient supplies for bad harvest years. Malnutrition can result and the next harvest is reduced because the debilitated cultivator and his family can no longer farm satisfactorily. Some governments even discourage the growing of hardy crops like cassava which could give rural people some security against famine if their grain crops fail, because they fear such 'famine crops' cause soil exhaustion.

Innovations may have both good and bad effects. In the drier parts of Africa in the 1950s and 1960s maize replaced traditional hardy cereal crops. Maize yields are seven or eight times those of traditional sorghum varieties when rainfall is between 300 and 600 mm/y, but maize is much less drought tolerant. It might be argued that the spread of maize cultivation has been a major cause of the present famine situation in Africa (Cross 1985). As well as being encouraged by governments and aid agencies, farmers also prefer to grow maize. Good maize yields with risks must be weighed against poor but stable yields, there may well be situations where the former is preferable.

The tendency has been to level the blame for agricultural problems and environmental degradation on population growth, climate or farmers' malpractices. However, some blame should be levelled at social, political and trade forces **outside** the drylands (Van Apeldoorn 1981: 75). In Ethiopia for, example, the government tried to resettle some 250,000 people on the traditional grazing lands of nomadic pastoralists; the result was conflict and environmental damage: the Tuareg pastoralist people of Mali have received similar treatment. In both Ethiopia and Mali the dryland people were blamed by the authorities (Pearce 1984a: 11). In North-East Brazil recurrent drought, famine and environmental damage are unlikely to be cured without socio-economic reforms affecting large landowners, businesses, that supply fertilizers, those marketing produce and water supply authorities, yet most time and money has been invested in trying to 'cure' the small farmers 'bad practices' (Hall 1978).

Drought

Over much of the tropics, especially the drylands, climate should be treated as a variable not a constant. At times a region may not have as much water as the people, crops and livestock need: this is a *drought* (Timberlake 1985: 28). The shortage of water may be due to poor rainfall or it may result from land use which increases evapotranspiration and/or reduces infiltration which recharges groundwater. Rainfall or groundwater supplies become inadequate

Table 1.4 Areas of land limited by drought

Region	Per cent of land limited by drought
North America	20
Central America	32
South America	17
Europe (mainly Spain)	8
Africa	44
South Asia	43
North and Central Asia	17
South East Asia	2
Australia	28

Source: Finkel 1977 (Table 2).

if human or livestock populations increase too much or if farming practices are intensified without alteration of techniques. It makes little sense, therefore, to define drought in terms of rainfall. A drought can be said to have arrived when crop or livestock yields diminish enough to affect people: in some regions people exist so near the limits of satisfactory subsistence that only a few per cent reduction is enough to do this (Wijkman & Timberlake 1984: 35; Glantz & Katz 1985).

Surprisingly it is often overlooked that drought and flood are interrelated: drought damages vegetation and soil so that precipitation penetrates the ground more slowly and streams are swollen with water; people disrupted by floods find it difficult to practise good husbandry or to make preparations to mitigate the effects of drought. Measures which make dryland agriculture more productive and secure (e.g. terracing or check-dams) are also effective in reducing floods.

Considerable efforts have been directed towards trying to establish whether climatic change or human land use or a combination of both are responsible for recent droughts in drylands. Some environmental scientists are convinced that carbon dioxide and/or particulate matter pollution of the Earth's atmosphere, largely due to the combustion of fossil fuels, markedly affects global temperatures (Kellogg 1978; Lamb 1982: 325; Harrison 1984). Carbon dioxide in the atmosphere, along with methane, water vapour, ozone and other gases, produce what is generally known as the 'greenhouse effect'. Put simply, the more carbon dioxide there is in the air the greater becomes the Earth's surface temperature. However, the temperature aloft is not raised anywhere near as much, so there is a marked thermal gradient. These changes might cause shifts in seasonal rainfall patterns, felt especially in lands fringing the Sahara and regions which depend upon 'monsoons'. Since 1900 there has been a roughly 10 per cent increase in global atmospheric carbon dioxide – whether this has significantly altered rainfall patterns has not been established.

There is a further complication, particulate matter in the atmosphere, natural and man-made dust and soot, especially ash erupted from volcanoes, reduces the intensity of sunlight reaching the Earth's surface. Because sunlight has to pass through a greater thickness of atmosphere at higher latitudes, assuming the particulate matter pollution is globally uniform (which is rather unlikely), temperatures should be reduced more away from the equator than at lower latitudes – this might cancel out any temperature changes caused by carbon dioxide. More research is needed to confirm if these speculations are in fact correct. It is as well, however, if planners bear in mind that irrigation may be needed in the future in some regions where it is not today, and that in some regions presently irrigated the need might diminish. There is also a possibility that increased levels of carbon dioxide, which lowers rates of evapotranspiration and increases production of plant tissues, might reduce plant water demands, improve drought resistance and increase crop yields – a doubling of present levels would probably raise crop yields about 5 per cent (Harrison 1984: 166). Unfortunately, global warming associated with such a rise in atmospheric carbon dioxide would probably melt polar ice and flood considerable areas of agriculturally productive lowlands.

In 1980 the World Meteorological Organization launched the World Climate Applications Programme with the goal of improving world food production by gathering and making better use of available information on past and present climate (Lamb 1982: 98). Such research is valuable because many of the records of drought are misleading, tending to reflect farmers' perceptions which are biased towards blaming climate. Climate is often blamed for drought by those who fail to appreciate land use – runoff relationships: 'The drying-up of streams and the wilting of crops are such convincing symptoms of drought that many an agricultural officer has spent long evenings plotting the rainfall records in search of evidence for a change in climate' (Pereira in Lal & Russell 1981: 6). The misconception that wet years are 'normal' and dry years are 'bad' has frequently been behind the failure of development projects in the tropical drylands, for both farmer and governments tend to make over-optimistic plans during wet phases which fail during dry phases (Biswas 1979a; Sirnanda 1979).

Desertification

Terminology is confused. *Desertification* has been used by some writers to describe the 'spread' of desert, paying little attention to cause. To talk of 'spreading' deserts is misleading, because it implies the movement of desert conditions outwards from deserts, whereas most desertification occurs *in situ*. A number of writers use desertification and *desertization* with the implication that both human and climatic causes are operative: the latter term should, however, really be restricted to situations where climatic causes predominate. Glantz (1977) used the term 'desertification' for degradation of environments with between 50 and 100 mm/y rainfall, and desertization for deterioration of environments with 300 to 600 mm/y rainfall. It probably makes best sense to use desertification to mean *environmental degradation as a result of human and climatic causes under any rainfall regime* (Hall 1978; Goudie 1981; Dennel 1982a; Grainger 1982; Leng 1982; Spooner & Mann 1982).

Probably about 20 per cent of the Earth's land surface and about 80 million people are affected by desertification (World Bank 1982: 164). An estimate 650,000 km^2 of arable and grazing lands have been affected along the southern margins of the Sahara (the Sahel – Sudanic Zone), and large portions of the fertile Nile Delta, the Sudan, Ethiopia, Chile, North-Eastern Brazil, Malagasy, Central Java, Kenya, Rajasthan (India) and Australia have been lost to agriculture or are presently threatened (Fig. 1.6 and Table 1.5). If desertification continues at present rates, probably around 60,000 km^2 (some would argue more) of useful agricultural land will be lost each year (Ruddle & Manshard 1981: 117).

Desertification is not restricted to the tropical drylands. It can even occur in the Amazon Basin, and experiences in the American Prairies and the 'Virgin Lands' of the USSR show that it is not only the farmers and administrators of developing countries who misjudge the agricultural capabilities of marginal land (Anon. 1977a; Chambers & Moris 1973: 517; Goodland & Irwin 1975; Franke & Chasin 1980).

It is reasonably easy to recognize human actions which lead to desertification: overgrazing, poor cultivation practices, poor irrigation management resulting in salinization, and collection of vegetation for fuel. Some of these activities may only become critical because of climatic changes, but establishing whether that is the case may be difficult. There are situations where socio-economic factors alone seem enough; for example, in parts of the Sudan the replacement of traditional village leaders with administrators appointed by the central government who had not the respect of local people resulted in the collapse of long-established restrictions on fuelwood collection and grazing.

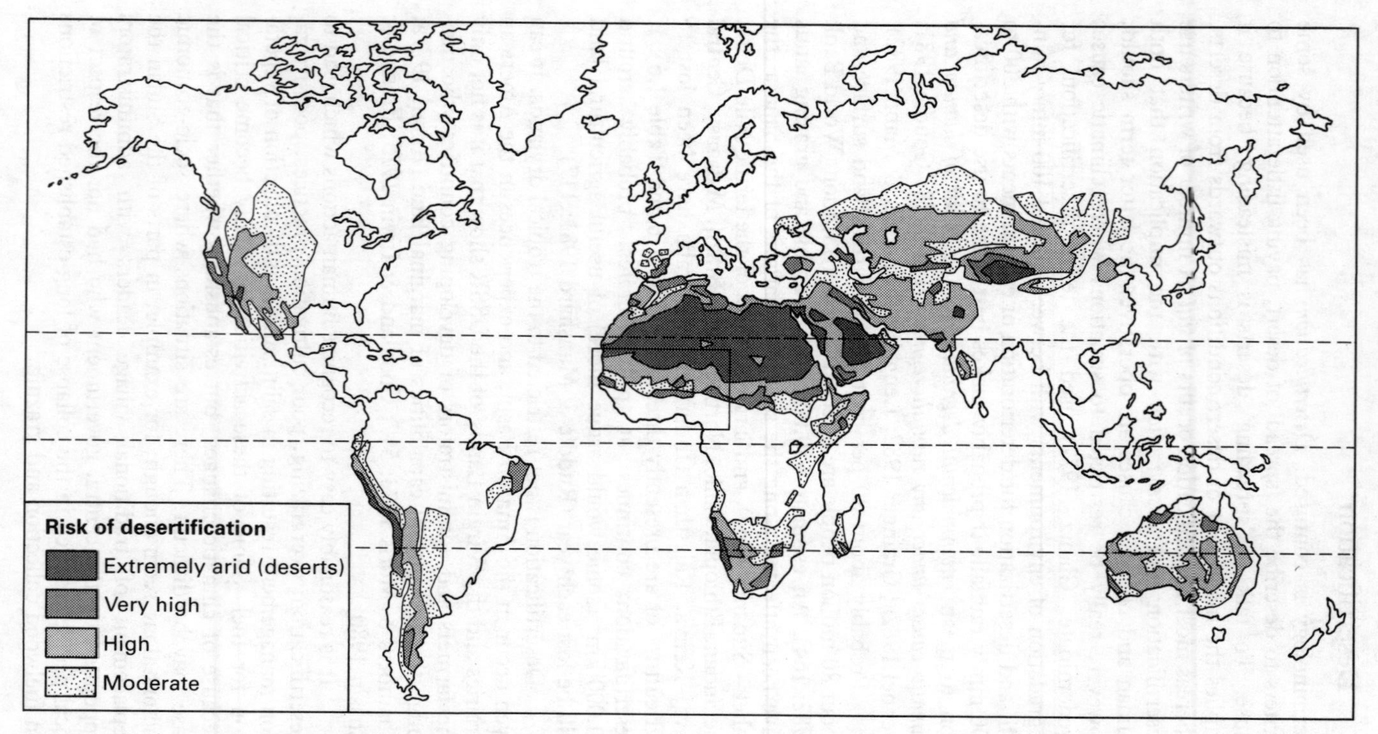

Fig. 1.6 Deserts and areas subject to desertification

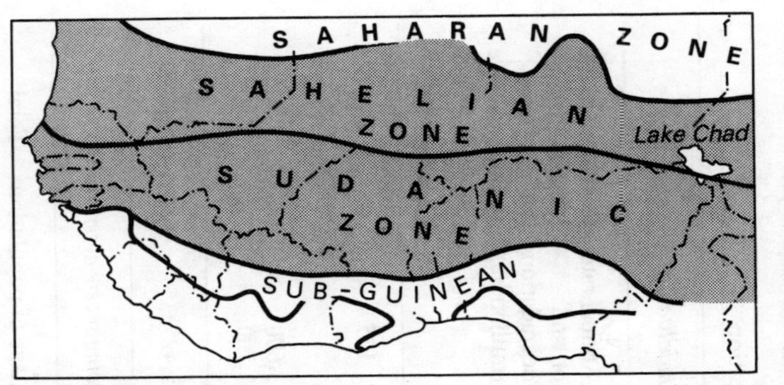

The Sahelian-Sudanic Zone

The subdivisions used in this inset map are based on those used by Van Apeldoorn (1981: 21):

Saharan – rainfall only in some years.
Sahelian – rainfall in almost every year, but with a rainy season that on average lasts less than 2.5 months, so that in most years and most places crops cannot grow without irrigation. Little problem with tsetse – flies attacking cattle.

*Sudanic** – with one rainy season of 2.5 to 7.0 months. In the southern portion of this belt the rainy season is longer (5.0 to 7.0 months) and tsetse – flies hinder pastoralist activity.

Sub-Guinean – with two rainy seasons totalling about 7.0 to 9.0 months' duration. Droughts do occur but are much less of a threat than they are to the north.

*In much of the literature, especially that published before the 1980s, the Sahelian – Sudanic Zone (Sahel) is the belt spreading from the Atlantic (including the Cape Verde Islands) roughly to Lake Chad, and bounded on the north by the southern 'edge' of the Sahara (Steeds 1985). This is the area shown by the inset map. Conditions of drought and desertification may be traced on across Africa, so that in effect there is a zone from the Atlantic to the Indian Ocean which includes southern Sudan. Recent literature and media coverage has concentrated on conditions in Ethiopia and the Sudan and there is thus a tendency to interpret the Sahelian–Sudanic Zone as meaning a cross-Africa belt. But as can be seen from the inset map, in the past 'Sahel' and 'Sudanic' have been often used when referring to north-south divisions of the Sahel west of Lake Chad.

Source: main map: IUCN, UNEP and WWF 1980 (map 2); inset map: Van Apeldoorn 1981: 20 (Fig. 3.1).

Table 1.5 Estimates of global desertification

A. Land affected, cost and net gains from rehabilitation

	Total area affected (million ha)	Annual rate of land degradation (million ha)	Cost of salvage ($US million)	Net gain from salvage ($US million)
Irrigated – waterlogged	25	125	106	112.5
– salinity problems	20			
Rangeland	3,600	3,600	180	180.0
Rainfed cropland	170	1,700	170	340.0
Total	3,815	5,425	456	632.5

B. Population living in areas recently undergoing severe desertification

	Area affected (million km²)	Percentage of population by livelihood type			Total population (million)
		Urban based	Cropping based	Animal based	
Mediterranean	1.3	31	60	9	9.8
Sub-Saharan Africa	6.9	19	37	44	16.2
Asia and the Pacific	4.4	27	54	19	28.5
Americas	17.6	32	56	12	24.1
Total	30.1	27	51	22	78.6

Source: UN Secretariat (1977) *Desertification, its Causes and Consequences*, Pergamon Press, Oxford (Table 18.3).

Attempts to legislate 'safe' limits for small farmers or herders, i.e. lines on maps beyond which people should not practice certain agricultural activities, have not generally been successful in controlling desertification, probably because the choice of the critical isohyet has usually been arbitrary and enforcement has been very difficult (Glantz 1977: 30). Preventing large-scale agriculture from causing desertification should be easier than controlling many small farmers or herders. The decision to develop marginal land should be based on the probable occurrence of drought, or on the poorest likely rainfall, **not on the average rainfall.**

Drought in the Sahelian – Sudanic Zone, the Sudan and Ethiopia

Between 1968 and 1974 the Sahelian or Sahelian–Sudanic Zone of Africa suffered the latest of a series of known droughts (Fig. 1.6 inset). This drought was 'triggered' by several years of less than average rainfall, rather than markedly low precipitation. Recovery from the drought is still at best only partial and in some parts of the Sahelian–Sudanic Zone things have got worse (Steeds 1985). There has been a tendency for those dealing with the Sahelian drought to consider that it would be quickly 'cured' when rains returned. However, in the Sahel, as in all regions suffering desertification, drought conditions may persist or return even after heavy rainfall. It has been estimated that well over 100,000 people had died as a result of the Sahelian drought by 1981 (Kates 1981: 71; Pearce 1984a).

Much of the literature on the Sahelian drought (and more recently parallel droughts in Ethiopia and Southern Sudan) presents it as an environmental disaster. However, it is as much a socio-economic phenomenon, 'drought': '. . . has become the scapegoat in arid parts of the developing world for basic failure in the agricultural production system' (Matlock 1981: 44). Put crudely there were too many people and too many animals for the environment to support. The Sahelian–Sudanic Zone (see Fig. 1.6 inset) could probably feed its present population, and certainly that of the mid 1970s, if more had been done to help rainfed cultivators. These farmers produce an estimated 95 per cent of all cereals in the zone, yet much of the funds which have got through to improve agriculture has been spent on developing commercially run, large-scale irrigation projects producing crops for export, which if anything further marginalizes poor rainfed farmers and hinders their recovery. A 1984 report suggested large irrigation schemes completed in the Zone between 1960 and 1979 barely made up for land lost through degeneration of such schemes in that period (*International Journal of Water Resources Development*, 2[1], p. 4).

Improving tropical agriculture

The People's Republic of China is nowadays apparently able to feed around 1,000 million people, while in the 1920s with a population of roughly 400 million famine was a common occurrence. Increased food production has been particularly marked since the 1950s: for example, in 1982 rice production was 2.8 times that of 1952; wheat production was 6.0 times that of 1952; potatoes 2.8 and meat 4.0 times that of 1952 (*China Facts and Figures* 1985). The major key to China's progress is generally acknowledged to have been the intensification of agriculture based primarily upon improved supply and management of water (improved transport and centralized administrative control also played a part). Could improved water supply achieve similar successes elsewhere? China's government has been able to organize groups of farmers and labourers to improve agriculture with minimal cash investment. This is encouraging, as many developing countries have difficulty finding funds. However, in China, under a communist regime, socio-political control by the government has been strong and thus people can be organized to pursue labour-intensive irrigation or drainage projects with very little cash reward. This has not been the case in many other developing countries. And in few developing countries are finished projects run with the levels of community spirit found in China. The development of water resources through 'self-help' could greatly increase the agricultural production of developing countries, but to do it in practice will require a 'work ethic' like that found in China or Cuba, Taiwan or South Korea. In some countries it may require time and a lot of effort to get it, in some countries it may come quite easily, for given adequate incentives small farmers can improve their output with little outside aid providing the pay-off or return is sufficiently high (Carruthers & Clark 1981: 16). In Japan during the 1920s and 1930s peasant farmers managed to raise rice crops four- or even fivefold in a few decades with little change in the traditional cultivation techniques or outside help by adopting artificial fertilizers and better crop mixes.

Agricultural development is commonly financed by loans or by aid. Loans can shackle a developing country as large proportions of funds become tied up for extended periods to meet interest payments and reduce debts. Sometimes projects financed by loans repay the investment too slowly or not at all. World recession in the 1980s has severely hindered agricultural development. Markets for crops like cotton, sugar or groundnuts have not been as buoyant as some planners hoped in the late 1970s. Another problem is *substitution*. This may be said to have occurred when an alternative supersedes an established commodity, for example, plastic fibres have severely affected jute sales; the use of artificial sweeteners, especially in soft drinks and confectionery in the USA and Europe, has had an impact on sugar sales. Similarly, tastes and fashions change and can upset established export markets. Substitution, changed tastes and fashions can create new opportunities, but on the whole '. . . never before have so many countries simultaneously reached their days of economic reckoning' (OECD 1982: 19). An increasing number of developing countries now have balance of payments deficits so bad that not only can they not find funds or attract foreign loans to expand or improve agriculture, they can no longer afford properly to run or maintain the projects they have already implemented. Even if a country has funds, supporting small-scale agriculture is a risky business for returns are slow and uncertain. Even when the economic climate is good, credit for small farmers is seldom easy to come by. The obligations to which recipients of aid are bound vary considerably depending on whether the support is purely humanitarian, no-strings-attached help or is tempered by political, commercial or moralistic motivations. Many aid agencies now have considerable expertise at their disposal and experience in the improvement of agriculture in developing countries, especially work with small-scale cultivators and herders in remote, marginal lands. Indeed there is more likelihood that a project will be satisfactorily planned, implemented and monitored after completion if it is managed by an aid agency than if it were in the hands of some developing country planning departments which suffer personnel changes, policy changes and shortage of funds, and which are generally unable to function with any continuity. Some governments still doubt the wisdom of spreading available funds thinly among many small farmers and have preferred to support large-scale agriculture or industrial development, hoping that the benefits generated would somehow 'trickle down' to the rural masses – a view which has often proved over-optimistic.

The question of what is needed to identify and encourage or aid small-scale farmers and herders who have the potential to 'take off' from unsatisfactory subsistence to 'self-sustained growth' has been much debated (Hill 1963; Schultz 1964; Wilkinson 1973). It seems likely that five 'essentials' are needed if small farmers are successfully to adopt innovations. A number of 'accelerators' can also help and speed up the process, but could be done without (Arnon 1981: 17). The essentials are: *markets for products* (for no small farmer will produce more than his/her family can consume unless it can be sold or bartered for a price attractive enough to reward the extra effort involved); *constantly changing technology; local availability of supplies and equipment; incentives;* and *transportation*. The accelerators include: *credit; education for development; group action by farmers; improving and extending agricultural lands; national planning; crop storage;* and *marketing facilities*. The degree to which accelerators help seems to depend on the novelty and complexity of innovation (new crops tend to spread relatively fast; new technology much more slowly); the cost of innovation (if it is costly credit will probably be more important); and profitability (the greater the returns the more likely are small farmers to adopt an innovation).

The majority of small-scale cultivators and herders in developing countries are illiterate; many of them live in regions with poor roads. Even if suitable cultivation techniques and crops can be developed most rural peoples will still need advice and some training if they are to be successfully adopted. Modern telecommunications, radio and increasingly television, broadcast via satellite to community television sets, will play a vital role in farm extension training, in altering attitudes which may inhibit progress, in improving hygiene and thereby health and productivity and in monitoring agriculture. In a few decades small farmers in remote regions may respond to nightly bulletins informing them of the need to apply water, fertilizer or pesticides to crops in the way farmers in the West today listen to weather forecasts (Clarke 1977; World Bank 1982: 74).

Governments or agencies wishing to develop agriculture may adopt either a *command approach*, wielding power to reorganize and manage improvement efforts; or a *market approach*, whereby economic incentives are used to 'tempt' and reward farmers for making improvements and maintaining standards. The approach to improving cultivation or herding can be merely technological, or it can be one more aware of the socio-economic needs and abilities of the people involved, or it can be a combination of both.

Between the early 1950s and the late 1970s a lot of effort went into promoting innovations which doubtless gave improved yields under optimum conditions, but which had been little tested under the conditions which most small farmers operated. Inevitably such efforts met with mixed success, often having replaced an existing production strategy, necessary inputs could not be supplied or became too costly, opportunities for marketing produce were not as good as had been expected or there were institutional or environmental problems. Failures in innovating were widely noted and slowly forgotten by rural folk who were anyway usually suspicious and reluctant to take advice from outsiders. Many improvement efforts might have worked if there had been more numerous, more suitable and better-motivated farm extension service officers to win the trust of the farmers or herders.

In many situations it will be better to improve what is already practised, rather than to try and transfer unproven strategies possibly evolved under different environmental, social and economic conditions. There are situations, especially in harsh tropical environments, where there is no proven modern agricultural strategy which can be beneficially and economically substituted for the traditional practices.

The green revolution approach

In the 1940s teams of scientists were sent by the Rockefeller and Ford Foundations from the USA to help the Mexican government improve wheat and maize production by use of improved seed, artificial fertilizer and other 'scientific farming' innovations. The Rockefeller and Ford Foundations initiatives were largely instrumental in establishing what became the first of several international agricultural research centres: Centro Internacional de Mejoramiento de Maiz y Trigo (CIMMYT) and the International Rice Research Institute (IRRI). These and other centres in the 1960s and 1970s played, and continue to play an important role in the improvement of tropical agriculture (Tables 1.6 and 1.7).

From the mid 1960s Consultative Group on International Agricultural Research (CGIAR) centres were disseminating *high-yielding varieties* (HYVs) of wheat and maize (developed at CIMMYT) and rice (developed at IRRI). Results in Mexico were so good that the catch-phrase 'green revolution' was used to describe the hopefully beneficial transfer of HYV 'packages' (seeds, fertilizers, pesticides and cultivation techniques) to developing countries. Green revolution implied peaceful, rapid improvement of agriculture through technological innovation. Improved yields of grain, it was hoped, would feed the hungry and raise farmers' incomes. Many developing country and Western statesmen welcomed the

Table 1.6 Developments in agricultural science some of which contributed to the 'green revolution'

Date	Development
1920s	Japan and Taiwan develop rice varieties which respond favourably to fertilizer application
1933	Hybrid maize varieties
1945	DDT, and later other organochlorine pesticides developed
1945–47	Advances in artificial fertilizers
1951–53	New herbicides
1956	Hybrid rice varieties first developed in Taiwan
1957	Hybrid sorghum varieties developed
1961	Dwarf hybrid (HYV) wheat varieties
1965	Dwarf hybrid (HYV) rice varieties, including IR-8 'miracle rice'
1969	Hybrid barley varieties

Note: the above dates are approximate. Research and development has continued since 1969.

Table 1.7 International agricultural research centres supported by the Consultative Group on International Agricultural Research (CGIAR)

Centre	Location	Date founded	Coverage	Crops/ activities
Asian Vegetable Research and Development Centre (AVDRC)	Taiwan	1971	Humid tropics, especially Asia	Beans, tomatoes, cabbage, etc., insect-resistant crops
Centro Internacional de Agricultura Tropical (CIAT)	Colombia	1967	World-wide, tropical lowlands, especially Latin America and Caribbean	Cassava, field beans, rice, pasture. Cattle, pigs. Modernizing shifting cultivation
Centro Internacional de Mejoramiento de Maiz y Trigo (CIMMYT)	Mexico	1966	World-wide	Maize, wheat, barley, triticale
Centro Internacional de la Papa (CIP)	Peru	1971	World-wide	Potatoes
International Board for Plant Genetic Resources (IBPGR)	Italy	1972	World-wide	Collection, documentation and conservation of genetic material of important crop species
International Centre for Agricultural Research in Dry Areas (ICARDA)	Lebanon, Syria, Iran	1976	World-wide, semi-arid /seasonal rainfall regions	Dryland farming systems for subtropics. Wheat, barley, pulses and irrigation research

Table 1.7 (continued)

Centre	Location	Date founded	Coverage	Crops/ activities
International Crops Research Institute for the Semi-Arid Tropics (ICRISAT)	India	1972	World-wide	Sorghum, millet, food legumes and farming systems for the semi-arid tropics
International Food Policy Research Institute (IFPRI)	USA	1979	World-wide	Food policy research
International Institute for Tropical Agriculture (IITA)	Nigeria	1968	World-wide, lowland tropics, especially Africa	Farming systems for the humid tropics. Root crop food legumes, maize, rice. Alternatives to shifting cultivation
International Livestock Centre for Africa (ILCA)	Ethiopia	1974	Africa	Livestock production
International Laboratory for Research on Animal Diseases (ILRAD)	Kenya	1973	Africa	Livestock disease research (trypanosomiasis and theileriasis in particular)
International Rice Research Institute (IRRI)	Philippines	1960	World-wide, special emphasis on Asia	Rice and rice-based farming systems

Table 1.7 (continued)

Centre	Location	Date founded	Coverage	Crops/ activities
International Service for National Agricultural Research (ISNAR)	Netherlands	1980	Developing countries	Assistance to national agricultural research
West Africa Rice Development Association (WARDA)	Liberia	1971	West Africa	Rice research

Note: CGIAR is an informal association of countries, multinational organizations and private foundations pledged to supporting and expanding research which will help solve agricultural problems common to many developing countries. In 1982 there were thirteen CGIAR centres and programmes. In addition to CGIAR research centres, there are other institutions involved in developing tropical agriculture, notably university departments. An excellent critique of the green revolution, especially CGIAR's role, is provided by Glaeser, B. (ed.) (1987) *The Green Revolution Revisited: Critique and Alternatives*, Allen and Unwin, London.

Sources: Ehrlich *et al*. 1977: 342; Griffin 1979: 247–8; Lal & Russel 1981: 4; Manassah & Briskey 1981.

green revolution, not least as a way of sidestepping 'subversive' demands for land reform, it was also welcomed by multinational agribusiness who sold fertilizers, pesticides and more recently HYV seeds (Brown 1970; Farmer 1977; Dahlberg 1979; Griffin 1979; Pearse 1980).

There has been endless, largely inconclusive, debate over the success or failure of the green revolution. In the 1960s there was a threat of severe and widespread famine, within little more than a decade over half of the wheatlands and about one-third of the padi-fields of developing countries were sown with HYVs (World Bank 1982: 69). In 1966 India was the world's second largest wheat importer; between 1970 and 1972 the country's wheat production was doubled largely through the adoption of green revolution technology, and after a brief setback India became self-sufficient in the late 1970s. In the Philippines, Taiwan and some other parts of South East Asia there were considerable successes in raising rice yields by growing HYVs; under favourable conditions two or three times the crop produced by traditional seeds and methods was possible.

The green revolution approach suffered from a number of weaknesses. It bypassed large parts of the tropics, for example sub-Saharan Africa, because in the 1970s only about 17 per cent of that continent's agricultural production was cereals, and much of that not wheat or rice. It would be better to call the HYVs, *potentially HYVs*, as their success depends upon appropriate management, proper application of inputs, especially irrigation, fertilizer and pesticides. Increased yields are given by HYVs through one or more of the following qualities: by responding better to fertilizers than traditional varieties, using water more efficiently (although some HYVs proved less tolerant of irregular water supplies), by requiring a shorter growing season, producing more grain relative to stem and leaf, better resistance to lodging – collapse due to wind or rain which hinders harvesting. Sometimes the HYV package did not fit the local environment. Unlike traditional crop varieties with considerable genetic diversity the green revolution HYVs lacked resilience and occasionally spectacularly succumbed to disease, drought or pests. Ideally the national research stations were to have adapted HYV packages to suit local conditions and needs, but they were often given insufficient time and funds to do so, and were too often sited in unrepresentative locations remote from the rural people and with much less harsh conditions than the lands the bulk of farmers worked.

The strongest criticism voiced against the green revolution has been that the packages were more easily or effectively adopted by the richer landowners and those with access to water for irrigation. The effect was to widen the gap between rich and poor farmers (Farmer 1977; Pearse 1980). Some have argued that the green revolution was an attempt by richer nations to apply their own conceptions of development to the South, working mainly through the local élites. Certainly in many regions the package were offered to richer more 'progressive' farmers and seldom were smaller farmers given aid to offset their disadvantages when it came to obtaining inputs. It is also worth noting that the production of food surpluses in no way gives any guarantee that it will be distributed to those who want it at the price they can afford.

Some of the green revolution problems arose in consequence of developments which could not have been foreseen in the 1960s. Particularly important were the 1973–74 Organization of Petroleum Exporting Countries (OPEC) petroleum price rises which pushed up the cost of inputs of fertilizer, fuel and pesticides. The international research centres, aid agencies and other bodies active in agricultural research are aware of the problem and increasingly innovations do not require petroleum-based technology.

Those who claim that the green revolution 'failed' overlook

the fact that the agricultural research centres which were active in the 1970s are still active. In effect the green revolution continues, albeit in an altered form. They also ignore the fact that without the green revolution there would have been much more famine in the 1970s than there has been. Nowadays the international research centres have diversified, research is in progress on new or improved varieties of presently used crops and livestock, better methods of tillage and crop storage, new ways of improving soil fertility and controlling pests and weeds. The problem now is that promising innovations must be pieced together into appropriate, sturdy, reliable farming systems which, hopefully, people will accept. To do that effectively, existing farming systems and the impacts of innovations must be studied.

The farming systems research approach

Agricultural extension has often been practised by cadres who instruct farmers with little discussion and little respect for or real understanding of their position, and not infrequently a grossly inadequate knowledge of environmental, social and economic conditions. Such an approach is highly unlikely to win the trust and backing of the cultivators or herders who are to be helped.

Some developing country small-scale cultivators or herders may already have evolved to the point where improvement of production is relatively easy. Others may have reached a 'plateau'; others may be far behind them in terms of skills, attitudes, capacity to innovate and/or they may be in remote and/or unfavourable environments. Ironically, it may be more difficult to increase the yields of poor, efficient farmers than poor, inefficient farmers (Schultz 1964). Bunting (1979: 1045) observed: 'The first challenge now before agricultural science and technology in the poorer countries is to understand the rationale of existing systems of production. The second is to devise innovations, appropriate to the environment as well as to the social and economic context which will enable producers to increase output from the resources actually available to them.'

The goal of the green revolution was to produce under irrigated agriculture more staple food grains: wheat, rice or maize. The goal of the *farming systems research approach* is to improve the aggregate production of the small farmer and his family's welfare. The farming systems research approach involves trying to economize on non-farm inputs, such as chemical fertilizer. An attempt is made to maximize the use of what small farmers have in relative abundance, for example family labour. Particular attention is paid to the farmer's risk-avoidance strategies and to understanding where their farming methods are weak or threatened. Put at its simplest, farming systems research attempts to understand the behaviour and needs of the small farmers and then agricultural research goes more than half-way to meeting their needs in full awareness of social, cultural and physical constraints (Redclift 1984: 112). To do this effectively involves the use of a multidisciplinary team to study the way a 'target group' of small farmers operates: their habits, abilities, needs and constraints. Farming systems research can either originate in the research station and then encompass the farmers, or it can originate in the farmer's fields and then stimulate appropriate work in the agricultural research stations. There is no reason why those involved in the green revolution approach cannot become involved in the farming systems approach; indeed much of the pioneering work on farming systems research was done at CIMMYT and IRRI (Clayton 1983: 137).

Land availability and the case for better use of water resources

The *Global 2000 Report* (Council on Environmental Quality and Department of State 1982: 16) noted: 'Land under cultivation is projected to increase only 4 per cent by 2000 because most good land is already being cultivated . . . In the early 1970s one hectare of arable land supported on average 2.6 persons; by 2000 one hectare will have to support 4 persons.' Hunger will affect many more people by the end of the decade if there is no improvement of agricultural yield in developing countries – in Africa alone thirty out of fifty-one states may be unable to feed their predicted populations at the end of the century (Harrison, 1983) (see Table 1.2). Land which is degenerating and becoming unproductive must somehow be protected and where possible made more productive; where practical, land which is presently unused or underused must be developed.

Much of the land which is degenerating could be rehabilitated with better water use, indeed a significant part of land damage in the tropics is caused by poorly managed irrigation. In many developing countries the most accessible water supplies have already been developed and domestic and industrial water demands will increase and compete with those of agriculture. The improvement of inefficient irrigation projects is therefore becoming very important (Falkenmark & Lindh 1976; Utton & Teclaff 1978).

Though there are unpleasant predictions for the future, there is without going to ecological extremes, plenty of useful land that could be farmed and much that could be better farmed *if invest-*

ment were directed to establishing suitable rainfed cultivation, adequate, reliable irrigation or effective moisture conservation strategies (George 1977: 57; World Bank 1982: 164). Some countries have been able to open up new land; for example, Indonesia has supported transmigration of Javanese settlers to Kalimantan; Brazil has settled poor 'Nordestinos' in parts of Amazonia; and Malaysia's Federal Land Development Authority has cleared vast areas of forest for land colonization projects. Parts of Africa are unfarmed at present because of the prevalence of diseases like trypanosomiasis or onchocerciasis. There is a chance that these might some day be controlled.

Many developing countries have little or no good unused land. To increase their agricultural production they will have to either intensify the use of already farmed land or crop marginal land. In either case little will be achieved without better rainfed cultivation strategies or irrigation. Developing nations face a dilemma: if they delay attempting to intensify their agriculture they will be unable to feed their growing populations and land degeneration may well occur; if they press ahead they may adopt current technology and methods which may in five or ten years compare unfavourably with later developments. Some of the presently expensive and complex but productive irrigation methods may become simpler to operate and cheaper to install.

Even for areas which have, or are soon expected to have, high population to food production ratios there is hope. Geertz (1963: 13) examining the interrelationship of physical environment, population growth and agricultural improvement in Indonesia suggested that although too rapid population increase may overtake food production a steady build-up even to a high density can actually trigger intensification of agriculture. A similar observation was made by Boserup (1965: 14, 117), who argued that a community undergoing sustained population increase has a better chance of modernizing agriculture than one with a stagnant population. In short, population increase might 'drive' agriculture. It is worth remembering that Japan (and similarly the People's Republic of China) increased agricultural production under conditions of severe overpopulation starting in the 1920s and in roughly thirty years had a production that outstripped her needs (Arnon 1981: 9). Those countries which have made progress with developing their agriculture, e.g. Japan, South Korea and Taiwan, have done so on a foundation of padi farming, which meant that the cultivators were from the outset quite skilled. In Africa and Latin America some have argued there is less of a tradition of irrigated farming, the intensification of small-scale agriculture through irrigation may therefore take more time and funds may be better spent on improving rainfed cultivation. Richards (1985: 53), however, cautions that such arguments based on Boserup's model of agricultural change (which implies that if land is in plentiful supply, cultivators are better rewarded by expanding the area cultivated than on changing to a more intensive system of cultivation) may be a useful stimulus for academic discussion, but might also be misleading as a basis for formulating agricultural development policies. Richards (1985: 86) suggests 'population pressure' is far from being a unique or especially significant stimulus to 'grass-roots' agricultural change.

Chapter 2

The water resource management system

Water as a resource

Roughly 97 per cent of the Earth's water resources are in the oceans and are saline. Only 2.5 per cent of the world's water is fresh water and of that a mere fraction is available to man (Fig. 2.1). The oceans, streams, lakes and groundwaters which are too saline to purify economically, polar ice and deep groundwaters, are

Fig. 2.1 The world's freshwater resources

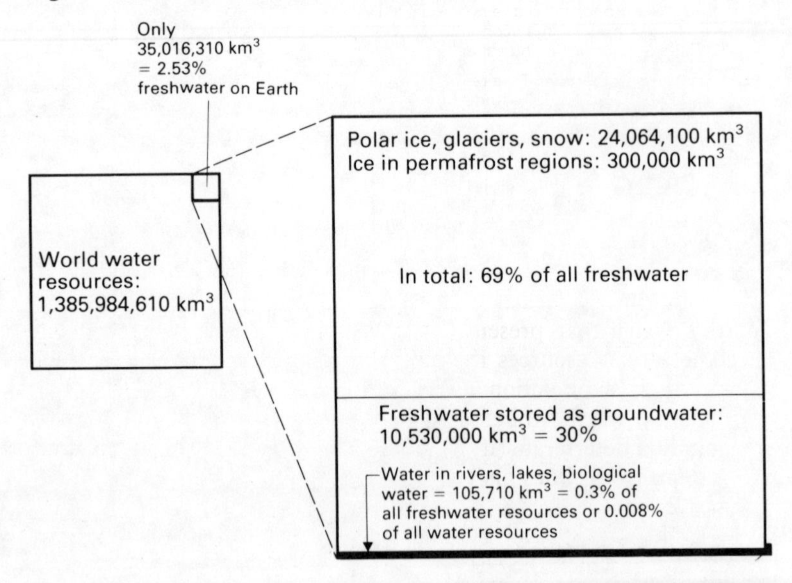

Source: based on Keller, R. (1984) 'The world's freshwater: yesterday–today–tomorrow', *Applied Geography and Development*, **24**, 7–23 (Fig. 1).

Fig. 2.2 Simplified classification of resource base, resources and reserves

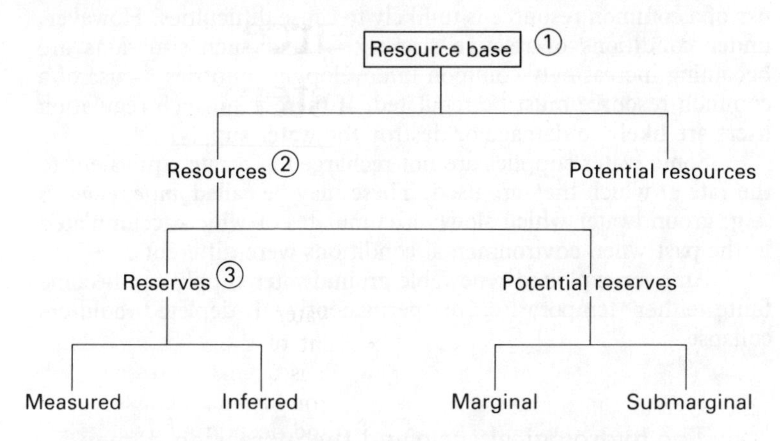

Notes:

1 The sum total of all components in the environment that would become resources as such if they could be extracted from the environment.

2 The proportion of the resource base that man can make available under given social and economic circumstances, within limits set by the level of technological advancement.

3 The proportion of the resource that is known (with reasonable certainty) to be available under prevailing social, economic and technological conditions.

Source: based on Park, C. C. (1980) *Ecology and Environmental Management*, W. M. Dawson, Folkestone and Westview Press, Boulder, Colorado. (Fig. 8.1).

potential resources; they are not presently usable and/or wanted. Water resources (*actual water resources*) are that portion of the Earth's water that presently has utility. Potential resources may become actual resources if there is technological innovation which makes their exploitation cheaper and/or if demand makes their exploitation worth while. On the other hand, if the economics of exploitation deteriorate (if, say, fuel costs for pumping rise sharply) or if demand declines an actual resource may cease to have utility (Fig. 2.2).

Some water supplies continue to be available for use so long as the rate at which they are consumed is less than the rate at which they are replenished and provided the environment is not altered to affect their quality or movement. These are *renewable resources* (streams, rivers, rainfall, replenished groundwaters). *Potentially renewable* would be a more accurate term, because renewal usually depends on satisfactory management. Poor management can lead to temporary or permanent, slight or total damage of water supplies.

Water is frequently a *common resource* (common-pool resource), not owned or controlled by any one individual or organization and used by many. Under conditions of relative abundance, use of a common resource is unlikely to cause difficulties. However, under conditions of relative scarcity — and such situations are becoming increasingly common in developing countries — use of a common resource must be regulated. If there is no such regulation users are likely to damage or destroy the water supply.

Some water supplies are not recharged at a rate equivalent to the rate at which they are used. These may be called *finite resources* (e.g. groundwater which slowly accumulates or which accumulated in the past when environmental conditions were different).

An over-exploited renewable groundwater supply can become finite either temporarily, or permanently if depleted aquifers collapse.

The hydrological cycle and the increasing demand for water supplies

The ultimate source of virtually all fresh water is the continuous distillation of the oceans by solar radiation and its condensation as rain (i.e. the sun 'drives' the hydrological cycle). Annually about 43,000 km^3 of water vapour are evaporated from the oceans and about 70,000 km^3 are evaporated or transpired (the emission of water vapour from pores in the leaves and stems of plants) by plants growing on the land surface, which keeps the amount of water vapour in the atmosphere more or less constant. Every year roughly 500,000 km^3 of precipitation falls, about 110,000 km^3 of which falls on land, which is significantly more than that which is evaporated and transpired – the result is *runoff* (Ambroggi 1980: 39).

The degree to which vegetation interacts with the hydrological cycle may be judged from the fact that of all the water returned to the atmosphere from the world's land surfaces, between 50 and 70 per cent is through *transpiration*, and only 30 to 50 per cent by *evaporation*. Much of the moisture left after evaporation and transpiration (roughly 40,000 km^3/y) runs to the sea as floods. Of this the maximum amount of water that might be used each year with the technological means presently available is probably somewhere between 14,000 km^3 and 25,000 km^3 (Ambroggi 1980: 39).

Taking the lower of the two previous estimates, 14,000 km^3/y, and assuming that roughly 5,000 km^3/y flows in uninhabited regions which are likely to remain so due to inhospitable environment, roughly 9,000 km^3/y are available to meet man's water demands world-wide (Ambroggi 1980: 39). In some regions use may be made of finite groundwater supplies in addition to water derived from precipitation.

Water is used for domestic consumption, i.e. drinking, washing and cooking, for industry and for agriculture. Domestic supply and industrial water demands are rapidly increasing in most countries, but by far the largest water user is agriculture. It should also be noted that much of the moisture used by agriculture is provided directly by precipitation, i.e. it is used before it becomes part of the 14,000 to 25,000 km^3/y streamflow to support rainfed crops directly (Ambroggi 1980: 40).

In the first century AD the total world population was roughly 200 to 300 million persons, in 1975 it was about 4,000 million, in AD 2000 it will be around 6,500 million. As population increases more and more water will be needed for agriculture and the ratio of available water to water demand will fall considerably.

So far most of man's manipulation of the hydrological cycle for his own ends has taken place at the streamflow/riverflow and, to a slightly lesser extent, groundwater storage phases. Attempts to modify weather to increase precipitation (mainly through cloud seeding) have had inconclusive or at best limited success. It may be possible in the future to collect fog, mist or dew where environmental conditions are favourable. Agricultural use of seawater is not feasible at present and extraction of fresh water from the sea (*desalination*) is too costly to be of general value to agriculture in developing countries. While technological advance may make the sea a practical water source one day there are no signs of this happening in the near future. In the foreseeable future most water will come from streamflow/riverflow or groundwater sources.

The watershed/river basin ecosystem

The term *watershed* is sometimes used to refer to the catchment boundary (or divide) which borders a *river basin*. However, it is more widely taken to indicate the entire catchment area of a river or drainage basin (Fig. 2.3).

The watershed/river basin is an ecological unit (*ecosystem*) in which the living and non-living environment function together. Water resources (with the exception of finite groundwaters) are seldom static: there is flow and water is an integrative element in the watershed/river basin ecosystem. The functioning of such an ecosystem can be monitored: the inputs of water (precipitation) and the outputs of water, transport of eroded material and dissolved nutrients can be established without resorting to impossibly detailed studies of interacting subsystems and their components. The watershed/river basin is therefore an ideal management unit (Laconte and Haimes 1982).

Fig. 2.3 Schematic diagram of a watershed/river basin ecosystem

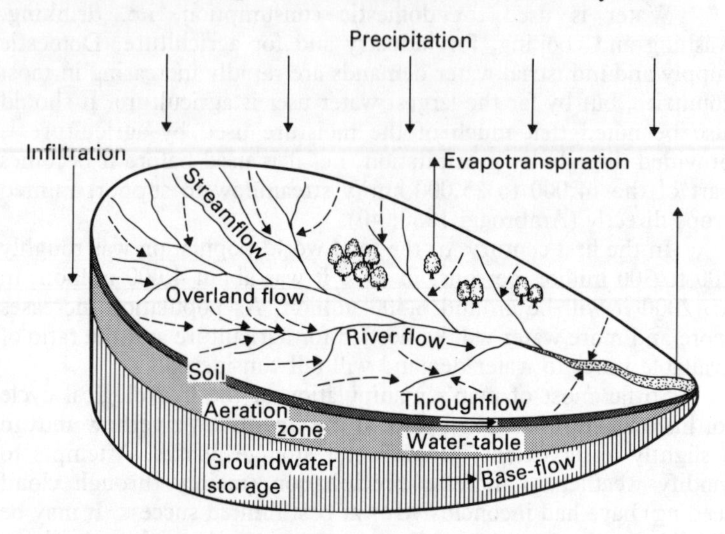

Source: Porter, E. (1978) *Water Management in England and Wales (2nd edn)*, Cambridge University Press, Cambridge.

Managing water resources

Watershed management

Most development within a watershed/river basin will have some effect on that ecosystem. Landuse changes, especially forest clearance, will affect the local microclimate and possibly regional climate (forests transpire a lot of water, keeping their immediate surroundings moist and to a lesser extent increasing humidity downwind). If the vegetation cover of a large enough portion of a watershed is altered to plants with either higher or lower evapotranspiration rates, groundwater accumulation may either decrease or increase respectively. Removal or change of the vegetation cover of a watershed also affects overland flow and infiltration rates altering quantity, timing and quality of streamflow and riverflow. Changes in streamflow and infiltration can affect groundwater levels and vice versa.

Location within a watershed/river basin affects land-use potential. Figure 2.4 illustrates the pattern of land-use which would be advisable in a tropical environment: in area (a) arable cropping would be ill-advised; in area (b) arable cropping would be possible

Fig. 2.4 Watershed/river basin: agricultural suitability

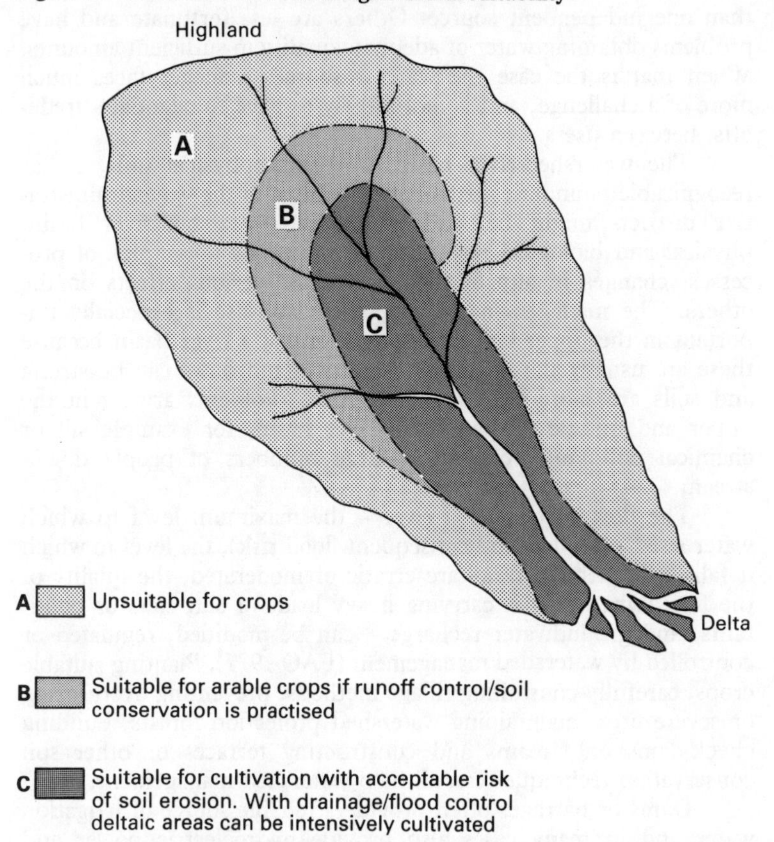

Source: Interim Mekong Committee 1982: 77 (Fig. 4.6).

provided suitable runoff control/soil conservation measures were taken; and in area (c) arable cropping could take place with less need for runoff protection and soil conservation.

Frequently more than one use is made of water: it may be needed for domestic supply, for livestock, for irrigation, for hydro-electricity generation, for dilution and disposal of wastes, (industrial or sewage), for cooling thermal power stations, for fishing, for recreation, for navigation. These different demands on the resource are not always compatible – if one user alters the quality of water supplies or restricts flows it may well affect another user. One of the tasks of the water resources manager is to try and ensure that various uses are compatible, and if not, to weigh up the costs and benefits and decide how to restrict use. Some localities are fortunate and have abundant water supplies, sometimes from more than one independent source. Others are less fortunate and have problems obtaining water of adequate quality in sufficient amounts. When that is the case the water resources manager faces much more of a challenge, and is more likely to have to negotiate 'trade-offs' between users.

The watershed/river basin is a biogeophysical unit, it has recognizable boundaries. It is not ephemeral in the way administrative districts might be, and within the watershed/river basin, physical and biological resources are linked by a complex of processes, changes in any of them can cause serious effects on the others. The management of watershed land-use is especially important in the upper and middle portions of a river basin because these are usually the portions where overland flows can be strong and soils are more easily eroded. Also, problems arising in the upper and middle portions of a river basin, for example silt or chemical pollutants, can affect large numbers of people downstream.

The flow pattern of a river – the maximum level to which water rises (*peak flow* and consequent flood risk), the level to which it falls, whether the flows are erratic or moderated, the quality of the flow (whether it is carrying heavy loads of silt, salts or pollutants) and groundwater recharge – can be modified, regulated or controlled by watershed management (FAO 1977). Planting suitable crops, carefully controlling tillage methods, preventing overgrazing or severe fires, maintaining watershed protection forests, building check-dams on streams and constructing terraces or other soil conservation techniques can all help watershed management.

Dams or barrages often provide a reliable source of irrigation water and in many cases also provide hydroelectric power and regulate floods, but they can be expensive, and in some cases their construction has been emphasized to the detriment of integrated development. Without adequate watershed management reservoirs are likely to become silted and land use in the catchment area will deteriorate. More water management could, and should, take place before the streamflow/riverflow phase of the hydrological cycle. Of the estimated 110,000 km^3 of precipitation which falls on the lands of the world 70,000 km^3 never reaches stream channels (much is returned to the atmosphere by transpiration). Despite its importance watershed management has had less attention and less funding than large dam or barrage construction, but as land use intensifies and competition for water resources increases it seems likely that developing countries will be forced to remedy this neglect. Successful watershed management requires close liaison between land-use planners and managers and water resources managers: it also requires constant appraisal. With modern developments in earth resources monitoring techniques and data storage this should become easier for developing countries.

Managing water resources systems

To be successful, water resources management must be aware of the linkages between land-use, streamflow and groundwater storage. Only in the last few decades has understanding of the structure and function of tropical environments developed to a level that is practically useful. This has largely been a consequence of work carried out during the International Biological Programme (1964–74), the International Hydrological Decade (1965–74), and more recently the UNESCO Man and Biosphere Programme. In spite of advances there is still far from adequate understanding of tropical ecosystems – their productivity, resilience to change and the subtle, close relationships between environment and man (Williams & Joseph 1973; Eckholm 1976; Furtado 1980; Balek 1983).

The water resources manager must be capable of assessing, exploiting and allocating supplies; it is also necessary to maintain the quality and quantity of supplies. The problems involved in developing and managing water resources are complex and extend across professional discipline boundaries: a multidisciplinary approach is therefore desirable.

The most important aspects of water resources management are the development and allocation of supplies (Fig. 2.5). Management of water resources for agriculture may involve developing and maintaining more than one water source. Some water sources may be affected by land use more than others, some may be available to 'back-up' others. By manipulating land-use and developing the appropriate 'mix' of sources it is often possible to ensure a satisfactory supply for agriculture. Having developed a water supply it must be delivered to the crops or livestock, often but not always this involves a *conveyance system* and arrangements for *allocation* and *distribution*.

If the maximum benefit is to be obtained from a water supply it must be used in the most appropriate manner, and after use it must be disposed of or reused satisfactorily. The linkages between these various subsystems in the water source – use – disposal/reuse system are illustrated in Fig. 2.5. For irrigation to give sustained, satisfactory yields it is not enough for the water resources manager simply to see that the particular set of subsystems shown in Fig. 2.5 functions satisfactorily. The water resources supply – use – disposal/reuse system interacts with other systems, the production – marketing system, the land management system and so on. It is

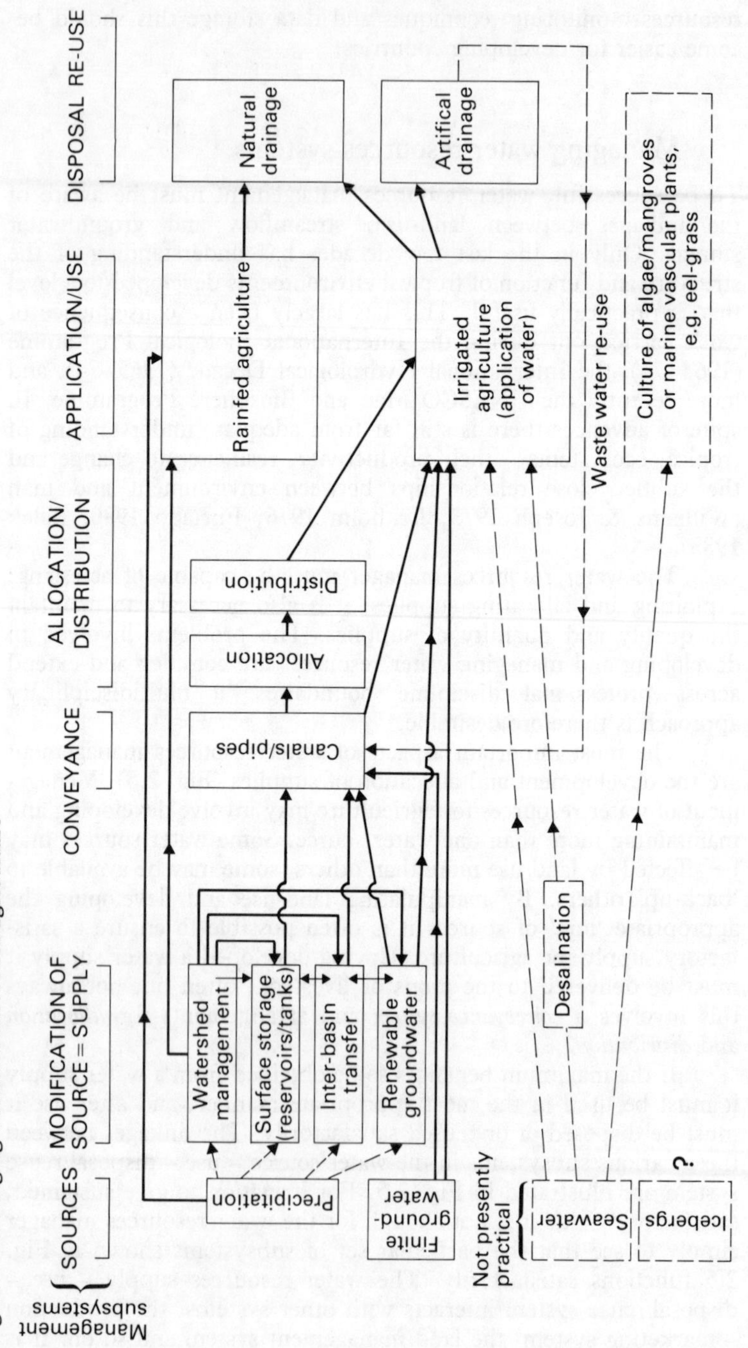

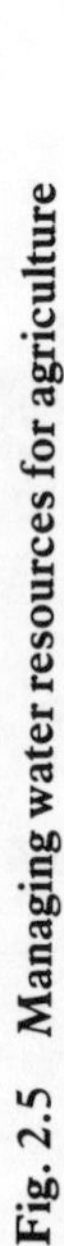
Fig. 2.5 Managing water resources for agriculture

Fig. 2.6 Inputs and processes in irrigated agriculture

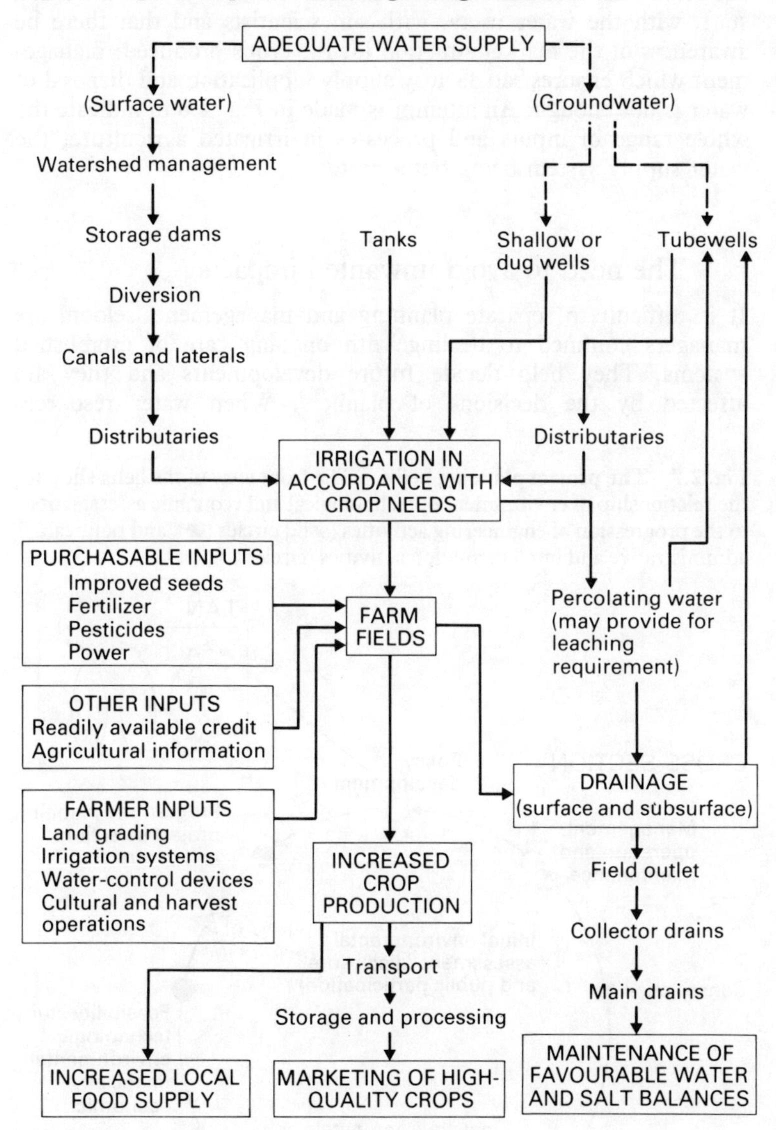

Source: Millikan & Hapgood 1967: 36 (Chart 1).

therefore vital that there be close liaison with agricultural extension staff, with the water users, with soil scientists and that there be awareness of the market situation for the crops produced: management which ensures satisfactory supply, application and disposal of water is not enough. An attempt is made in Fig. 2.6 to indicate the whole range of inputs and processes in irrigated agriculture, the water supply system being but a part.

The need to avoid unwanted impacts

It is difficult to separate planning and management, seldom are managers confined to dealing with ongoing care of established systems. They help decide future developments and they are affected by the decisions of planners. When water resources

Fig. 2.7 The project planning helix. The front view of the helix shows the relationship of environmental, technological and economic assessments to the progression of engineering activities (solid circles (●), and political, administrative and implementation activities (circles ○).

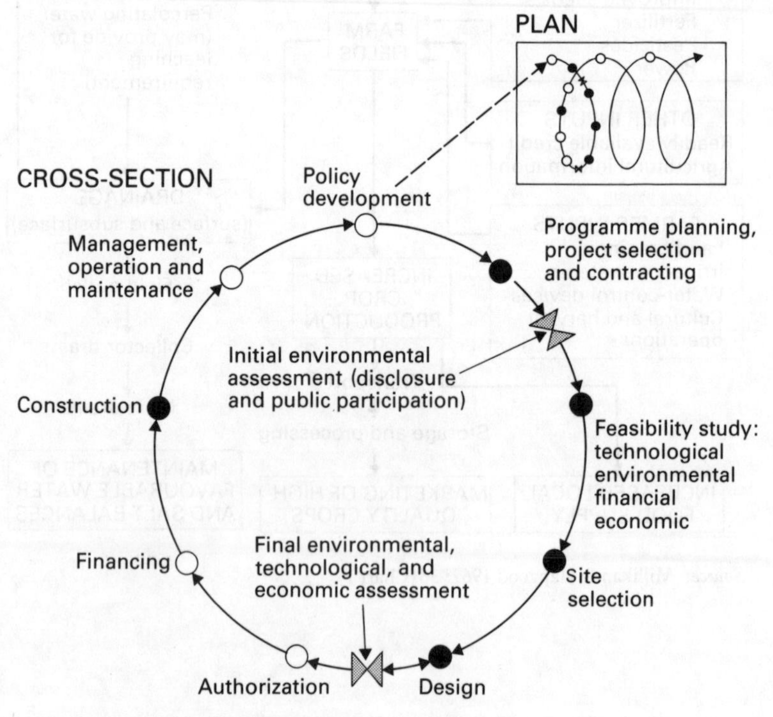

Source: Gunnerson & Kalbermatten 1978: 242 (Fig. 12.1).

management identifies the objective(s) which are to be met by water resources development there should be adequate pre-project/ pre-programme appraisal sufficiently in advance of proposed developments and sufficiently thorough to ensure that problems which might arise are identified and can be dealt with. Such an *impact assessment* should consider environmental, socio-economic and socio-cultural costs and benefits of the strategy(ies) available to achieve the objective(s). Efforts should also be made to research and evaluate relevant hindsight experience and there should, wherever possible, be consultation with the people who are likely to be affected by the development. This should ensure that they are aware of the situation, are able to adopt improvements and innovations and to identify socio-economic or socio-cultural factors which might hinder or aid management, and to establish what they want and need. A rather idealized view of how the policy development – project appraisal – management – policy development spiral should function is shown in Fig. 2.7.

Once a particular strategy is selected it has to be implemented and then maintained. To do this satisfactorily requires an ongoing monitoring of environmental and human impacts. Without adequate monitoring it is difficult to identify whether it is necessary (and practical) to alter plan or programme should conditions or demands change or unforeseen problems arise. And without monitoring it would be difficult to assess the degree of success of the project or programme.

Managing shared water resources as demands upon them increase

Increasingly, water resources development involves the use of river systems or groundwater shared by more than one user, region, state or country. It is worth noting that roughly one-third of developing countries are federations, and that federal and state authorities often do not see eye to eye. Even within one region different, often conflicting, interests may wish to develop water, for example, hydroelectric generation may compete with fisheries or irrigation needs. To develop such resources fully and to minimize harm, to the interests of any of the parties involved, to maintain, and where necessary improve, flows and quality may require interregional, interstate or international cooperation. Since the 1930s a number of countries have sought to use a *river basin planning approach* to resolve these problems (UN 1970; 1975; OAS 1978; Faniran 1980; Saha & Barrow 1981; Godana 1985: 169–249, 298–325).

The river basin can be developed and managed by a suitable team of specialists and administrators – a *river basin development commission or authority*. If the basin is shared by more than one region, state or country the commission or authority could be composed of representatives from the interested regions, states or countries. River basin development commissions can oversee water resources development and, if wished, pursue comprehensive or integrated planning and management using water resources development as a 'tool' to achieve much wider regional development and/or socio-economic goals (OAS 1978: 5).

In 1978 there were over 200 internationally shared river basins, a number of which are now managed by international river basin commissions (Fig. 2.8). For example, in West Africa there are four important river basin authorities: in Mali the Office du Niger; in Burkina Faso the Volta Valley Project; Chad and Nigeria manage the Chad River Basin Commission; and Senegal, Mali and Mauritania manage the Senegal River Valley Development Authority (Organization pour la Mise en Valeur du Fleuve Senegal). Some of these, and other international river basin authorities have been successful; however some, despite grandiose plans, seem to have gone only a little way towards achieving them.

Laws controlling the use of water resources shared by more than one country

Law is an instrument which can be used to smooth out conflicts of interest generated in the sharing of water resources. It also provides guidelines for ordering future conduct. Law can be determined by court action(s) which set a precedence that becomes a 'guideline' for future cases (the process of *common law* in the West). Law may also come from legislation where an administrative body, for example a government, passes a statute when it sees a need. In many countries a constitution affects water rights and water management, because it binds legislation and common law or its equivalent. The present international law concerning use of water resources is the product of centuries of endeavour towards formulating a set of substantive principles and procedural instruments to balance and harmonize divergent national interests. While there is no formally ratified rule of international law prescribing that a basin state must have the prior consent of other basin states in order to use and develop the waters of an international drainage basin within its own territory, there is a legal duty to give notice in the cases where such use is likely seriously to affect the rights or other interests of another basin state (Caponera 1983: 182).

The International Law Association has introduced the con-

Fig. 2.8 International river basins

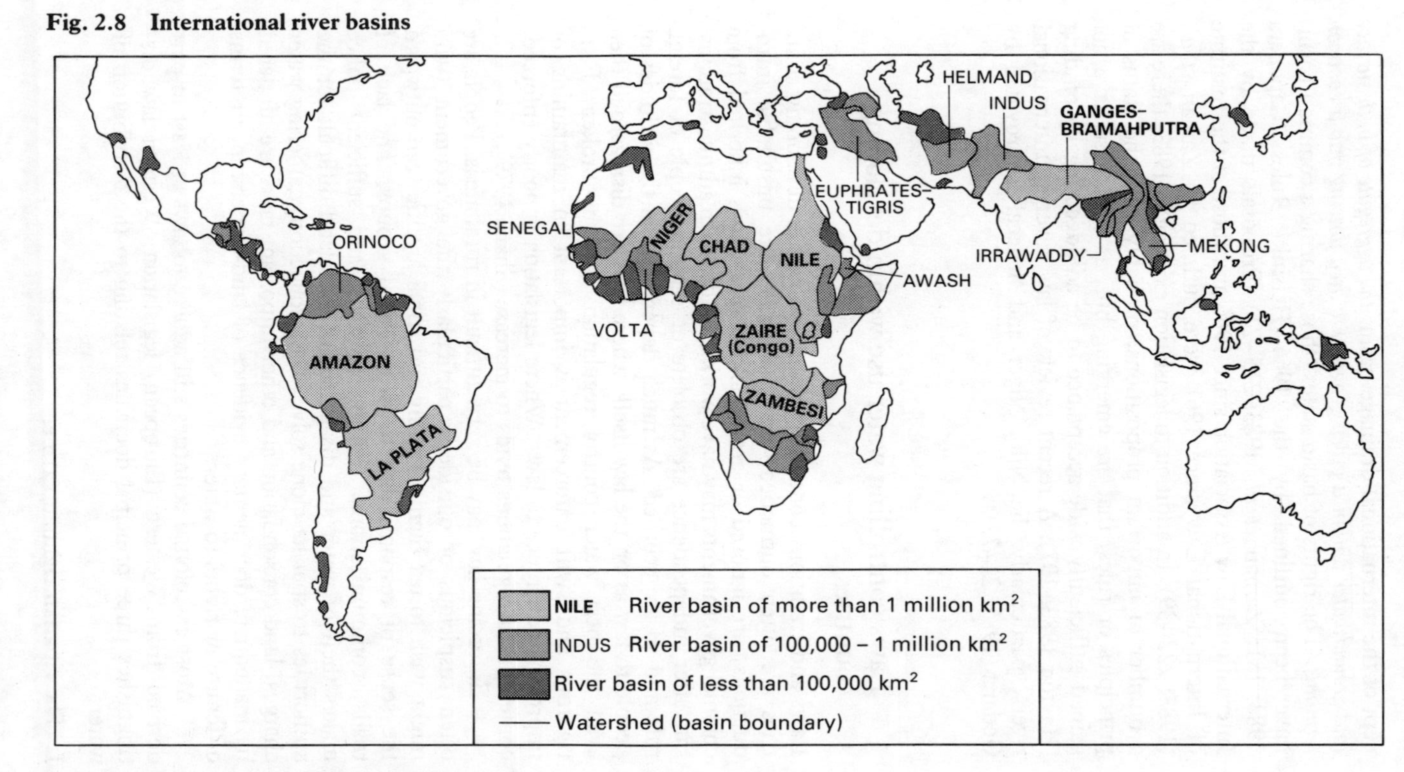

Source: IUCN, UNEP and WWF 1980 (Map 4 – no page number).

cept of the international drainage basin: *the aggregate of both surface and groundwater within a given geographic area flowing into a common terminus*. The rights of basin states (those sharing a common basin) have been outlined by the 1966 Helsinki Rules (Caponera 1983: 181; Zaman *et al.* 1983: 215–21). Proposals made by the International Law Association since 1966 and those of the Institute of International Law since 1961 were outlined by Zaman *et al.* (1983: 222–39). In addition to legislation existing in 1983, there are a number of important propositions, probably with the status of principles or rules, that are emerging but, as yet, these have not gained sufficiently wide acceptance to be adjudged legally binding (Hayton 1983: 197). A recent review of law covering international rivers, particularly the Nile, Niger and Senegal is provided by Godana (1985: 21–77).

Laws controlling water use within developing countries

Laws and traditions controlling water use within developing countries are often outmoded, are inadequate, are unsuitable introductions, are ignored or are unenforceable. Laws inherited from colonial governments may have worked under such administrations, but after independence are obsolete or remind people of foreign rule and so are scorned. As much depends upon the attitudes of people to law as on the law itself, where laws are disregarded then even the best water control regulations will be broken. Fair, rigorous and swift enforcement is important in maintaining or improving adherence to laws. Where legislation is poorly enforced, contempt for regulations tends to increase (Shane 1979).

Enforcing law may be very difficult in rural areas. People are often suspicious of outsiders, particularly the government (who might wish to tax them). Family and tribal loyalties can outweigh the sense of responsibility to wider social groups. The lack of roads, communications, lighting at night and sufficient police makes theft of water and disagreement over land difficult for the authorities to spot let alone solve. Data collection, gathering water charges, land consolidation and other important tasks are hindered by inadequate documentary evidence of land-ownership, or terms of tenure or rights to water.

Most ex-colonial countries still adhere to laws at least in part derived from Western (European) legislation. Western law distinguishes three principal doctrines relating to the distribution of water:

1. The riparian rights doctrine.

2. The correlative rights doctrine.
3. The prior appropriation rights doctrine.

Table 2.1 lists the main features of the riparian rights and prior appropriation rights doctrines.

Riparian rights involve basically the rights to use water as a consequence of owning land bordering a watercourse; the doctrine has spread from England to the USA and Australia and from France to parts of Africa. The major features of riparian rights are that: (1) it should give equal rights of use to owners of land which borders on or touches a stream or across which a stream flows; (2) a riparian right is 'attached' to land-ownership – a user can take up the right to use water at any time even if he/she had not done so before and to do so affects existing users. The right is 'usufructuary' i.e. the owner does not own the water (the resource itself can belong to the state or some other authority) only the right to use it. Thus one who enjoys riparian rights should receive flows from upstream landowners(s) without material change in quality or quantity and should ensure that downstream owners enjoy the same. Use of water for anything other than household consumption plus a few livestock is likely to be regarded as excessive, and nowadays in most countries would be controlled – licensed by the state.

Table 2.1 Riparian and prior appropriation rights to water

Components	Riparian rights*	Prior appropriation rights†
Water ownership	Inherent in the land	Only from prior use or statutory rights
Location of land to which water rights accrue	Must adjoin stream	Need not adjoin, i.e. can be distant from stream
Use of water	Diversion for 'natural uses' only; non-consumptive	Consumptive: prior rights may claim whole
Amount of water used	No limit in theory	Fixed – by prior right – by licence
Duration of water right	Infinite; does not lapse if not used	Finite; lapses if not used

* Derived from British common law. † Derived from Roman or canon law.

Source: Heathcote 1983: 270 (Table 17.1) (Hamming 1958).

Riparian rights allow an owner to extract groundwater from under his land without restriction irrespective of its effect on others or on the conservation of groundwater resources. At its most basic the riparian rights doctrine is archaic and really suited only to humid lands where water is not especially scarce. Because riparian rights go to those who have the watercourse on their land, and those away from it have no rights, it can hinder water transfers (Trelease 1964; Rodda 1976: 315–30; Hodges 1977: 3; Anderson 1983: 228).

Some countries have adopted the correlative rights doctrine. This is basically a modification of riparian rights to include some consideration of parties other than the holder of the rights. The doctrine was developed in California after a 1903 court case and was an improvement on the situation that was then operative which allowed landowners to exploit groundwater with no consideration for others.

Under the correlative rights doctrine: (1) if groundwater demand exceeds supply, all overlying landowners must reduce use on a coequal basis; (2) where water supplies are in excess of reasonable needs of those overlying them, water may be put to non-overlying uses, i.e. piped or led away (Anderson 1983: 224). Where correlative rights are enforced, access to water can be restricted by the state in times of water shortage – a considerable improvement over the situation with riparian rights.

Prior appropriation rights may be summed up as 'first in time first in right'; the earliest appropriator has a right superior to later appropriators (i.e. historical precedence). Later appropriators get what is left by the earlier, and at times of low flow may be deprived. Under prior appropriation rights a person occupying land can take groundwater irrespective of others because water is regarded under the prior appropriation rights doctrine as a saleable commodity separate from the land. The doctrine is suited to situations where there are water shortages or if the water is to be diverted, for example, transfer from one river basin to another. Prior appropriation rights have been blamed for encouraging excessive use of water resources, because if flows are not being used, rights to it may be forfeited – in effect 'use it or lose it'.

Administrative disposition of water-use rights is tending to supersede the three doctrine just described. Under administrative disposition, water is controlled fully or partly by the state which then grants permits or licences to users. In some countries the state now controls all surface waters; for example, in Peru the 1969 General Water Law made water and the channels containing it, without exception, the property of the state, so there are now no individual owners of water in the country (Urbina 1975). Many writers, but by no means all, argue the best water resources management is exercised where water is the property of the state and all utilization is controlled by permit for beneficial use (Anderson 1983: 13–43 debates the advantages of private versus state ownership of water resources).

Many developing countries have non-Western legal systems. Some follow Islamic law and common sense born out of years of experience of using water in drylands. According to Islamic theory, water is divided into three categories: that coming from (1) rivers; (2) wells; and (3) springs. Rivers are further subdivided into greater and lesser rivers. Great rivers, such as the Tigris, the water of which is deemed sufficient for all the needs of cultivation and from which anyone can lead off water without much affecting neighbours, are owned in common by all Muslims. Lesser rivers are further divided into: (a) those which have sufficient water to be diverted without the need for storage dams to irrigate the land situated along their course (and from which canals can be led off to water more distant plots), provided such actions do not prejudice the position of lands situated along river banks; (b) those rivers across which barrages or dams have to be built. When the latter is the case, lands situated higher up the watercourse have a prior right to water. This prior right (chirb) makes sense because any other distribution would involve passing limited water supplies further and thereby incurring greater wastage through leakage, evaporation and theft. The amount of water users can extract from smaller rivers depends upon circumstances, local needs and custom. Islamic law regards canals dug to bring water to otherwise barren land as belonging to those who dug them (Caponera 1954; Spooner 1974: 46).

There are some developing countries with indigenous water laws; for example, in Indonesia traditional adat (law) is administered through village or 'parish' units (marga) which lease water rights to individuals (McAndrews & Chia Lin Sien 1979: 132). Early in European history legislation and rights were evolved to control the use of certain common resources and prevent damaging over-use. However, in many countries (especially in Africa) there is little or no legislation to control the development of common resources. Groundwaters in particular have proved difficult to manage in a number of countries largely because they are considered a common resource.

Land and water are often treated as separate 'entities'; ideally there should be integrated planning and management of land and water resources development. To achieve that, legislation controlling water resources development must be coordinated with laws relating to land tenure, land use and inheritance. Unfortunately, water resource managers often have little control over anything other than their water source(s) and supply system (Anderson 1983; Falkenmark 1984).

Hindrances to water resources development due to management faults

Choice of strategy, timing and administration of water resources development are more often than not subject to troublesome political and bureaucratic pressures. Such pressures may well cause more hindrance and be more difficult to overcome than environmental and technical problems.

Water resources planning and management has generally been dominated by politicians, economists and engineers, and 'success' has commonly been judged on economic or technological criteria. Prediction of environmental and socio-economic impacts may even be seen as the work of 'soft science' and therefore get minimal attention. Concern for the aptitude, attitudes and needs of the people affected by development has often been inadequate, yet consideration of such factors is critical if water resources development is to succeed.

Developing countries for a variety of reasons are frequently unable or unwilling to find suitable expertise for what have often been slow, costly and complex pre-project, in-project and post-project studies. Studies must be speeded up and funds must be made available to pay for them, and whenever possible indigenous planners and managers, rather than foreign consultants, should be used to do the job. Getting such personnel may not be easy; professionals in developing countries are often lost in a 'brain drain' to richer nations, and those who do remain tend to concentrate on 'safe' issues or prestigious issues which may not be very important or even relevant to their country's development (Myrdal 1968: 15, 939). A consequence of the shortage of skilled manpower in developing countries is that two problems commonly arise; firstly there is often inadequate data for the planner and manager, and secondly there is a shortage of trained personnel to interpret and use what is available. Cheap, rapid, reliable methods of data collection and project appraisal may help to overcome these difficulties.

Too often key management positions are held by persons without sufficient experience or who are in some other way unable to make satisfactory decisions. Even where managers may have the ability to make decisions they may not have the authority to do so – decisions are seldom made by officers on the spot who can act appropriately, sensibly and swiftly when action is needed. Too often general administrators and/or civil engineers who may be based in a city many kilometres distant make decisions and, frequently there is little consultation with agronomists, ecologists or social scientists (Bottrall 1981a: 6).

Management in many developing countries is not only blighted by lack of funding, it suffers from lack of security because of inadequate experienced manpower and the mandate to act without reference to superiors who are slow to respond. Bureaucrats in many developing countries have a rapid turnover: an election defeat or career ambitions frequently mean replacement of entire planning or management teams, or sudden abrupt changes in the terms of reference and policy guidelines. The time horizon for planners and managers of developing country projects and programmes is rarely greater than five years, commonly much less. Inevitably planning, implementation and management reflect this uncertainty and need for haste. For example, in 1970 the Mahaweli Development Programme was begun in Sri Lanka. It is one of the world's biggest hydroelectricity and irrigation schemes and originally it was envisaged that the five dams and associated works would be constructed over a thirty-year period. However, in 1977 a change of government took place and the decision was made shortly after to accelerate the implementation to completion in six years – a reduction of seventeen years – despite dire warnings against such haste from various well-informed bodies and experts, including the World Bank and the International Monetary Fund (Madeley 1983: 18).

Projects which are implemented are often large and inflexible and planners and managers can do little to adapt to unforeseen difficulties. As funds commonly run out on completion of the infrastructure there is little available to pay for the rectification of faults. Furthermore, expatriate expertise has by this time left the country. Money can often be found for infrastructure, but for less 'concrete' things, for example training managers, appraising management strategies and developing new and improved ones, there may be a reluctance to provide funds. Engineering structures, for example, a new dam or canal, give physical testimony to a backer's support, but even very successful management has far less publicity value. It is hardly suprising, therefore, that management gets neglected.

A wise course of action before the implementation of a large and costly project or programme would be to proceed with a pilot project/programme. However, pilot projects/programmes have been rather unpopular in developing countries despite their value in 'trouble-shooting', training personnel and gathering data. The Kafu Project in what is now Zimbabwe, the Medjerda Project

(Tunisia) and the Tana Basin Project (Kenya) are some of the relatively few cases where pilot projects were used prior to the main implementation. Pilot studies are especially important if new techniques are being introduced to a region, for example, microcatchments should be tested for at least two years to see how they fit local rainfall, slope and soil conditions before widespread installation is pursued.

From farmer up to senior agricultural administrators, irrigation planners and water resource managers there is a great need to stimulate more experimentation and to draw away from the presently widespread tendency towards unquestioning adoption of Western ideas or resignation to accept the *status quo*. Cautious innovation – the adoption of, and where need be, the adaptation of new ideas – should be encouraged.

Even the best, most intelligent development/management strategy can suddenly be upset by external forces. For example, irrigation projects are now far more expensive to run since the 1972–74 and 1979 OPEC petroleum price increases than could ever have been envisaged by planners before 1972. Few could have foreseen the recent deceleration in growth of the world economy and with it reduced demand for many primary agricultural commodities, yet this has severely hit the profits of many irrigation projects built since the 1960s, which would otherwise have been quite viable.

In large part because of the recently apparent external difficulties, those planning water resources and agricultural development are now placing greater emphasis on less expensive, easily available input, 'appropriate technology' projects which produce a greater range of products (rather than one, possibly vulnerable, export crop) or food. Ironically, one of the major threats to irrigation development from the mid 1970s to the present – the OPEC petroleum price rises – has forced planners and managers into promoting more sensible projects and programmes.

The nature, causes and possible relief of water resources problems in developing countries are summarized in Fig. 2.9. Three main problems are recognized (shown on the left of the figure: data incomplete/uneven; inadequate use of data; data unsatisfactory when applied to water resources development. The various causes of each of these problems are suggested in the middle of the figure, and the possible means of reducing the problems are suggested on the right.

More than 85 per cent of the world's cultivated land was used for rainfed cropping in 1980, and only about 12 per cent was irrigated; by the end of the century irrigation will still only make up about 13 per cent of the world's total cultivated area (Ambroggi

Fig. 2.9 Tropical water resource development: problems, causes and possible relief

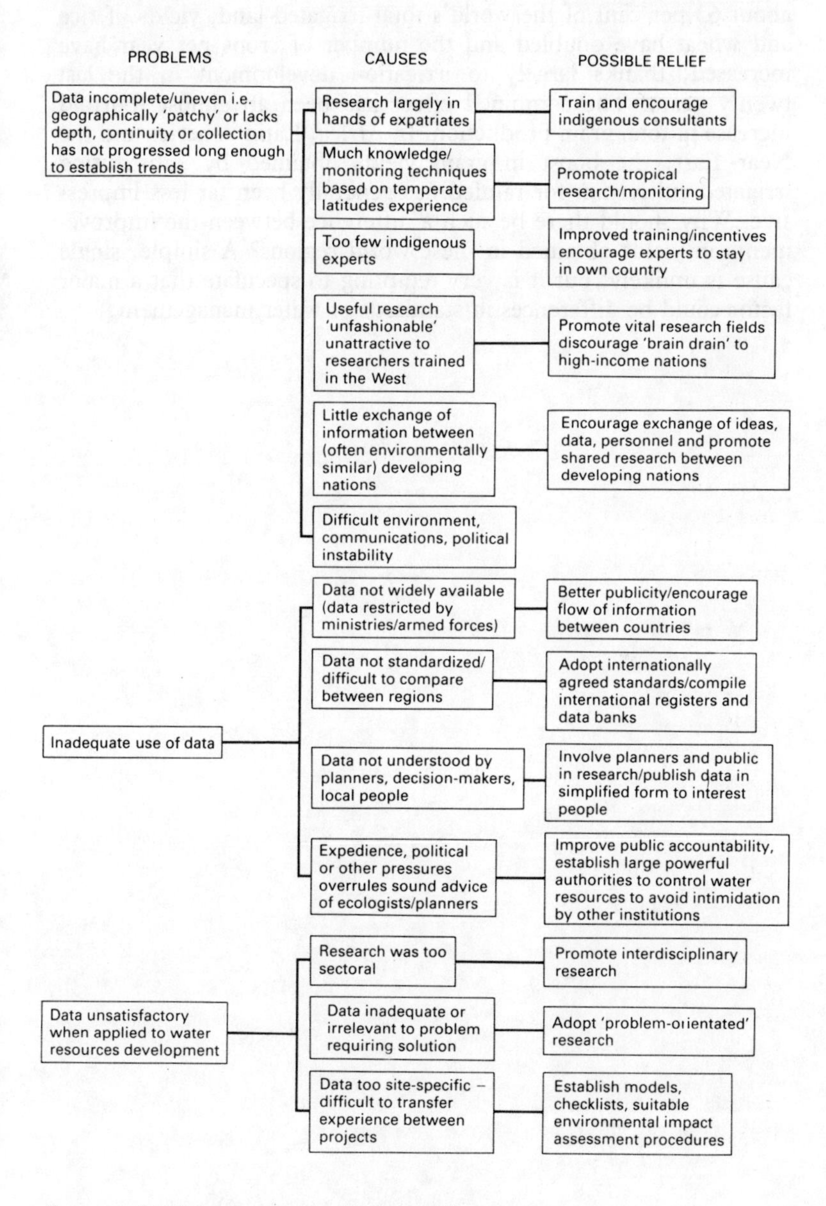
PROBLEMS
CAUSES
POSSIBLE RELIEF
Data incomplete/uneven i.e. geographically 'patchy' or lacks depth, continuity or collection has not progressed long enough to establish trends
Research largely in hands of expatriates
Train and encourage indigenous consultants
Much knowledge/ monitoring techniques based on temperate latitudes experience
Promote tropical research/monitoring
Too few indigenous experts
Improved training/incentives encourage experts to stay in own country
Useful research 'unfashionable' unattractive to researchers trained in the West
Promote vital research fields discourage 'brain drain' to high-income nations
Little exchange of information between (often environmentally similar) developing nations
Encourage exchange of ideas, data, personnel and promote shared research between developing nations
Difficult environment, communications, political instability
Data not widely available (data restricted by ministries/armed forces)
Better publicity/encourage flow of information between countries
Data not standardized/ difficult to compare between regions
Adopt internationally agreed standards/compile international registers and data banks
Inadequate use of data
Data not understood by planners, decision-makers, local people
Involve planners and public in research/publish data in simplified form to interest people
Expedience, political or other pressures overrules sound advice of ecologists/planners
Improve public accountability, establish large powerful authorities to control water resources to avoid intimidation by other institutions
Research was too sectoral
Promote interdisciplinary research
Data unsatisfactory when applied to water resources development
Data inadequate or irrelevant to problem requiring solution
Adopt 'problem-orientated' research
Data too site-specific – difficult to transfer experience between projects
Establish models, checklists, suitable environmental impact assessment procedures

Source: article by author Barrow 1983: 440 (Fig. 13.1).

1980: 41). The relatively limited extent of irrigated agriculture belies its real or potential importance. In South Asia, which has about 63 per cent of the world's total irrigated land, yields of rice and wheat have doubled and the number of crops per year have increased, thanks largely to irrigation development in the last twenty years. The combined effect has been an almost fourfold increase in total grain production. In Africa, Latin America and the Near East, the boost in grain yields obtained by substituting irrigated cultivation for rainfed has generally been far less impressive. Why should there be such a difference between the improvements in yields obtained in these world regions? A simple, single cause is unlikely, but it is very tempting to speculate that a major factor could be differences in standards of water management.

Chapter 3

Water resources in the tropics

The hydrological cycle and its variability

There are regions where it is reasonably certain that the right amount of suitable quality water will be available at the right place at the right time. However, this is seldom the case within the tropics and is becoming even less common as agricultural, industrial and domestic demands for water supply increase, and as deforestation, land development and possibly climatic fluctuations alter water supply availability.

Tropical rainfall and evaporation

In Chapter 1 the tropics were broadly subdivided on the basis of moisture availability into: humid, where precipitation is usually abundant; subhumid, where precipitation is sometimes problematical; and tropical drylands where it is very often scarce. Excess water can be as much as problem for agriculture as shortage; in the humid tropics disposal of overland flow is periodically necessary, often almost year-round. In the subhumid tropics safe disposal of overland flow can be a problem, mainly in the wet season(s), but also during intense storms which may punctuate the dry season(s). Even in the tropical drylands there can be intense storms, and with sparse vegetation cover overland flow can be considerable. In many parts of the tropics floodwaters are an essential element in the agriculture, in some they are a threat.

Over a large part of the tropics precipitation varies from season to season and from year to year. The difficulties caused by temporal variation are amplified by other rainfall characteristics. In particular, rainfall intensities tend to be high in the tropics, and a considerable proportion of the rain is concentrated into a comparatively small number of heavy storms. Much of this type of rainfall does not become available to agriculture. Instead of contributing to soil moisture reserves which can be used by plants during dry weather, stormwater runs over the ground surface causing soil erosion and flooding. Rainfed crops can be drastically affected and the timing and depth of flooding also varies affecting floodwater-dependent cultivation. The problems caused by rainfall variability are compounded by high rates of evaporation and transpiration which can cause a soil to dry out very soon after rainfall, especially where soils are free-draining.

The decision to adopt a new or improved crop variety or cultivation practice is commonly made after referring to a locality's mean annual rainfall statistics. This can be a mistake because a high mean annual rainfall, well above theoretical crop needs, can be inadequate if it is comprised of irregular, infrequent downpours. Even short-lived droughts if they coincide with some critical stage in crop development or vital cultivation practice, especially where soils are free-draining, can be disastrous. Although at first inspection average annual rainfall over wide areas of the tropics may seem more than adequate for cultivation or safe grazing, in reality it may be too erratic or ill-timed.

Various researchers have developed 'moisture indexes' to try to assess whether precipitation can adequately support crop growth and to calculate the water supply requirements of irrigated crops; however, none are fully satisfactory. The best known are those of Köppen, Thornthwaite and Meigs (for a review of these see: Barkley in Hall *et al.* 1979: 73–97). Meigs's refinement of Thornthwaite's index which relates potential evapotranspiration to actual transpiration may be used to subdivide the tropics (Grigg 1970: 158). If evapotranspiration balances actual transpiration (a Meigs index of 0) active crop growth should be possible. An index of 0 to −20 indicates subhumidity, −20 to −40 semi-aridity, −40 to −56 aridity and below −57 extreme aridity (Heathcote 1983: 16).

Average annual rainfall values and length of dry or wet seasons, being based on measures of central tendency, have limited value. Tinker (1977: 583) commented: '. . . average is an almost meaningless term in the Sahel where it may not rain significantly for three years and then rain torrentially in one place, while five kilometres away little falls at all.'

Rainfall variability may be expressed as the ratio of maximum annual rainfall to minimum annual rainfall; as a percentage deviation of annual rainfall from mean rainfall; or as a ratio between standard deviation and the mean rainfall (Morales 1977). If a region's rainfall data is heavily skewed, which is commonly the case in the subhumid tropics and tropical drylands, rather than normally distributed throughout the year, the median (which is unaffected by extreme values) might be a more useful statistic for assessing precipitation. If the mean must be used it should be accompanied by an indication of variability.

Irrigation planners often rely on plotting annual rainfall against the probability of occurrence. This can be related to the minimum estimated growing season water requirement of the crop(s) under consideration, and to the ability/willingness of farmers to bear a crop failure. Some agricultural planners prefer to consider deciles of rainfall, so that they can judge in how many years out of ten the rainfall exceeds the critical lower limit for successful growth of the crop they are considering, and how often they may expect severe drought or flooding.

Interception and depression storage

Before discussing moisture infiltration and storage or runoff it is necessary to consider *interception* and *depression storage* (see Fig. 3.1). The amount of precipitation actually reaching the ground surface largely depends upon the nature and density of the vegetation cover (if any), and on the type of buildings, roads and other structures if there has been construction in the locality. Whether natural or man-made the cover intercepts part of the falling precipitation and temporarily stores it on its surfaces (leaves, branches, roofs, pavements and so on). Initially the interception capacity of vegetation which is dry is quite high, and a large part of the precipitation is prevented from reaching the ground. In due course, provided the precipitation exceeds the rate of evaporation, sufficient water accumulates to overcome the surface tension by which it is held and, from that point further additions of moisture are almost completely balanced by water droplets falling from leaves or running down stems (Ward 1967: 63). When precipitation is light and brief and the rate of evaporation is high, little or none of the moisture may reach the ground. When showers are very light and evaporation rates are high, it may require only very sparse vegetation cover to cause total interception losses. Where there is a thick cover of vegetation even quite heavy rainfall may yield little surface flow and do little to recharge groundwater. The ability to intercept moisture varies with plant variety; some shed water fast, some retain it. Atmospheric humidity, amount of sunlight and windspeed also conspire to make interception losses a rather variable phenomenon.

Sometimes vegetation intercepts moisture which might otherwise not be precipitated in that locality. Some tropical coastal vegetation and highland forests survive largely on intercepted mist and cloud, in some cases they may even exceed their interception capacity and contribute to streamflow and groundwater recharge. Urban areas and surfaced highways can either intercept virtually 100 per cent of precipitation if they have no outlet to the ground,

Fig. 3.1 The water–soil–crop system

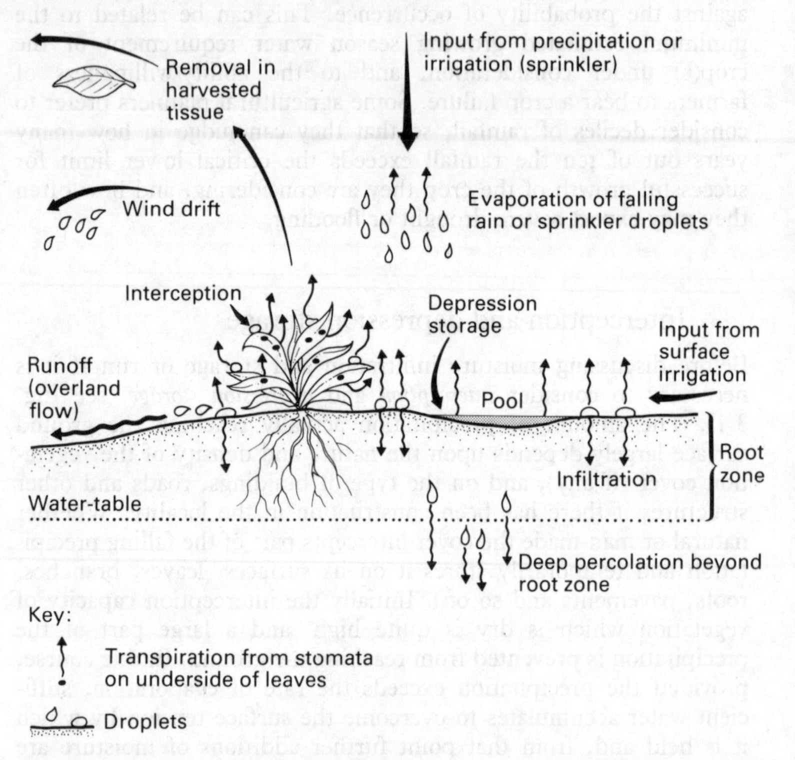

or if there is a drainage system water may be channelled to rivers far more quickly than if the landscape were unmodified. In general though, if precipitation is heavy or light but sustained, water finally begins to reach the ground. Depression storage (or surface storage as it is sometimes called) is that water retained in hollows and depressions in the ground surface during and after rainfall. This water is either evaporated back to the atmosphere, used by vegetation or infiltrates into the soil. None of it appears as runoff. Low-intensity rainfall may simply lead to depression storage, possibly leading to some infiltration of moisture into the soil, but once the accumulation of moisture exceeds the infiltration capacity of the soil, the depressions fill up and a continuous undelayed movement of water takes place over the ground surface to streams (*overland flow*).

Infiltration

The rate of entry of water into a soil under the action of gravity

Table 3.1 Typical infiltration rates for various soil textures/particle size

Soil	Infiltration rate (mm/h)
Clay	1–5
Clay loam	5–10
Silt loam	10–20
Sand loam	20–30
Sand	30–100

Source: Stern 1979: 88 (Table 24).

(*infiltration* or surface intake) is greatest when the soil is dry at the start of precipitation or earliest irrigation applications, and usually decreases, sometimes considerably, as the topsoil becomes saturated. Infiltration stabilizes if the precipitation or irrigation application is sustained at a roughly constant rate for a given soil – the *infiltration rate* (basic intake rate), which indicates how much water a soil can soak up in a given period of time. Typical infiltration rates for various soils are given in Table 3.1.

The infiltration rate determines how long it will take rainfall or irrigation to water a plot of land, for example, a soil with a 20 mm/h infiltration rate will take four hours to absorb 80 mm of water per unit area. When infiltration is slow, rainfall or water applied to the land may be wasted by flowing away (possibly causing soil erosion as it does so), or forms puddles from which evaporation losses are high. Some soils, particularly those clay soils which crack when dry, initially allow rapid infiltration, but this soon decreases to stabilize at quite a slow rate (because soil cracks have sealed or soil particles have swollen after wetting).

Soil moisture

Between the solid particles of a soil are a complex of interstitial channels and voids which provide space for moisture, air and soil organisms. These interstitial spaces may be temporarily or permanently filled with water during and for a while after precipitation, flooding or irrigation; when they are filled, the soil is said to be *saturated*. The maximum amount of water a soil can hold when saturated is known a its *saturation capacity* (in actuality it is rare for all soil space to be filled with water, as some air is inevitably trapped, Wiesner 1970: 4; Stern 1979: 83). Where drainage is impeded leading to saturation a soil may be said to be *waterlogged*.

Soil moisture (i.e. what is present above the *water-table*, see Fig. 3.2) can be divided into three categories (this is a simple but rather dated classification): *gravity*, *capillary* and *hygroscopic*. Gravity (or vadose) water can remain in the soil above the water-table for only a short time before it drains away under the force of gravity; in sandy soils this may take a day or less, in clay soils three or four days. Even when drained of water the spaces in soil still contain considerable amounts of moisture. Capillary water is held by forces of surface tension around soil particles and in smaller pores and fissures. There is also water vapour in the 'empty' voids and a film of hygroscopic water adhering to soil particles by molecular attraction (see Table 3.2). Capillary water can 'creep'

Fig. 3.2 Soil moisture

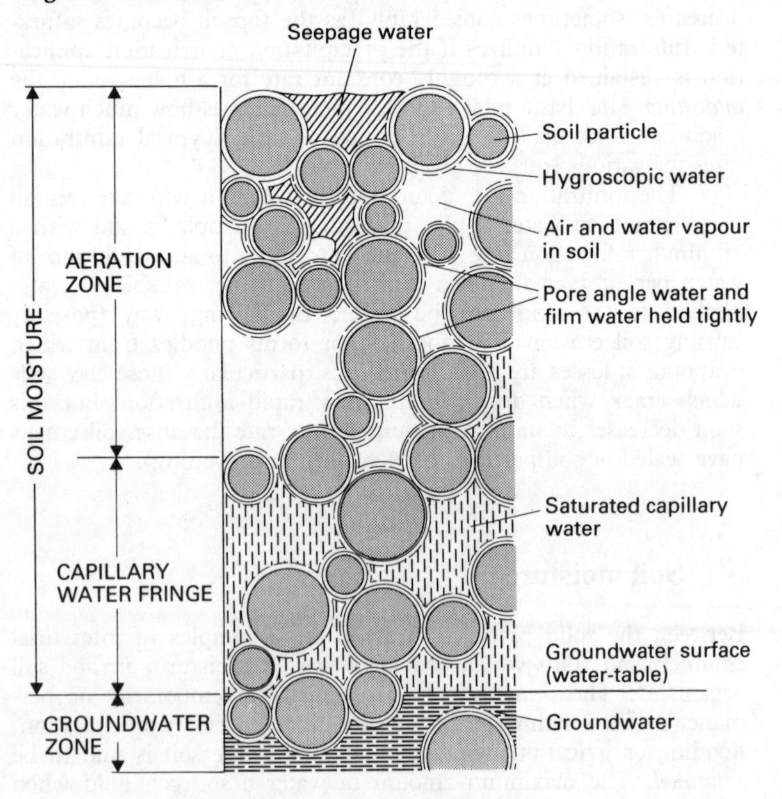

Source: based on Wiesner 1970: 39 (Fig. 8).

Table 3.2 Available moisture for various soil textures/particle sizes at field capacity

Soil type	Available water	
	%	mm/m
Fine sand	2–3	30–50
Sandy loam	3–6	40–100
Silt loam	6–8	60–120
Clay loam	8–14	90–210
Clay	13–20	190–300

Source: Stern 1979: 85 (Table 21).

above the water-table to a height determined by the texture and composition of the soil. The zone between the water-table and the height to which capillary rise occurs is known as the *capillary fringe* (or zone). When the water-table is shallow the capillary fringe may be bounded by the soil surface and an upward movement of water will take place as evaporation removes what reaches the surface. This upward movement of water may carry salts from the groundwater and/or leach salts out of the soil itself, and these are concentrated at the soil surface. In practice the capillary fringe is likely to be of uneven depth as the soil texture and composition is seldom uniform; any wetting of the ground surface or salt deposition is therefore likely to be uneven and patchy even on a well-levelled field.

Because *hygroscopic* (pellicular) moisture is firmly held to the surface of soil particles, plants may have difficulty using it, some *xerophytes* (plants able to withstand arid conditions) might, however, draw upon it *in extremis* (Wiesner 1970: 38).

It is usual to refer to the maximum amount of water remaining after gravity water has drained away as the *field capacity* of the soil. A typical definition of field capacity would be: *the moisture content of a deep, permeable, well-drained soil a few hours to several days after a thorough wetting* (usually field capacity is expressed in millimetres) (Kovda *et al.* 1973: 210). As a soil constantly drains – and soils differ in the rate at which this happens there – is not really a constant amount of water held against gravity. To attempt to standardize definitions the field capacity is nowadays usually defined as the *quantity of water held at a particular suction pressure forty-eight hours after wetting* (Carruthers & Clark 1981: 81).

Groundwater

There is probably about 3,000 times as much water stored as groundwater within 800 m of the Earth's surface as in all the world's rivers and streams (Arnon 1972: 146). *Groundwater* may be defined as water which accumulates in porous rock at moderate depth, its upper surface being the water-table. Commonly, groundwater is at a pressure greater than atmospheric pressure. Most groundwater is derived from precipitation (and is termed *meteoric water*); some has been trapped in rock since that rock was formed and is known as *connate water*. Frequently the groundwater has a lower limit, a stratum of dense rock which hinders further downward seepage. Alternatively the water-bearing rock (*aquifer*) may be compressed with depth so that at a level depending on the resistance of the local rock to compression there are no voids for water to occupy (Bouwer 1978: 2). *Soil moisture* (but not true groundwater) can be held above the water-table against gravity by capillary force; the quantity held and the distance it rises above the groundwater depends upon the rock/soil texture (Fig. 3.3). Water-tables are rarely static. Usually they rise and fall seasonally or as a result of withdrawals of water from wells or boreholes. Groundwater is the source of water in wells and springs, and entering streams it provides the stable *base flow* which largely maintains the supply of water in rainless periods.

Plant roots require oxygen as well as water, consequently most plants root between the ground surface and the water-table in the aerated, water-vapour-rich but unsaturated, vadose water *root zone*. Should the water-table rise, saturating this zone for any significant length of time, plant roots and soil organisms will be damaged or killed. If the soil permits capillary rise there is a very great danger when groundwater lies near the surface that an upward movement of water will deposit salts in the root zone or on the soil surface (salinization).

Groundwater storage, recharge and movement

Aquifers (rock formations which store, transmit and yield water) may be divided into:

1. Unconsolidated rocks – sands and gravels (these predominate);
2. Open-structured rocks – sandstones, breccias, conglomerates, which have structures permitting water storage and movement;
3. Limestones which store water and allow it to flow in fissures and joints;
4. Shattered crystalline rocks (and other broken-up consolidated rocks).

Fig. 3.3 Occurrence and distribution of subsurface water

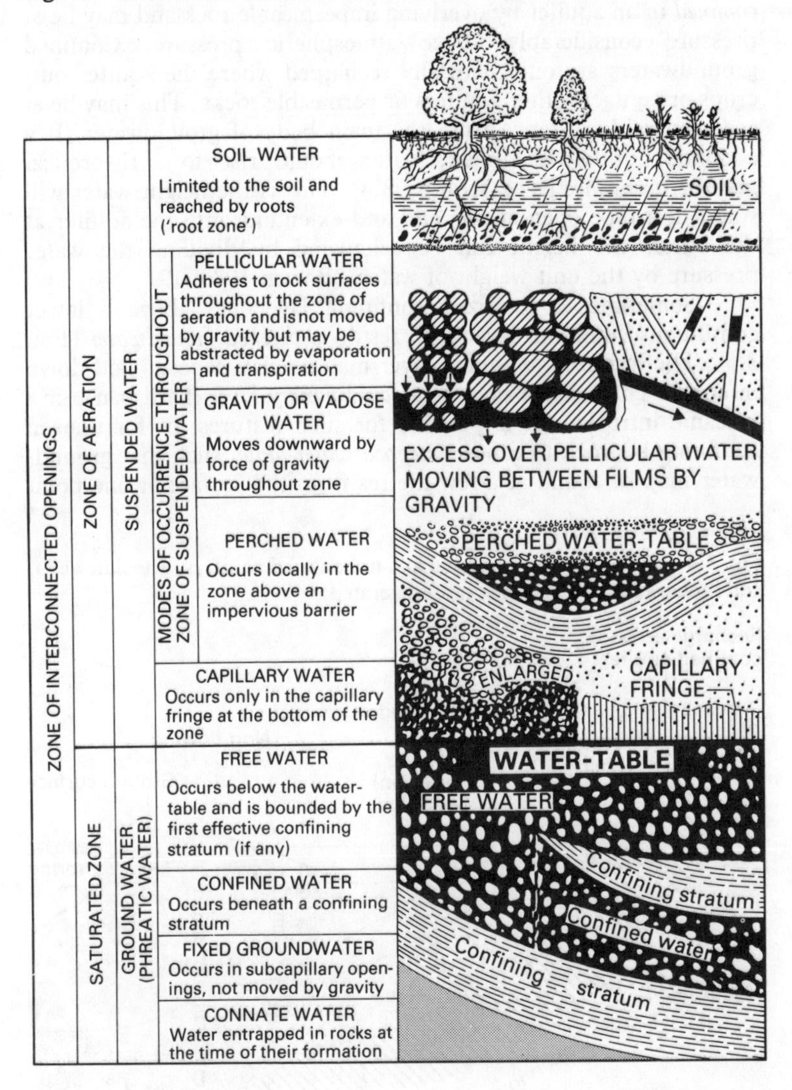

ZONE OF INTERCONNECTED OPENINGS
ZONE OF AERATION
SUSPENDED WATER
MODES OF OCCURRENCE THROUGHOUT ZONE OF SUSPENDED WATER
SOIL WATER
Limited to the soil and reached by roots ('root zone')
PELLICULAR WATER
Adheres to rock surfaces throughout the zone of aeration and is not moved by gravity but may be abstracted by evaporation and transpiration
GRAVITY OR VADOSE WATER
Moves downward by force of gravity throughout zone
PERCHED WATER
Occurs locally in the zone above an impervious barrier
CAPILLARY WATER
Occurs only in the capillary fringe at the bottom of the zone
SATURATED ZONE
GROUND WATER (PHREATIC WATER)
FREE WATER
Occurs below the water-table and is bounded by the first effective confining stratum (if any)
CONFINED WATER
Occurs beneath a confining stratum
FIXED GROUNDWATER
Occurs in subcapillary openings, not moved by gravity
CONNATE WATER
Water entrapped in rocks at the time of their formation
SOIL
EXCESS OVER PELLICULAR WATER MOVING BETWEEN FILMS BY GRAVITY
PERCHED WATER-TABLE
ENLARGED
CAPILLARY FRINGE
WATER-TABLE
FREE WATER
Confining stratum
Confined water
Confining stratum

Source: Tolman 1937: 39.

Where an aquifer is *unconfined* by overlying rock, the water-table is free to rise and fall as the stored water volume changes (under such circumstances groundwater may be described as *free groundwater*). Such unconfined groundwater can be locally recharged by precipitation, seepage from streams, canals, reservoirs, floods, or by deliberate recharge. Groundwater may, however, be *confined* in an aquifer by overlying impermeable rock and may be at pressure considerably above atmospheric pressure. Confined groundwaters are only naturally recharged where the aquifer outcrops or contacts other aquifers or permeable rocks. This may be at a considerable distance from the main body of groundwater. If a confined aquifer is tapped, water should rise to a theoretical *piezometric surface* level (see Fig. 3.4). This is the height water will stand in a pipe open at one end and extending into the aquifer at the other (the height can be calculated by dividing the water pressure by the unit weight of water) (Bowen 1980: 19).

Sometimes a localized confining layer lies above a lower continuous confining layer, the result is *perched groundwater* (Fig. 3.5). The localized confining layer may be clay deposits laid down in stream channels, it may be a *concretionary layer* (hard pan), or a volcanic intrusion. It is possible for such features, or for uneven confining bedrock to cause a *stepped water-table*, such that groundwater may be found at a few metres from the surface at one point

Fig. 3.4 Groundwater: confined and unconfined aquifers, artesian wells and recharge area (vertical scale exaggerated)

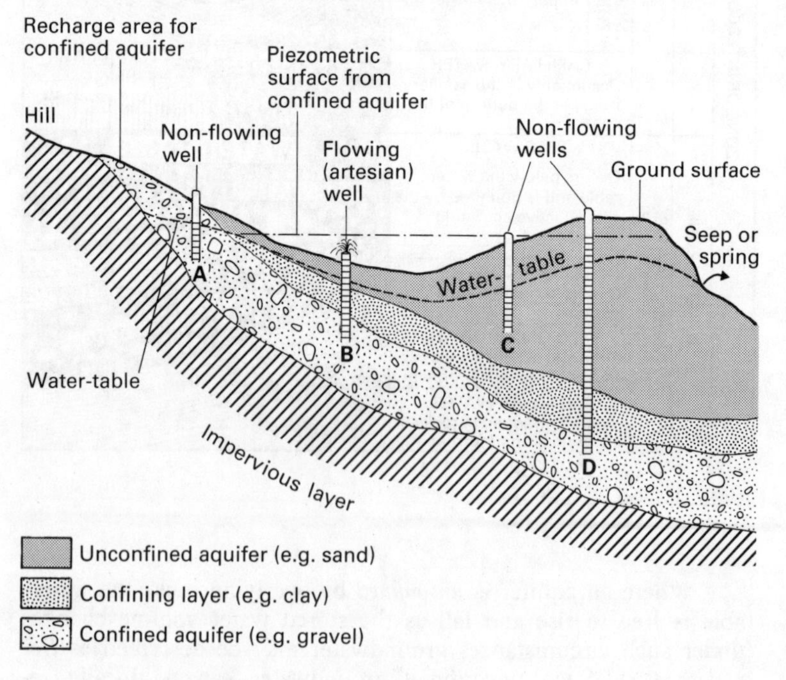

Source: based on Israelsen & Hansen 1962: 43.

Fig. 3.5 Perched groundwater and variation in depth of water-table below ground surface (a) *Perched groundwater* above an impervious clay layer deposited in a stream channel and buried. (b) *Perched groundwater* above impervious clay lenses (typically deposited on the bed of a seasonal lagoon) buried in alluvial deposits. (c) *Variation in depth of water-table below ground surface* due to uneven impervious bedrock.

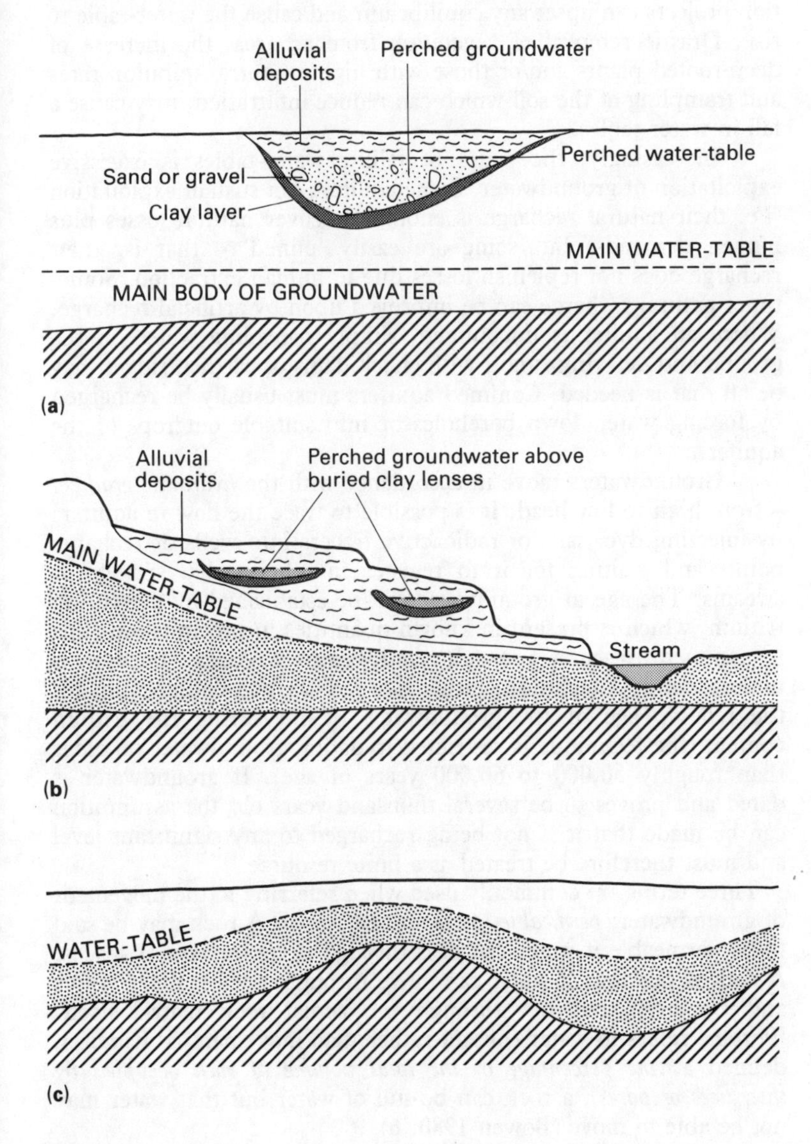

Sources: (a) and (b) very roughly based on Ward 1967: 247 [Fig. 7.3]; (c) author.

say and at 50 m elsewhere, even though the ground surface is generally level.

Shallow aquifers are often roughly in balance with recharge more or less equivalent to losses through drainage and withdrawal by deep-rooted plants. Removal of deep-rooted vegetation on a sufficiently large-scale, seepage from reservoirs, canals and irrigation projects can upset any equilibrium and cause the water-table to rise. Drastic removal of vegetation from an area, the increase of deep-rooted plants and/or those with high evapotranspiration rates and trampling of the soil which can reduce infiltration, may cause a fall in water-tables.

Increasingly, the cause of falling water-tables is excessive exploitation of groundwater. Some aquifers can sustain exploitation (i.e. their natural recharge is enough to cover natural losses plus human demands) but some are easily 'mined' – that is, their recharge does not replenish losses due to human extraction. Sometimes natural recharge can be improved upon by artificial recharge. Where an aquifer is unconfined, simply spreading water on the ground and delaying runoff with check-dams or earthen ridges may be all that is needed. Confined aquifers must usually be recharged by forcing water down boreholes or into suitable outcrops of the aquifer.

Groundwaters move in accordance with the *hydraulic gradient* – from high to low head. It is possible to trace the flow in aquifers by injecting dye, salts or radioactive tracers into wells or recharge points and waiting for it to register in other wells, springs or streams. The age of groundwater can be established by measuring tritium, which is present in known quantities in precipitation (this is assumed to be a constant) and begins to decay once it is underground. A technique rather like radiocarbon dating estimates the relative age of the groundwater by measuring the tritium content (the technique is restricted to dating groundwater no older than roughly 50,000 to 60,000 years of age). If groundwater is dated and proves to be several thousand years old the assumption can be made that it is not being recharged to any significant level and must therefore be treated as a finite resource.

Three terms are commonly used when referring to the movement of groundwater: *permeable, pervious* and *porous*. A rock may be said to be permeable if water can pass through it via interstitial spaces. A rock is pervious if water can move through it via joints, fissures and fractures, but not through the actual rock matrix itself. 'Porous' refers to water content rather than movement (it has been defined as *the percentage of the total volume of rock occupied by interstices or pores*); a rock can be full of water but that water may not be able to move (Bowen 1980: 6).

The availability of groundwater

Before groundwater can be exploited it first has to be found and, if it is to be rationally used, its extent, quality and rate of renewal must be assessed. So far the exploration for and mapping of developing countries' groundwaters has been patchy: a review of the efforts by the United Nations (UN) to rectify this was presented by Taylor (1979).

The mapping of groundwater may be either of *proven* (carefully researched and assessed) or *estimated* reserves. Springs, seeps, existing wells, mines or boreholes sunk in search of oil, coal, metals or for general geological survey purposes often indicate the presence of groundwater, but more detailed geophysical and geological studies are needed to provide useful estimates of the extent and character of reserves. Water-divining (*dowsing*) has been traditionally used in many countries and, despite much official scepticism, still helps to site wells (Bowen 1980: 137). In alluvial lowlands it is quite common to find former stream channels with sand or gravel infill buried beneath less permeable deposits. Often water still moves through such channels which may be recognized by remote sensing or ground surveys and tapped by shallow wells or boreholes.

To make rational exploitation of groundwater resources it is necessary to know the *yield* (the volume of water discharged per unit time either by pumping a well or borehole or by natural flow into a well, sump or springs). What is termed *safe yield* is the maximum rate of withdrawal which had no undesirable effect on the groundwater or aquifer. Figure 3.4 illustrates four wells/boreholes and a spring: wells A and C do not flow naturally (they are *gravity* or *water-table wells*), B and D are *artesian* or *confined groundwater wells* and the water they produce rises above the local water-table (in the case of B exploitation would require no pumping). The rate at which groundwater can be exploited or replenished depends on the type and size of well or spring and the rate at which water can move through the aquifer to the well(s) or spring(s). When a well or borehole is left sufficiently long without water being drawn from it inflow will fill it until it stabilizes at what is known as the *static water level*. Withdrawal of water will usually lower that level, and the distance between static water level and the level after an extended period of withdrawal is termed *drawdown*. Around a well or borehole or group of wells/boreholes from which water has been withdrawn for an extended period, the water-table will be distorted – depressed downward – to form a funnel-shaped *drawdown cone* centered on the well(s)/borehole(s). Diameter, depth and length of the well/borehole passing through the aquifer(s) (the percolation rate of which is also important) and the suction to which a pump can subject the walls of the well(s)/borehole(s) determine yield (initially, when withdrawal begins there will also be some water stored in the well).

Runoff

Runoff (streamflow, stream or river discharge, or catchment yield) is the movement of water over the surface of the earth in response to gravity, in channels which range from that containing the smallest ill-defined trickle to those containing rivers as large as the Congo or Amazon. In general, runoff represents the excess of precipitation over evapotranspiration when allowance is made for storage in or under the ground surface. Much of the excess precipitation comprising runoff reaches streams and rivers only gradually, having been intercepted and delayed by vegetation, loose debris or stored as groundwater. In effect runoff in channels is 'smoothed' compared to the irregular receipt of precipitation over a region. Runoff is normally expressed as a volume per unit of time. Commonly used units are: the *cumec* (1 m^3/sec), *cusec* (1 ft^3/sec), *millions of gallons per day (mgd)*; or *unit depth on a catchment area per day, month or year* (Ward 1967: 309).

Runoff may be subdivided into several component parts: *direct precipitation* on to stream channels, swamps or other water bodies (this is seldom large and is difficult to estimate so is often ignored in runoff calculations); *surface runoff* comprises water that finds its way over the ground either as *sheet flow* or in trickles and streams to main stream courses. The term *overland flow* is sometimes used as an alternative to surface runoff, but is better applied to sheet flow of water from the point at which precipitation makes contact with the ground to the point at which it enters a defined channel. *Interflow* takes place when part of the precipitation infiltrates into the soil, meets a relatively impermeable layer and spreads until it escapes into a stream channel. Interflow moves more slowly than overland flow or surface runoff and is likely to be greatest where thin, permeable soil overlies impermeable rock or hardpan layers. Where such soil conditions exist, interflow can be an important component of the total runoff. *Groundwater runoff* (base flow, dry-weather flow or effluent seepage) is the contribution to runoff made by groundwater when it reaches stream channels. It is the overflow from the slowly changing reservoir of moisture below ground. Groundwater runoff helps to smooth out total runoff regimes by contributing water when there has been little precipitation. In some parts of the tropics, streams originate from highland snowfields or glaciers, and have a *snowmelt component*. Sometimes snow and ice melt slowly and like groundwater runoff helps maintain runoff when there is little precipitation. Sometimes, however, there is rapid melting and the streams it feeds have a brief period of high flow.

When precipitation takes place over a catchment which is not already thoroughly wetted by previous storms, interception, and depression storage and infiltration combine to prevent much water reaching stream channels. Often surface runoff does not become significant (unless there is an intense storm which can be common in the tropics) until most of the soil moisture has been replenished; once it starts, however, it increases quite rapidly. Simplifying things somewhat, moderate rain over a sparsely vegetated, thin soil/impermeable bedrock watershed will quickly lead to massive increases in streamflow and a similarly rapid decline between storms. Moderate rain over a deep soil and/or permeable bedrock watershed will yield a gradual and more moderate increase in streamflow and one which declines more slowly after the rainfall and, provided stored groundwater can reach the stream, flows are likely to be maintained. In short the relative importance of the two main components of runoff, surface and groundwater runoff, determine the pattern of streamflow. If the former dominates it is likely to be relatively erratic, with a greater risk of floods and dry channels.

Water quality

Practically all runoff and groundwater, indeed even rainwater, contains some dissolved salts and other impurities which may greatly hinder their use for irrigation, human and livestock consumption or industrial use. Groundwaters are more likely to be rich in dissolved impurities than runoff, and it is reasonable to generalize that high clay content rocks tend to yield more saline waters than sandstone and grit aquifers (Bowen 1980). Demand for water is increasing in most tropical countries, and existing water supplies may also decline as a consequence of poor watershed management and/or pollution. In the future users will almost certainly have to manage with poorer-quality water. Yaron (1981: 40) summed up the situation: '. . . the question is no longer one of "how good is the water?" but rather of "what can be done with this water?"'

More often than not discussion of water quality centres on salt content. However, there are other factors which affect water quality: physical impurities (sediment and silt); non-organic pollutants (pesticides, herbicides, industrial pollutants especially heavy metals, artificial fertilizers, detergents, natural toxic trace elements, radioactivity); organic pollutants (bacteria, viruses and other disease organisms, nutrients derived from sewage or decomposing agro-industrial waste or decaying weeds, algae, antibiotics); and heat pollution. Water management strategies considerably influence water quality. Poorly designed and operated water conveyance systems may suffer from sediment and silt problems; better-designed and managed systems fed with the same water may not. Similarly, skilful, appropriate irrigation management can utilize water which would ruin an inappropriate and/or badly managed system.

Physical impurities

Sediment (detrital material carried and, when flow conditions permit, deposited by flowing water) and *silt* (fine sediment with particle size between 0.002 and 0.006 mm) in suspension are not usually very harmful to crops or livestock, indeed they may settle on the land and help sustain soil fertility. It should be stressed, however, that not all sediment is fertile. Problems may arise if silt and sediment settles in channels or reservoirs; it can also erode pump parts and irrigation spray jets, clog pipes or trickle irrigation equipment. Heavy sediment loads are the norm for many tropical streams, especially where there has been extensive deforestation, overgrazing or poor husbandry in the watershed (Balek 1977; Dobby 1973).

Non-organic pollutants

Some groundwaters, streams and lakes contain excessive amounts of non-organic pollutants due to natural causes. Groundwater is often contaminated as rainwater percolates through the soil or when it is stored in the aquifer, either from the aquifer itself or by contamination by saline groundwater or seawater seeping in. Oil-well drilling, mining and irrigation development may also contaminate groundwater or streams with salts, sulphur compounds or heavy metals. There is an increasing frequency of industrial spillages and in some countries airborne particulate pollution from industry and vehicles is causing contaminated precipitation – 'acid rain'. Some of these pollutants can affect very wide areas of watersheds and some may even be considered to have worldwide effects. Usually polluted water supplies are more of a problem in the mid and lower portions of watershed/river basin ecosystems, where streams, groundwater lakes and reservoirs have become contaminated as a result of human industrial and agricultural activities.

Some chemical pollutants can be very harmful to plants, animals and man, even in minute quantities. In many aquatic ecosystems lower organisms tend to concentrate certain pollutants: for example heavy metals (mercury, selenium, cadmium, lead, tin, arsenic), polychlorinated biphenyls, certain pesticides like DDT and radioactive compounds. Higher organisms preying on lower organisms may further concentrate pollutants in their tissues, in effect a 'biological magnification' operates and renders 'safe' levels of pollution in the environment a danger to higher organisms. Even non-toxic compounds may cause problems, for example, small quantities of detergent in irrigation supplies may gradually remove lubricant from pump bearings leading to increased maintenance costs.

Organic pollutants

As with non-organic pollutants the problem is greatest in the mid and lower portions of watershed/river basin ecosystems. Organic pollution may result from excess nutrients leaching from farmland, from agro-industrial waste, human or livestock effluents. Increasingly, cultivators are obliged through scarcity of uncontaminated supplies to use water that is heavily contaminated with organic pollutants, for example, sewage effluent or irrigation return flows.

Organic pollution problems may be broadly subdivided into those due to contamination with pathogenic organisms, which affect farmers or contaminate produce, and those due to excessive nutrients. The quantity of certain indicator organisms, for example the bacteria *Escherichia coli*, is often used as an indication of the degree of pathogenic contamination. Excessive nutrients commonly give rise to *eutrophication*, which may be defined as excessive biological productivity caused by excess phosphorus, and nitrogen, which causes micro-organisms, algae and aquatic weeds to grow to such a degree that they deoxygenate the water. Contamination with water enriched with phosphatic artificial fertilizers, certain detergents and sewage are the usual cause of eutrophication. When a body of water is enriched with nutrients, micro-organisms multiply and consume dissolved oxygen. *Biological oxygen demand* (BOD) is generally used to indicate the degree of organic pollution. It may be defined as *the molecular oxygen needed to stabilize, by aerobic biochemical processes, the decomposable organic matter present in the water*. Microbes will degrade (and in so doing deoxygenate the water) wastes until there is no oxygen left – there may therefore be a cessation of microbial purification before all wastes have been consumed. Ideally a water supply, to be useful for irrigation should have a BOD of less than 3 mg/litre, a BOD above 10 mg/litre indicates that the water is virtually free of dissolved oxygen.

Waste water as a resource

Waste water can be treated to remove physical, chemical and organic pollutants. Simple settling may be sufficient; a well-designed canal intake or system of silt traps can remove sufficient sediment and silt to avoid a water supply system choking up. It is usual to refer to the screening out or settling out of sediments as *primary* or *physical treatment*. *Secondary treatment* (which involves the activity of micro-organisms to break down wastes) commonly using the techniques of slow filtration, activated sludge treatment or settling lagoons, requires more expensive facilities, space and time. Secondary treatment reduces the BOD risk and reduces the number of disease-carrying organisms in the effluent, but does not remove all bacteria and viruses, nor does it remove toxic chemicals or all nutrients. Eutrophication may still take place after secondary treatment. *Tertiary treatment* involves the use of chemicals, activated carbon or electrochemical techniques to neutralize chemical pollutants and kill pathogenic organisms. At present the costs are too high and the equipment too complicated for widespread use in the developing countries.

Land disposal has many advantages over the previous treatments, not least it is cheap. Sewage effluent or agro-industrial waste is applied directly around crops. Provided these are non-food crops (e.g. cotton) or are food crops that are heat-treated before consumption this may be a viable way of using wastewater and reducing the risk of eutrophication of streams or other water bodies. Crop growth may be dramatically raised due to the nutrient content of the effluent. However, there are disadvantages: crops can be contaminated with pathogenic organisms heavy metals, high levels of nitrates, salts and even antibiotic, drug and hormone contaminants, and, if care is not taken during the application of effluent, there may be a drift of pathogens or other contaminants on the wind or into groundwater where they can harm people or livestock. Other problems associated with use of sewage or farm effluent is that there may well be aesthetic or religious objections to the use of unclean water – these can be avoided if timber or fibre crops, rather than food crops are irrigated. In many parts of the tropics townships presently use septic tanks for sewage disposal or simply discharge effluent into streams. As populations increase this means of disposal will have to be changed and this may force the consideration of land disposal. Some cities already use land disposal, for example, Mexico City and Melbourne (Australia) both irrigate pasture land with sewage effluent. The risks of using effluent are discussed in Finkel (1983: 47–53), while significant, they are not insurmountable.

Heat pollution

Some groundwaters emerge well above the ambient temperature and in the tropics this is more likely to be a nuisance than an advantage. Contact with hot rocks not only raises groundwater temperature it generally causes mineral enrichment. Industrial activity or electricity generation by thermal power stations can raise the temperature of streams used for cooling water and this may render the supply unsuitable for irrigation until it has sufficiently cooled. Hot, mineral-rich groundwater rapidly corrodes ordinary steel pipelines and other irrigation equipment. Algal and bacterial growths tend to build up in warm, mineral-rich water and both organic and chemical pollution may have greater impact on organisms if the water temperature is raised.

Sometimes the return flow from an irrigation project has been sufficiently warmed by the sun as it ran through shallow channels to raise its temperature several degrees, and it may be that this could affect crop yields if it were reused for irrigation. Certainly many farmers in India hold the view that cool irrigation water raises the yield of padi rice, but this has not been proven or disproved.

Salinity problems

One or a number of salts may be present in water in sufficient quantities to cause problems. Plants, livestock and man can be harmed by high concentrations of a single salt (i.e. the toxic effect of a particular ion) but more often, by high total concentrations of more than one salt (i.e. the total volume of dissolved solids). There are a number of commonly occurring salts including chlorides, sulphates, sodium salts, calcium salts and magnesium salts. Problems may also arise if the balance between sodium and calcium plus magnesium is upset; when this happens soil structure generally suffers. The actual concentration of salts which may be harmful in any particular situation depends very much upon the chemical characteristics of the soil, on the type of crop grown and on the management of the water, not just on the quality of the water supply. Salinity in a soil is not a constant, it varies from place to

Table 3.3 Expression of salinity: direct and indirect means of assessing total dissolved solids

Parts per million (ppm)	Concentration in terms of TDS related to weight of water in which they are dissolved; 1 ppm is equivalent to 1 mg of solute in 1 kg of solution, 1 ppm = 1 mg/litre. In the irrigation literature unit tons of dissolved solids per acre foot of water (taf) or grains per gallon (gpg) are sometimes used (1 taf = 735 ppm)
Milligrams per litre (mg/litre)	Concentration in terms of weight of solute per volume of solution
Equivalents per million (epm)	Concentration of substance in solution in terms of its chemical equivalence
Electroconductivity (mhos/cm or micromhos/cm)	Indirect determination of TDS by measuring electrical conductivity (EC) of solution, as salinity increases so does conductivity. Calculated for a solution at 25 °C and generally measured by reading resistance between two electrodes inserted into saturated soil. An EC of less than 250 is low, 750 to 2,250 medium and 2,250 to 4,000 high. The actual electroconductivity of water depends not only on the total quantity of dissolved salts, but also on the ionic composition of the solution

Sources: Yaron *et al.* 1973: 75; Bowen 1980: 81; Hoffman 1981; Yaron 1981: 21.

place even within one small field depending on local variations in drainage, soil qualities, tillage pattern and height of the ground surface above the water-table. Salinity problems may also vary in time, during dry seasons when there is little rain to leach salts out of the soil, salinity may rise, falling after wet season rains. In drylands salts may build up over a period of years when the land is cultivated and/or when irrigation is practised, and may decline slowly if the land is left fallow.

Salinity is commonly expressed in terms of *total dissolved solids* (TDS or TDS/litre), or as *parts per million* (ppm) or milligrams per litre (mg/litre – which is now more generally preferred). There are also indirect methods of assessing salt content of water (Table 3.3).

Typical salinity levels of fresh water, brackish water, seawater and so on, an indication of the salt tolerance of plants, livestock

Table 3.4 Some selected salinity levels (ppm)

Fresh water	0 to 1,000
Brackish	1,000 to 10,000
Salty	10,000 to 100,000
Brine	Over 100,000
Rainwater	2.7 to 26.5 (varies largely with distance from sea)
River water:	
Niger	60 to 80
White Nile	174
Indus	250 to 300
Colorado	795
Some rivers exceed 1,000	
Groundwater	200 to 3,000 (some groundwaters fluctuate, being most saline at the end of the dry season before rains dilute them)
Seawater	35,000 (average)
Maximum palatable level for man (short term) 3,000 (Adams *et al.* 1978 claim man can withstand 8,000)	
Recommended long-term maximum level for man 500 to 750	
In extremis sheep and camels may drink 15,000 to 25,000 (short term)	
Maximum for sheep and cattle long term? less than 3,000 (poultry are sensitive to salt). Young animals and breeding stock are less tolerant of high salinity	
Irrigation	
– Provided drainage is adequate salinity below 750 should pose no problems	
– Salt-sensitive crops give restricted yield between 750 and 1,500	
– Most crops give restricted yields somewhere between 1,500 and 3,500	
– Salt-tolerant crops give reasonable yield between 3,500 to 6,500	
– Salt-tolerant crops yields fall somewhere between 6,500 and 8,000	
– Some salt-tolerant wild species (halophytes) withstand 8,000 and above	

Sources: Adams *et al.* 1978; Gupta 1979; Goudie 1981; Yaron 1981.

and man and an indication of the utility of saline water for irrigation use are given in Table 3.4.

Plants and salinity

Excess salts damage plants in a variety of ways, but the most common problems are caused by the salts affecting the osmotic relationship between roots and soil moisture. Plants may not be able to extract adequate moisture from the soil if there is *osmotic interference* – stress due to the total concentration of soluble salts in the root zone. (Salts are selectively absorbed by plants from the soil. If osmotic potential of soil water is greater than that of the plant tissue, water will flow out of the roots into the soil.) Those salts which produce large numbers of ions per molecule and remain most completely dissolved into individual ionic components, produce the greatest osmotic effects (for a fuller discussion see Yaron 1981: 22–44).

In general, any excess of salts will affect the extraction of nutrients from the soil by plants and may alter the soil's ability to retain nurients – the effect is supression of plant growth. The relative concentration of sodium to calcium and magnesium is important, because high levels of sodium (as sodium chloride, sulphide and sulphate), which can be present in alkaline soils, depress plant growth by upsetting nutrient uptake (particularly uptake of calcium) and/or damaging the soil structure. Fine-textured, clay-rich soils are most vulnerable to sodium damage, generally becoming less permeable as a result (Yaron *et al.* 1973; Hoffman *et al.* 1980; Breckle 1982).

Salinity can hinder the conversion of ammonium salts (in artificial fertilizers) to nitrate by nitrifying micro-organisms in the soil, which plants can use for growth. Rising salinity may therefore negate application of fertilizers. Excessive concentrations of some specific ions, *trace elements*, can be toxic to plants, livestock and man. Some of these are listed in Table 3.5.

Most plants are more sensitive to salts during seed germination, seedling growth and when flowering or fruiting. The seed and seedling stages are vulnerable not only because the plant structures are immature and delicate, but also because tiny root systems draw moisture and nutrients from near the soil surface where salts tend to concentrate. Most woody perennials are sensitive to sodium, and perennial fruit plants tend to dislike chloride more than most annual crops. The tolerance of various crops to salts, the ways in which planting may reduce the risks of salt damage and the use of saline water for irrigation are discussed in Chapter 7.

Vast reserves of water are to be found in the sea, in estuaries and lagoons, underground, in lakes and saline streams, irrigation return flows, sewage and industrial effluent. Stream salinity can change as a consequence of land development and agricultural water use. Also, impoundment and water abstraction, can lead to seawater penetrating inland into deltas and estuaries where it may hinder farming, and contaminate aquifers.

With the correct strategy and skilful management, plus appropriate crops, cultivation can be sustained with quite poor-quality water. As good agricultural land and good-quality water supplies become scarce the development of techniques and crops

Table 3.5 Elements which can be toxic to plants, livestock and man

Boron	Essential for plant growth, but toxic if concentration exceeds an optimum level (over 1.0 to 2.0 ppm affects livestock and man, over 2.0 ppm affects plants. Levels have been known to reach 5.0 ppm). Boron is more difficult to remove from soil than other soluble salts
Selenium	Occasionally groundwater can be sufficiently rich in selenium to be toxic to plants. Livestock can be affected by as little as 0.05 ppm
Lithium	Some groundwaters have enough lithium in them to harm citrus fruit (which are susceptible)
Fluorine[a] Chlorine Copper[a] Zinc Lead[a] Cadmium Manganese	Occasionally there are sufficiently high levels of one or more of these elements to affect plants
Arsenic	At levels above 0.2 ppm affects livestock
Aluminium	May result from leaching of soil

[a] Sometimes affect livestock or man.
Note: Little is known of the effects of trace elements on soil micro-organisms, especially those which fix nitrogen (Bernstein 1962).

Sources: Cox & Atkins 1979; Gupta 1979; Goudie 1981.

that can allow the utilization of even very poor quality water must receive support. Investment in developing salt-tolerant crops and salinization avoidance/control strategies is more likely to benefit the farmers of poor countries than investment in desalination, at least in the foreseeable future.

Chapter 4

Systems of water management in the tropics

'. . . Some model builders look like cyclists pedalling along elegantly and powerfully on a chainless bike. . . . their influence on policy formulation is not at all in proportion to the intellectual effort devoted to water management modelling over the last ten years' (De Donnea 1982: 18).

Introduction

An increasing number of people involved with water resources and agriculture management in developing countries hold the conviction that such complex real-life problems cannot be solved by a single discipline and that an interdisciplinary approach is needed. A meaningful interdisciplinary approach is one which breaks down problems into distinct *subsystems* (elements) in such a way that each belongs to a well-defined discipline but links between them are clearly understood. Each subsystem should then be analysed and discussed using techniques and terminology of the appropriate discipline. The relevant points which emerge should then be disseminated and understood by other disciplines involved in the assessment (Yaron 1981: 1–20).

A typical water resource/agricultural system might comprise of six subsystems (Fig. 4.1). 1) *Goals and objectives*; 2) the *core* of the system, or the *input-output mechanism*; 3) *decision making*, and *controllable inputs* to the core; 4) *exogenous factors* or uncontrollable inputs (eg. weather, the state of the world economy, price levels); 5) *outputs* or *outcomes*; and 6) *evaluation of outputs* or a *feedback-and-control* unit. The figure is a gross oversimplification, for example, the water resources/agricultural system is dynamic, whereas Fig. 4.1 is describing things at one given point in time. The six subsystems could also be subdivided: for example, the *core* (2) could be divided into *demand* and *supply components*. The demand component could be further subdivided into agricultural, domestic, industrial. The supply component could be subdivided into sources

Fig. 4.1 Stages in operation of a water resource – agriculture system. Each link between boxes (subsystems) represents flows of **(a)** information; **(b)** orders and instructions; **(c)** water; **(d)** other materials; **(e)** funds. Exogenous factors include uncontrolled inputs such as weather, state of world economy, prices for produce, etc.

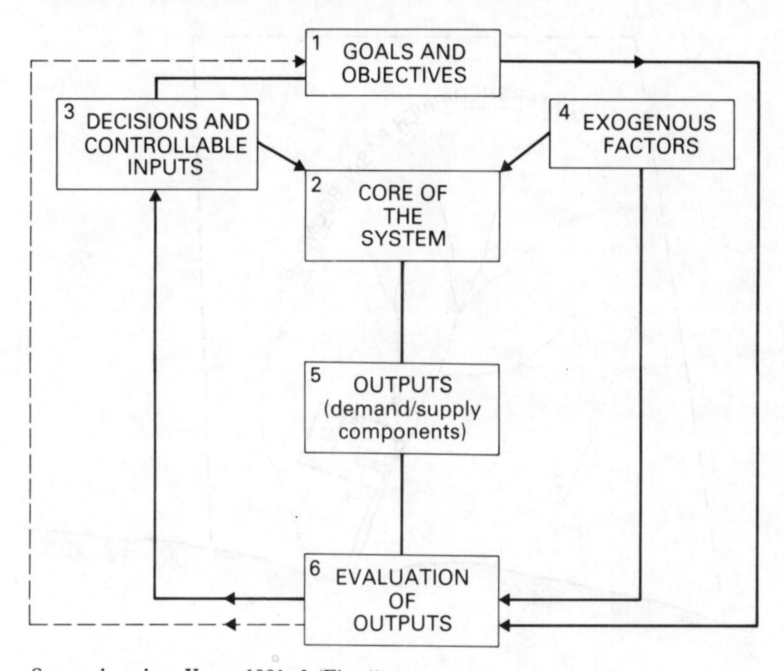

Source: based on Yaron 1981: 3 (Fig. 1).

(of water and salinity), the conveyance system and the disposal system for removing excess water and salts.

The water resources/agriculture management continuum has a number of levels: the national (Fig. 4.2 A); the regional (watershed/ river basin) (Fig. 4.2 B); the subregional (project) (Fig. 4.2 C); the subproject (water users' group, irrigation district, village unit and so on) (Fig. 4.2 D); the individual water user (Fig. 4.2 E); the field or plot level (Fig. 4.2 F).

In irrigation management or development of rainfed cultivation *decision variables* and *environmental conditions* are generally considered first, then, *exogenous variables* are related to *target variables* (Fig. 4.3). Two phases of management can be recognized, phases I and II in Fig. 4.3.

Water resources and agricultural development may take five broad forms and the management demands of each are different:

Fig. 4.2 The watershed/river basin ecosystem: the national–regional–subregional–project–subproject–individual holding plot continuum (a) National. (b) Regional (watershed/river basin). (c) Subregional (irrigation project). (d) Subproject (user group/irrigation district). (e) Individual (user). (f) Field or plot level.

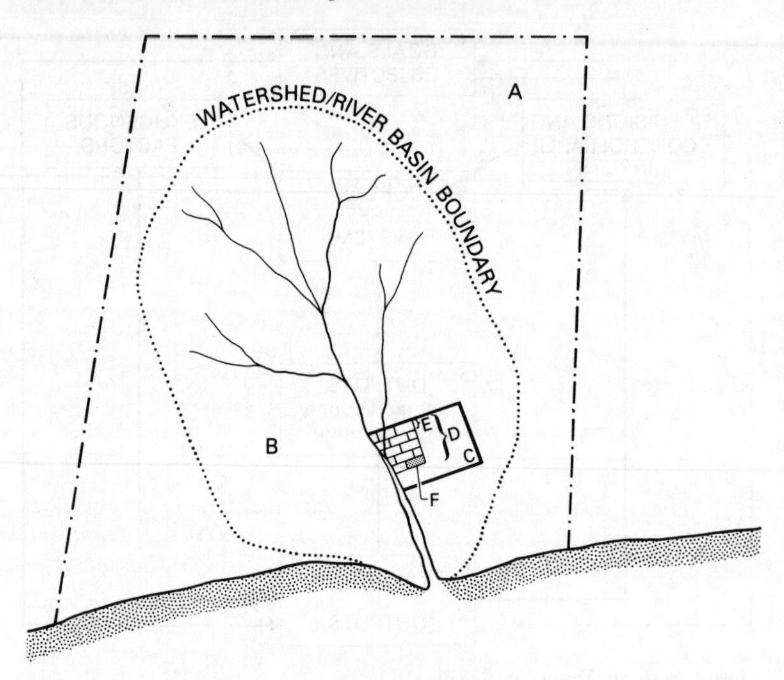

1. Rainfed, non-irrigated cultivation and/or livestock herding.
2. Irrigation independent of a supply system.
3. Irrigation dependent on a supply system.
4. Irrigation dependent on a finite, shared groundwater supply.
5. Irrigation dependent on groundwater which is unlikely to be 'overdrawn'.

In a given locality more than one of the previously mentioned forms of water resources and agricultural development may be found, for example, in a region with predominantly rainfed cultivation there may be some irrigation projects. Water resource and agriculture development of forms (1) and (2) depend on natural precipitation and natural runoff. Supply, allocation, distribution and disposal of runoff are unlikely to be important management considerations; management will mainly involve improvement of moisture collection, reduction of erosion, reduction of evapotranspiration losses and optimization of fertilizer use, cultivation

Fig. 4.3 Phases of irrigation management

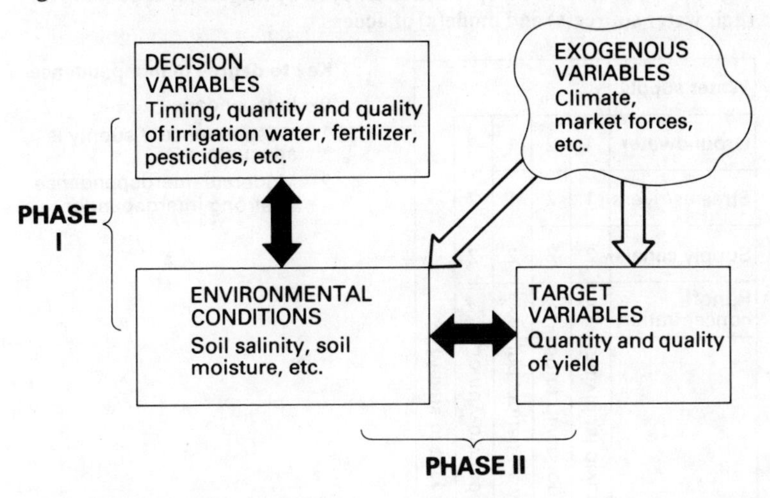

Source: Based on Yaron 1981: 6 (Fig. 2).

techniques and selection of appropriate crops. The management demands of water resources and agricultural development form (3) include control and maintenance of supply/conveyance system(s), allocation and distribution of water and, if necessary, disposal of excess water. The primary management consideration with water resources and agricultural development form (4) is the control of groundwater extraction. Forms (4) and (5) may well demand close management of disposal of excess water.

Although a large part of funds spent on developing water resources to improve agriculture has been spent on irrigation projects supplied by canal conveyance and distribution systems, many of the farmers of developing countries depend on natural rainfall, shallow groundwater or the diversion (and sometimes small-scale storage) of water from springs, streams or rivers flowing through or past their land. Cultivators with direct access to a stream, spring, river or well have little or no dealings with supply authorities and enjoy considerably more independence than those who do (Fig. 4.4).

Some users acquire water by contract or agreement, i.e. they agree supplies from a supplier in exchange for payment, goods, services or an undertaking to market produce through a specific channel. When users have direct access to water, authorities can have considerable difficulty monitoring and controlling water use. They may be able to tax produce, tax the farmer on the hectarage he sows or irrigates, restrict size and number of pumps owned,

Fig. 4.4 The level of independence enjoyed by irrigators, depending on their water source(s) and mode(s) of access

Water supply	Individual diversion	Joint diversion*	Individually-owned pump	Jointly-owned/hired pump
Groundwater	1	2	1	2
Streams/rivers	1	2	1	2
Supply canals	2	2	2	2
Runoff concentration	0	2		

Means of access

Key to degree of independence
0 – Independence
1 – Independence *if* supply is adequate
2 – Moderate interdependence to strong interdependence

Note:
* Where for example, water runs from one farmer's field to the next.

Source: based, with modifications, on Bromley 1982: 6.

issue licences and/or control fuel and electricity supplies, but more often than not water consumption is loosely controlled. However, where a supply authority conveys water to a group of users by canal or pipeline, or where groundwater is deep and can only be exploited with government-sunk boreholes, control of water use is possible.

Canal supplies in particular can be adversely affected by poor management and by the misbehaviour of water users when management is slack.

Managing irrigation supply/conveyance systems

Except where water is directly available to farmers, irrigated cultivation involves two groups of protagonists – water users and supply authorities. It is the job of the supply authority to manage the following tasks:

1. *Water delivery* – ensuring supplies to users via the conveyance system(s) are maintained. This involves repair, servicing and day-to-day running of a canal or pipeline system.
2. *Water allocation* – the distribution of available water to users. This often involves *scheduling* (the arrangement of rotation of supplies; in practice at any given time only a portion of the total users supplied by a canal or pipeline are actually getting water).

The delivery and allocation of water has two dimensions, each with a different set of problems and possible remedies: a technical dimension, which relates whether the methods to secure an optimal match between supply and demand are appropriate; and a social/political dimension, which concerns the capacity and will of officials to manage, maintain and repair canals or pipelines and allocate/ration/distribute water, often in the face of powerful pressure to misallocate (Bottrall 1981a: 13; Replogle & Merriam 1981).

Water delivery

Conveyance is by main canal, pipeline or aqueduct and then by *distributaries* (secondary channels, lateral channels or laterals), or even tertiary channels or pipes to the point where the user's responsibility begins – the *turnout* (farmgate), (Fig. 4.5). The system linking water source to crops may be divided into two subsystems: the *conveyance subsystem* (Fig. 4.5, A to B), and the 'on-farm' *distribution/application subsystem* (Fig. 4.5, B to C).

Referring to Fig. 4.5, the flow of water at B must be sufficient to cover losses during on-farm conveyance to the crops (losses do not cease at the turnout, on-farm losses are commonly considerable). Hopefully the flow at B is reliable and not subject to reductions or cessation. Between A and B water supplies suffer *conveyance losses* – seepage, spillage, evaporation, and theft. Conveyance losses are least from pipes or enclosed aqueducts; from well-maintained and lined canals in the tropics losses might well be 20 to 30 per cent (between A and B of Fig. 4.5). Poorly maintained, unlined canals, especially in dry environments, can have far higher losses.

Low *conveyance efficiency* (i.e. poor transfer of water from A to B) can be a problem even in richer nations; for example, Carruthers & Clark (1981: 86) cited an Australian irrigation project where 97 per cent of water supplies were lost between surface and farm turnouts. Great care must be exercised when one is presented with estimates of conveyance efficiency to ascertain whether this refers to transfer from (see Fig. 4.5) A to B, A to C or B to C (Stern 1979: ii; Zonn 1979; Biswas *et al.* 1983).

Fig. 4.5 Water supply stages and common measures of efficiency of supply (a) Flow at source. (b) Flow at turnout. (c) Flow into field. (d) Return flow (surface) – this is excess, 'used' or waste water from scheme, some water will leave the irrigation scheme by subsurface flow.

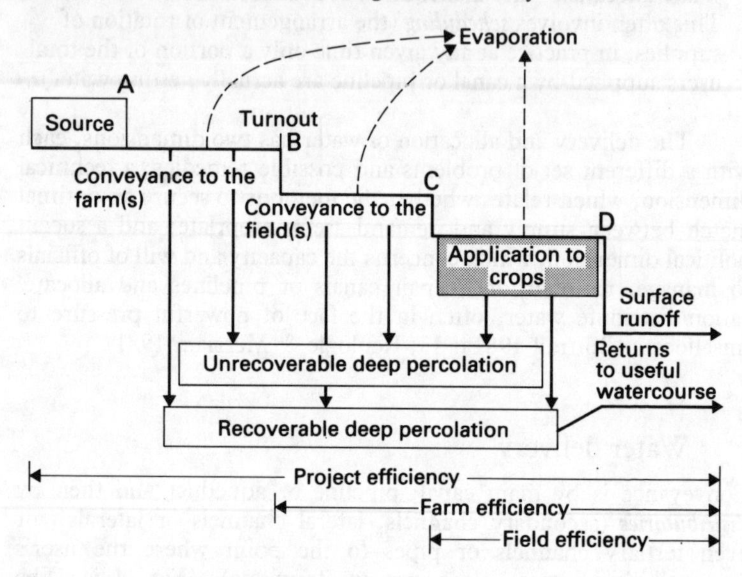

Source: modified from Howe & Easter 1971: 113 (Fig. 10).

The area of land which can be irrigated with water from a particular source is the *irrigation potential* (for discussion of procedures for calculating the irrigation potential of a water supply, see Hazelwood & Livingstone 1982). If land and other factors of production are in abundant supply, the irrigation potential of a water source depends on the demand for water and the capacity of the supply system to meet it. Demand and supply are seldom static, usually they both fluctuate during the year and from year to year. *Peak demand* is rarely at the time of peak supply. Irrigation potential depends on supply in relation to demand at a particular time – the crucial period when supply is lowest in relation to demand.

Scale of irrigation management

Authorities which choose to improve agriculture can adopt a variety of approaches. They can assist traditional irrigators or rainfed cultivators to modernize, but allow them to retain their autonomy, they may incorporate traditional irrigators or rainfed cultivators into large government schemes or they may establish new small-scale irrigation or improved rainfed cultivation schemes. A given supply of water may be distributed to many small-scale irrigators or to a few large-scale users. According to Hazlewood & Livingstone (1982: 8) large-scale users tend to demand water most when it is scarce, and small-scale users are more likely to require it when it is more abundant (at least in Africa). Therefore it may be possible to establish and sustain a mix of smallholder and larger-scale irrigation or a mix of permanent and seasonal irrigation, provided their peak demands are at different times.

Where there is a difference in elevation between water supply and farmland such that the former lies at a higher level, or if there is a supply of groundwater under pressure (artesian water), the irrigator enjoys considerable advantages – supplies can be fed to crops under the force of gravity or artesian pressure. When water can be fed to crops under pressure without need for pumping it is termed *gravity irrigation* and the area which is irrigated is known as the *command area*. Some methods of irrigation need water at such a pressure that pumps may be needed even if there is a head of water, this is often the case with sprinkler and some types of trickle irrigation.

Much of the irrigation development effort in the South (over 50 per cent) has concentrated on large-scale gravity-flow irrigation, usually fed by a canal system and producing export crops or food for urban populations. The costs of establishing this type of irrigation are high (e.g. in India and the Sudan they are similar, on average a little over $US2,000 per hectare; and in East and West Africa $US10,000 per hectare is not unusual) and, if the management is not good, much can go wrong (World Bank 1982: 62; Mather 1984: 182). Not only are the costs high, once the decision to invest has been made there has usually been little flexibility to adapt to the unforeseen – money is tied up for long periods and more often than not the developing country is committed to continuing heavy interest payments on loans used to finance the schemes.

The performance of large-scale irrigation projects in developing countries has generally been poor (Wade 1982a: 8). Large-scale by no means guarantees efficiency; for example, between 1951 and 1965 India established seventy-two major irrigation projects with a total command area of supposedly 13.4 million ha. By 1966 only 25 per cent of that hectarage actually received water (Falkenmark & Lindh 1976: 97). Large projects have commonly failed or have been very slow to repay the investments made on them. For example, the Kufra Scheme (Libya) seems to have had little hope of paying for itself – as in 1983 investment costs were $US8.8

million for 51,000 irrigated hectares and annual maintenance costs were $US3.0 million (Heathcote 1983: 201).

Heavy investment in large-scale irrigation has done little to alter gross statistics; most of the world's food is still produced by small-scale rainfed cultivators and small-scale farmers using irrigation. In view of the fact that those countries which have progressed from low levels of agricultural efficiency to 'agricultural development', e.g. Japan, Taiwan, South Korea and the People's Republic of China, have done so by increasing small farmer agriculture yields through improving their irrigation (Pearse 1980: 225). The present relative neglect of small-scale irrigation would therefore seem ill-advised.

One reason for the neglect of small-scale irrigation and the improvement of rainfed cultivation is the commonly made presumption that large-scale projects, especially irrigation projects, are somehow more efficient than small-scale. This seems to stem from several beliefs; firstly that large is somehow 'modern'; secondly, that there are 'economies of scale' – both assumptions have little strength. If technical efficiency is a goal, and one that is better met on large-scale projects, it should be realized that it is a rather unimportant achievement if it merely means saving labour because many developing countries desperately need to generate employment and labour is often cheaper than technical inputs. It has been suggested that large-scale irrigators or rainfed farming schemes are usually prepared to accept a one-in-five risk of crop loss, small-scale rainfed producers can probably withstand greater risks (Hazlewood & Livingstone 1982: 24, 35). This is because it costs the latter so little to prepare a hectare or so extra rainfed cropland – even if they only occasionally get a good harvest from it the gamble is worth while.

Small-scale irrigation can be successful; an example is the Matam Self-help Small Padi Irrigation Scheme (Senegal). This used local labour and materials for construction and is now self-sustaining, profitable and costs less than one-third of most large-scale irrigation projects (Carruthers & Clark 1981: 11). Although it is more expensive and difficult to provide water and services to scattered small-scale irrigators the problems are not insurmountable. In Tanzania for example, there has been some success in establishing government-aided, self-help (ujamaa village) settlements which have brought scattered homesteads together as villages at the centre of four large blocks of padi land (each block farmed by about 50 families). These 200-family villages can be provided with health care, schools, blacksmiths, agricultural extension services and so on far more easily than can scattered homesteads; the farmers can be coordinated fairly easily and do not have to walk great distances to get to their land.

To summarize, small-scale has many advantages which include: provision of more employment; an ability to withstand more frequent crop failures; it need not tie up such large sums of money as large-scale irrigation; it tends to obtain satisfactory results because those working are the direct beneficiaries and are therefore highly motivated; and it economizes on scarce management. On the negative side, attempts to supply water to numerous small farmers can be very wasteful and costly if not carefully planned. Because water wastage can be high and because it is difficult to monitor and control many small irrigators, salinization and human disease problems are more likely to arise. Many developing countries are beginning to promote small-scale irrigation (Eicher & Baker 1982: 135; Blackie 1984).

An important observation has been made by Bottrall (1985: 6) that most of the literature on irrigation concentrates on two groups: (a) the water managers, and (b) the water users (the farmers or tenants). There is, Bottrall argued, a third, numerous but neglected group – the landless rural poor – in Bangladesh experiments were under way in 1985 (with promising initial results) to see whether this group could maintain large-scale irrigation projects in return for using spare land for crops or agroforestry and irrigation ponds or channels for pisciculture.

The distinction between large-scale and small-scale may not always be clear cut. An authority can lay out an irrigation system and subdivide it into many units for cultivation by individual smallholders – usually retaining some degree of coordination to minimize wastage of water and ensure that crop production is efficient and maintenance is attended to. The Gezira Scheme (Sudan) adopts such an approach, although the overall coordination is probably stronger than is usual. It is also possible for small farmers to irrigate small plots which coalesce into large-scale aggregate 'irrigation systems'. Because such aggregates have not grown in a coordinated manner they are technically rather different from true large-scale irrigation schemes.

Sometimes governments decide to convert large-scale irrigation schemes to smallholder use, primarily for political motives, and if the infrastructure is not suitable this can be a catastrophic mistake. Water supply/irrigation managers must balance desire for equity against the need to get maximum crop returns per unit of water (Johl 1980: 8). Scale of operation may not be the only important criterion, Spooner (1974: 48) felt that: '. . . rather than distinguishing according to size (small-scale and large-scale or single village and multi-village systems) it would seem more feasible to make economic distinction based on complexity.' Systems in which the individual operator can control and maintain his own part, whether or not he knows the details of the total system,

Spooner called 'simple systems', whereas those which require engineering and maintenance beyond the abilities of individuals or groups involved in actual cultivation he termed 'complex systems'.

Whether or not large-scale irrigation has given worse or better returns on investment than small-scale, the latter will have to have more funding in the future because it is cheaper to establish and because it is so widespread. For example, in India about half the total irrigated land (roughly 22 million ha) is in the hands of smallholders (Carruthers & Clark 1981), and establishing such irrigation is on average one-fifth as expensive as establishing large-scale systems (Arnon 1981: 67). Not surprisingly, India has given aid to smallholders to introduce or improve irrigation. In Pakistan, the People's Republic of China, Southeast Asia, Iran and parts of Latin America there has also been government help, but in Africa south of the Sahara the potential for smallholder irrigation has only recently begun to be realized. Although there are extensive areas of fertile land in Africa overlying shallow, good-quality groundwater which could be easily reached with hand-dug wells (Arnon 1981: 68), the abundance of land makes it easier for small farmers to extend rainfed farming than start irrigation (Eicher & Baker 1982: 134).

Interaction between government and farmers

Efficient, reliable water supply is important, although in itself of little help if the user fails to grow the right crop(s), or fails to get adequate yields due to poor husbandry, inadequate inputs or pests, or if he is unsuccessful at marketing the harvest. Supplying water is just one important part of achieving satisfactory crop production.

Three broad types of situation may be encountered by those charged with managing water supplies: there may be some form of irrigation already practised (traditional methods or modern methods in need of improvement or expansion); there may be a non-irrigated mode of production (e.g. pastoralism or rainfed cultivation); or the land may be completely unused. The opportunities and challenges presented by each situation are quite different. Where there is some existing form of irrigated or non-irrigated crop production this must be carefully examined to see if it is feasible to change it, and if so, how. Where there is already some form of agricultural production any innovation, any attempt to improve existing practices is an intrusion. Unless the agents of the government or aid agency are aware of how agricultural development takes place at the farm level, any intrusion they might make, however well intentioned, is likely to yield disappointing results. Attempts to introduce irrigation, improve existing irrigation or uprate rainfed cultivation or pastoralism must be carried out by those who have a knowledge of the technical circumstances of tropical peasant agriculture and an understanding of the determinants of farmer behaviour and of the decision-making process of the farmers – in short a thorough understanding of the factors which influence agricultural development at 'grass roots' level (Clayton 1983).

The links between irrigation and society are often strong, yet until recently there were surprisingly few attempts by social scientists to understand and map out these links. Wittfogel (1957) put forward a thesis about the pervasiveness of centralizing authoritarian tendencies in irrigated agriculture-using societies, and more recently there have been studies of the social organization, especially of canal-supplied irrigation, like that of Hunt & Hunt (1976). There is, however, a need for much more research, for, in the words of Levine (Coward 1980: 51): 'Knowledge of the interrelationships between water and plant growth far exceeds our knowledge of the interrelationships between water and the human element in [water] delivery and utilization. . . .'

Sometimes the cause of unsatisfactory irrigation development is poor design or engineering (i.e. technological faults), more often it is due to poor management, occasionally both. In practice poorly designed irrigation systems which are well managed seem to fare better than well-designed systems which are under poor management (ASAE 1981: 103).

The potential benefits of improved management of water supply and irrigated cultivation in developing countries can hardly be stressed strongly enough. For example, at a conservative guess a saving in water wastage could be made in **most** South Asian irrigated rice schemes (Bottrall 1981a: 24). In the Mwea Irrigation Settlement Scheme (Kenya), which is generally acknowledged to be one of Africa's few successful projects, Chambers & Moris (1973: 7) concluded that managerial organization was the key ingredient.

Managerial improvements are usually considerably cheaper to make than physical changes to water supply systems; they can also usually be implemented much faster, yet the response to water supply problems is often to try to uprate the infrastructure at considerable expense. Physical infrastructure is generally more visible and impressive to voters and visiting dignitaries, so this attracts more funding and enthusiasm than management improvement (Johl 1980). Management improvements also tend to be bypassed because initiating new schemes is more attractive to administrators and engineers than 'rectifying the problems caused by others'. In the late 1970s the then President of the International Bank for Reconstruction and Development, R. McNamara, noted: '. . . the drama of harnessing a major river may be more exciting than the prosaic task of getting a steady trickle of water to a parched hectare, but to millions of smallholders that is what is going to make the difference between success and failure' (George 1977: 237).

Improved water supply management aimed at more efficient and just distribution could achieve a 'water revolution', i.e. a widespread, rapid boost in crop yields, but unlike the green revolution which has often accentuated income disparity between richer and poorer farmers, a water revolution would if anything improve equality (Bottrall 1981a). Unfortunately, as discussed above, money for improving irrigation management is likely to be more difficult to come by than money for building infrastructure.

Better management of irrigation not only requires funding, it also demands capable managers. This means there must be reform and management training; neither tend to be easy to institute (Skogerboe *et al.* 1982: 331). It is not uncommon for there to be poor contact, poor coordination and even rivalry between various departments and ministries involved in water resources/irrigation development. This is especially troublesome when a number of bodies are involved who have no agreed common objectives or plan of action, and worse, may not be of equal status or power. In South and Southeast Asia, for example, water supply and drainage are often controlled by irrigation and drainage or public works departments, and these are frequently older and 'outrank' departments responsible for agricultural extension, marketing and so on. In South India one irrigation department maintained a rigid seven-day water delivery schedule while the agriculture department extension service advised farmers to water at five-day intervals – clearly an impossibility. Coordination may be hindered if irrigation departments are organized on the basis of canal command areas, but agriculture departments use administrative boundaries. Effective planning, monitoring and management are hindered if key data are not collected or are held by government agencies who restrict or delay access to it (Bottrall 1981a: 77). Conflict between states and nations and insurgency can greatly hinder irrigation development or improvement of rainfed cultivation or pastoralism.

The first step in improving the management of existing water supply/irrigation systems, and to improving the design of future ones, is to assess where the inefficiencies lie. Useful evaluation guidelines, and an excellent review of the management of large (Asian) irrigation schemes are give by Bottrall (1985). Common weaknesses of irrigation organization and management and suggested remedies may be found in Table 4.1. A recent attempt to develop an analytical framework for monitoring and evaluating irrigation project management in developing countries pays considerable attention to water distribution problems (Bottrall 1981a).

Israel has demonstrated what can be achieved with better water management plus updating of irrigation technology: over the last few decades while considerably increasing agricultural production the Israelis have markedly reduced the amount of water used to produce those crops. In Sri Lanka strict water supply management on one irrigated rice scheme led to a 50 per cent increase in rice production within a single season, possibly because it led to reduction in waterlogging and better water rationing throughout the growing season (Bottrall 1981a: 14).

Communications between water supply authorities and water users

Getting good contact between various authorities is by no means the only requisite for satisfactory irrigation development, there must be adequate communication between the water supply, agricultural extension, marketing and other authorities and the cultivators. The management of large irrigation projects is too often in the hands of officials who are too far removed from the 'grass roots', who lack incentive to make the effort to find out what the cultivators want and/or are bound by inflexible operating rules (Coward 1980). There are even cases where officials deliberately cultivate poor communications; for example in the Punjab (Pakistan) irrigation officials have been reported to distance themselves from farmers in order to acquire an 'image' of being 'important' (this is an attitude by no means restricted to the Punjab – Merrey 1982: 99). The inertia of bureaucracy can cause poor communications: for example, largely illiterate settlers in North-East Brazil faced a twelve-month delay between attending a two-week demonstration course on irrigated farming (for most of them the only introduction they had) and actually getting their first plot of land, by which time they had forgotten much of the information given to them (Hall 1978: 92).

Poor communications between irrigation managers and cultivators generate a range of commonly occurring scenarios:

1. Where maintenance and repairs are a communual responsibility with no one person delegated to the task, problems may not get reported to the authorities and so remain unremedied.
2. Officials or technical experts in the field may hesitate to report problems to the administration for fear of casting a shadow over their own careers (only 'successes' tend to get reported).
3. The administrative structure often ensures that the least effective officials are to be found in the most remote and problem-ridden areas (precisely those areas in need of capable

Table 4.1 Components of organizational and management reform – implications for sequential action

Common weaknesses	Remedies (by sequence)[a]
Structure	
Poor horizontal coordination	1. Unified planning agency at provincial, national levels 2. Unified project structure
No specialist water distribution cadre	1. In-service courses for existing engineers 2. New university curricula for training new engineers 3. Establishment of new cadre
No watercourse management extension cadre	1. In-service training for existing field staff 2. Recruitment and training of new staff (where necessary)
No planning, research and monitoring unit	1. Design of procedures 2. Recruitment of new staff 3. Some in-service training
Overcentralized responsibilities within project organisation	1. Changes in work scheduling, job descriptions 2. Major changes in definition of responsibilities
Excessive staff stratification and barriers to upward mobility	1. Changes in promotion policy 2. In-service training
Excessively differential salary structure	1. Changes in salary structure
Overcentralized location of field offices	1. Better local facilities and/or bonus payments for 'hardship' posts
Insufficient powers to project managers	1. Improved management system 2. Unified project structure 3. Bonus payments to project for good performance
Process	
I. Overall project management	
Confused objectives	1. Clarification at government level 2. Clarification at project level + incorporation into written procedures

Remarks[b]
Major policy decision
Major policy decision/L
S
M
L
S
Negotiation with Finance Ministry
P
Minor expenditure
S
P
Policy decision
Policy decision
S
Major policy decision
Policy/financial decisions
P
Major policy decision
Major policy decision
Policy/S
P

Table 4.1 (continued)

Common weaknesses	Remedies (by sequence)[a]
Unsystematic budgeting and work programming, without participation	1. Improved procedures
Departmentalism	1. Improved procedures (interdepartmental committees, etc.) 2. Structural change
Authoritarian behaviour	1. Procedural reform 2. Management training 3. Structural change
Poor monitoring	1. Improved procedures
II. Specialist activities[c]	
Deficient system design*	1. Physical rehabilitation
Inappropriate water distribution methods*	1. (Change in system design, if necessary) 2. Research into new methods + incorporate into procedures
Deficient procedures for planning, implementation, monitoring	1. Develop new procedures, handbooks 2. In-service training
Inadequate skills	1. In-service training for existing officials and staff 2. Long-term training for new engineers
Inadequate resources (finance, staff, equipment)	1. Improved procedures 2. Reassessment; detailed case to be made for new resources 3. Increased water charges and/or bonus payments to project for good performance
Poor motivation	1. Improved procedures 2. Changes in promotion policy, plus training 3. Bonuses for good performance
Ineffective legal framework	1. Implementation of existing legislation 2. Changes in existing legislation

[a] Sequence is sometimes determined solely by the importance of the initial remedy (in the absence of which no other action can be of any assistance) but more frequently by the likely ease and speed with which it can be introduced.
[b] L = long-term; M = medium-term; S = short-term; P = procedural reform.
[c] Items with an asterisk apply only in the case of water distribution.

Remarks[b]
P
P
Major policy decision
P
S
Policy decision
P
L
(L)
S/M; P
S/M; P
S/M
S
M
P
Negotiation with Finance Ministry
Major policy decision
P
Policy decision
Major policy decision
S
M

Source: Bottrall 1981a: 183.

staff): anyone well-regarded by higher superiors is likely to be transferred to more comfortable postings nearer a city.

4. Officials tend to be transferred too often, moving just as they get a 'feel' for a region or project.
5. Farmers or local officials cannot get a response from decision-makers with sufficient power to help them (because those with such power are remote from them in the city); Lees (1978: 48) dubbed this the 'my superiors have not yet answered my letter syndrome'. Rarely is decision-making in developing countries delegated to officers on the site who have the ability and power to act appropriately, sensitively and quickly.
6. Administrators do not wish to get their feet muddy. Literally, they are reluctant to forsake the comforts and/or security of the city, main roads and developed regions, sometimes for fear of terrorism, more often because they dislike discomfort and sometimes because they suspect colleagues may plot in their absence.

Administrative structures

Bureaucratic management at its most rigid delegates little power to the water users. It is a common style of administration, frequently used to run large-scale, canal-supplied irrigation schemes. An alternative is for the supply authority to delegate much more power and operate *community management*. Between these two extremes there are a wide range of part-bureaucratic/part-community management approaches (Coward 1976; Chambers 1977; 1981; Bottrall 1981a).

Bureaucratically-managed water allocation

It is not uncommon for water supplies to be allocated by either a government agency (an irrigation department, a regional authority or a body responsible for a particular canal or irrigation project) or an aid agency. Where there is fairly contiguous, large-scale irrigation, management by, say, a canal command area authority can be effective. Unfortunately, a common characteristic of such agencies is that employees who administer the allocation of water supplies usually do not farm themselves, indeed they may know very little about the practice of cultivation and have only a limited understanding of the farmer's irrigation needs (and sometimes a lack of interest in those needs). Commonly there are poor communications between farmers and supply/irrigation development authorities, and often poor liaison between water supply, irrigation development and other managers (Bromley 1982: 46). Efforts have been made to improve the management of water allocation in many countries. For example, in India although canal irrigation is widespread and has had a long history, in the 1970s it was often (and in many places still is) inefficient. Anxious to ensure that existing canal-supplied irrigation systems functioned better, and new developments fared better, the Indian government decided in 1973 to create *command area development authorities*. Establishment of these authorities began in 1974; they were to be powerful organization and management bodies each with a single line of command able to integrate the various functions involved in canal-supplied irrigated agriculture. In effect an integrated area development approach was to be adopted for all irrigation projects over 100,000 ha.

Command area development authorities grew in number from thirty-eight in 1978 to sixty in 1983 (Syed Hashim Ali 1983: 1). Where there are areas of canal-supplied irrigation interspersed with areas of rainfed cultivation, integrated area development with management by a combined water supply/irrigation/agricultural improvement authority is probably best (Commonwealth Secretariat 1978: 9). It is not unusual for farmers on canal-supplied irrigation schemes to hold some rainfed farmland – sometimes there is under-exploitation of the irrigated land because farmers spend too much time on their rainfed crops, though sometimes the opposite occurs; integrated rainfed/irrigated land management would help avoid this problem. An indication of the range of variation in delegation of responsibility to individual farmers or groups of farmers is given in Fig. 4.6.

Often the various authorities involved in irrigated agriculture development have been reluctant to surrender their functions and change their ways to facilitate better management. The take-over by command area development authorities of existing large-scale irrigation schemes has therefore been fraught with problems, and heated debates still occur over the responsibilities of such authorities when new schemes are implemented today. In India at least, finding an appropriate method of management is not enough, long-established interests and institutions tend to resist change and nullify the effects of innovation. As one writer noted in 1980, command area development authorities **should** work but in practice seldom do (Syed Hashim Ali 1980: 16).

In Malaysia, the Philippines and Indonesia there has been a similar trend towards establishing unified irrigation authorities to improve contact between water supply authorities, agricultural extension services and the water users. For example, the Muda Scheme (West Malaysia) or the Kembu Scheme (Indonesia).

Fig. 4.6 Generalized pattern of responsibility distribution: a comparison between selected projects

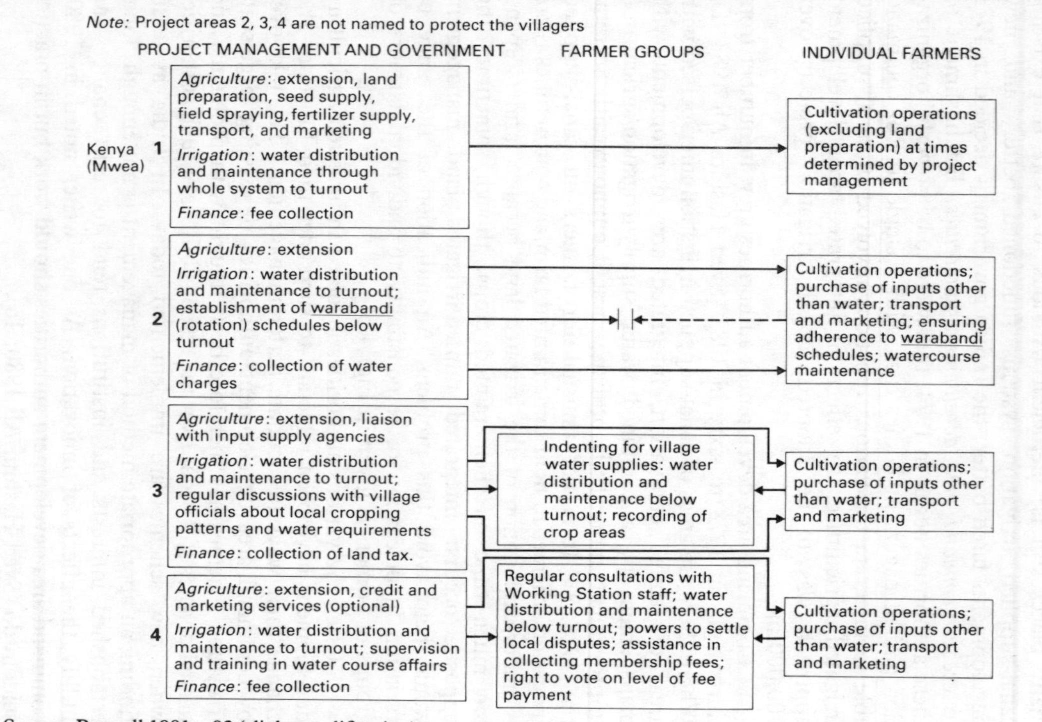

Source: Bottrall 1981a: 92 (slight modification).

In northern India and Pakistan bureaucratic management often adopts quite a *laissez-faire* approach, although the farmers may be obliged to sell produce through a marketing wing of the water supply/irrigation development authority (which deducts from their profits to pay for supply infrastructure and maintenance) there is little other obvious control. In such situations the water supply manager is not in direct control, but is more in the position of the manager of a service industry and the user is the client – who can complain and/or refuse payment if dissatisfied.

Strongly paternalistic management does not necessarily seem to be a disadvantage. The Gezira Scheme (Sudan), before about 1970 often cited as one of Africa's few successful projects (and probably one of the world's largest with over 800,000 irrigated hectares) is run with virtually military discipline, and has proved to be one of the cheapest to establish, costing roughly £UK8 (*c.*$US20) per hectare in the late 1970s (Gaitskell 1959; Hall 1978: 10). The farmers in the scheme seem able to bear the paternalism, probably because the lure of higher than average income outweighs resentment at being ordered what to do. Of considerable significance, the management of the Gezira Scheme has a good grasp of what is happening at the 'grass roots' level.

Similarly, the Mwea Scheme (Kenya), which is generally held to be another of Africa's few successful irrigation projects, is administered by a semi-state authority which exercises rigid control over many tenant smallholder irrigators who grow rice. The authority is responsible for supplying water and marketing the produce. It has the power to evict those who perform poorly, decides what the farmers should grow and when they should plant, weed and harvest. Tenants of the scheme have to hold at least 4 ha, and can expand their holding, but only if they are deemed to have performed well. Under such a management regime quite inexperienced peasant farmers have achieved relatively high levels of technical performance. Tenants are free to invest their profits as they wish and various incentives encourage good rice production; one key to the success is that efficient farming is rewarded and inefficient is punished. Tradition and traditional laws seem to have been largely retained and management discourages excessive mechanization to keep employment levels high. Like the Gezira Scheme, the Mwea Scheme management maintains close contact with tenants, using field assistants as intermediaries (each responsible for 125 farmers and issued with a motor cycle to enable them to make regular rounds), (Chambers & Moris 1973).

One problem with bureaucratic management is the risk that the irrigator will be too little involved in a project. The following case-study illustrates how such alienation can occur and how management can fail to identify problems in advance.

Case-study of the Senegal River Delta

Between 1946 and 1949 a large-scale, mechanized padi rice scheme was established in the Senegal River Delta at Richard-Toll (Senegal) by the colonial authorities. By 1960 this scheme was producing about one-tenth of Senegal's total rice needs from about 5,400 ha, but it cost too much and employed too few people. Following independence in 1961 Senegal set up an organization to oversee a renewed attempt at large-scale rice production in the delta, the Société d'Amenagement et d'Exploitation des Terres du Delta du Fleuve Senegal (SAED).

Between 1965 and 1975 30,000 ha were to be brought under padi rice cultivation (Adams 1981: 328), and SAED were to closely supervise peasant farmer cooperatives, provide seed and fertilizer, hire out machinery and market the rice crop. The region that SAED had set out to develop was peopled largely by nomadic Fulani herdsmen with only one sedentary village. It was therefore decided to bring in and settle 9,000 families from outside the region to work the padi. Six new villages were built between 1964 and 1967 and all went reasonably well until 1968 when the 8,000 ha of rice then in cultivation yielded the abysmal crop of 500 tonnes.

The low rice harvest of 1968 was due to low river levels and to a number of other factors; for example, it was apparent by 1968 that of the 30,000 ha planned for development, 19,000 ha were too saline. Much of the land scheduled for padi was not levelled and the simple flooding irrigation techniques drowned the rice in places and in others let the soil dry out, which reduced yields and led to localized salinization. On the roughly 11,000 ha which were brought into cultivation, crop losses averaged 30 per cent or more.

In an attempt to get rapid expansion of large-scale rice production SAED had invested heavily in mechanization, it also controlled most of the services and inputs – seeds, fertilizers, credit, roads, housing, water supply, marketing. The result was that the peasants had little to do apart from sowing and threshing (roughly forty days' work a year), and because they were not allowed to grow subsistence crops the peasants consumed quite a lot of the rice produced, depressing already poor yields. It was found that SAED was doing more and more and getting less and less return, so the cost of services and inputs rose as a consequence of this, the peasants found it harder and harder to make worthwhile profits and so after 1968 people began to leave the Delta in increasing numbers (drift which was aggravated by the opportunities for unskilled labourers to find work in France at that time). Another problem, at least upstream, was that rice fetched only two-thirds of the price that millet (a rainfed crop) could command.

Since 1972 SAED has attempted to improve irrigation by installing pumps and levelling land, and has tried to divide up the large unwieldy cooperative structure through which it had tried to run things. Efforts have been made to diversify the cropping and to give the peasants a more active role in management and cultivation.

Community-managed water allocation

Social scientists tend to see the poor performance of irrigation schemes as stemming from the unresponsiveness and inefficiency of over-centralized management bureaucracies, and the solution to such difficulties as lying in developing new organizational structures and management methods which encourage a greater degree of participation in decision-making by the cultivators (Oxby & Bottrall undated: 2). Frequently, greater delegation of responsibility to local communities for day-to-day organization and the creation of support organizations results in more efficient water use and better crops. Support groups should be capable of combining responsiveness to community needs with the ability to exercise sufficient impartial authority to minimize and settle conflicts between individual users or groups of users. Provided conditions exist or can be created which are conducive to local initiative, water users often organize themselves remarkably well and are capable of assuming substantial responsibilities, including the management of local conflicts (Lowdermilk 1985). However, it should not be forgotten that the weaknesses of irrigation projects may stem from poor planning, poor design and/or poor overall project management. Without reform of the water supply authority and irrigation management authority, increased user involvement in a scheme will achieve little. It is fashionable to back community involvement and the 'election' of some community management groups is simply 'cosmetic', the involvement comes too late to contribute to design of infrastructure and does little to improve management.

There may well be a 'threshold', i.e. a certain size of scheme or number of water users, which if exceeded renders management by the water users inefficient. It is likely the effectiveness of user management depends on average farm size, the number of farms per lateral channel with canal-supplied irrigation, degree of social cohesion/social stratification, the level of farmer education and experience with irrigated agriculture, the technical characteristics and size of the command area and the degree of supervision by an agency. In practice such a threshold has not been adequately studied.

Water users themselves can organize and manage their supplies through representatives which they elect, appoint or have appointed for them. Some of these *community-managed* systems are traditional, sometimes they are recent introductions promoted by governments or aid agencies. In a number of countries, for example the Philippines, where over 30 per cent of all irrigated land is community managed, experience has shown that involving water users from the start of irrigation planning is far better than the common practice of planning and constructing a scheme and **then** trying to establish water user associations to manage and maintain it (Korten 1982). Water users, if involved at the start of planning, can offer considerable local knowledge which helps planners avoid mistakes. Also, and perhaps more important, the water users can identify with the scheme and therefore are better motivated.

In Indonesia, Malaysia, the Philippines, Thailand, Japan, Sri Lanka, in much of India, Pakistan and Latin America community water management systems traditionally predominate. There is considerable diversity; in Japan communities tended to form water supply management cooperatives (Coward 1980: 127); in Taiwan and South Korea water users established irrigation associations to allocate supplies, maintain canals and control other aspects of irrigation (Wade 1982a: 9); in Himchal Pradesh State (India) private or communual khuls (farmer management groups) control water supplies; in Senegal, halpulaar groups perform in a similar manner; in parts of Indonesia, notably Bali, water allocation is managed by the subak, a complex, autonomous, legal corporation-cum-agricultural planning unit and religious community in which locally elected leaders make decisions on behalf of the water users. Anywhere from 25 to 700 individuals may be managed by a subak (Coward 1980: 70; Bottrall 1981a: 223). In north-western Luzon (Philippines) indigenous cooperative irrigation societies (zanjeras) control units ranging in size from 1 ha to 1,000 ha (but typically nearer 40 ha). An analysis of the development, structure and function of zanjeras was provided by Siy (1982). Who suggested that building upon these and similar indigenous institutions might result in better community management of irrigation than would be the case if new institutions were introduced.

Most community-managed water allocation organizations tend to be channel (canal) based, rather than community (village) based, i.e. a supply canal may serve a group of users who may well come from more than one settlement. A three-tiered structure of management has been successfully applied in various countries. Typically this would take the form of a *water district* with a staff responsible for overseeing the whole irrigation project, *water federations* which manage the laterals, and *water user organizations/associations* which manage groups of users along a lateral (Reuss *et al*. 1979: 418). An example of the successful adoption of a tiered system of irrigation management which involves farmers in decision-making may be seen in the following case-study.

Case-study of the Minipe Irrigation Project

The Minipe Irrigation Project (near Kandy, Sri Lanka) is a fairly typical large South Asian scheme, comprising about 6,000 ha of padi. The thousands of farmers on the Project all have individual water requirements which have to be met by drawing supplies from a complicated system of main canals, distributary channels (*laterals*) and field channels (fed from the laterals). Poor water management practices were prevalent from the main supply canal level right down to individual farmer level. There was a fairly typical maldistribution of water, whereby those nearest the main channel received greater supplies of water than those further down the supply network. Field officers of the water supply authority ignored the farmers, and supplies of seeds, fertilizers and other inputs were often delayed or unavailable. The supply system was poorly managed and the farmers were undisciplined, disorganized and had no effective share in overall management. The net result was that as well as inequity in water distribution the full irrigation potential was never realized due to water wastage.

Reform was needed which would give the farmers greater participation in the management of the Project, which would improve the initiative and motivation of field officers and various state agency staff involved in irrigation development, and which would improve the supply of inputs. Experiments were begun to try and develop an approach which would meet these needs on an area of about 800 ha. For nearly two years the farmers on this pilot scheme were involved in management through elected representatives (farmer leaders) who took responsibility for operation and maintenance of the supply system. This responsibility motivated the farmer leaders actively to seek the full participation of individual farmers in their charge. The state officers responsible for water supply, farm extension and so on, became part of sub-Project-level and Project-level water management committees composed mainly of farmer leaders. The result was that, answerable directly to the people, they exercised their authority with care and caution.

In practice a system of management was arrived at for the Minipe Project in which those officers who were not really keen on helping the farmers would usually transfer away at their own request. Officers were in effect forced to work with the farmers and to work well without any additional financial inducement (a real advantage for developing countries).

Once the pilot scheme was seen to be a success the Minipe Project was divided into six sub-Project areas of between 200 and 1,000 ha. Each of these sub-Project areas are managed by sub-Project water management committees. Activities related to the

Project as a whole are managed by a Project water management committee. Both the sub-Project water management committees and the Project water management committee are composed mainly of farmers.

Recently the Minipe Project has begun to experiment with farm-level water management committees serving areas of 150 to 200 ha. These consist of two or three farmer representatives and an officer of the Irrigation Department and of the Agriculture Department. There have also been efforts to motivate school children in water management matters, as they are seen to be the next generation of farmers – they arrange voluntary labour gangs to repair channels or agricultural roads and to weed padi-fields.

The reorganization of the Minipe Project has involved minimal expenditure, has encouraged the best use of available resources and above all seems to have greatly improved water management. The Project has shown what community participation in management can achieve. Another hint of the potential of community management may be had if one contrasts canal-supplied rice irrigation in India which is largely under bureaucratic management, with similar irrigated rice schemes in South Korea where management is practised through community associations. Though the environmental conditions and infrastructure are broadly similar, South Korean yields are roughly four times those of India. It is tempting to seek an answer to this discrepancy in the management/organizational structure differences, and Wade (1982b: 6) observed that in India: '. . . canals are managed by a centralized, governmental hierarchy . . . the irrigators' prescribed role in canal operation and maintenance is confined to "below the outlet" [i.e. below the farm turnout] . . . and administrative staff who make decisions . . . move from one project to another during their careers . . .' Whereas in South Korea: '. . . each canal system is managed by staff paid for, and appointed by, the beneficiaries . . . staff normally stay with one association [water user association] for most of their careers.'

Bureaucratic–community-managed water allocation

Pasternak (in Bottrall 1981a: 201) observed: '. . . there is a threshold of complexity in irrigation systems at which cooperation must give way to coordination; at which those served by the system relinquish decision-making power and their direct role in settling disputes.' When that threshold comes varies, but depends largely on technology, availability of water and the social cohesion of the community or communities involved. When community management becomes impractical there is an alternative to bureaucratic management: *bureaucratic–community management* (joint management). Under such management, as the name suggests, there is a compromise, some management functions are performed by the water supply/irrigation agency: delivery along the main canal(s) or pipe(s) to the lateral channels/pipes and the settling of disagreements or disputes between water users/groups of users. One or more water user associations/groups then manage water distribution along the laterals to their individual members.

Most bureaucratic–community-managed irrigation systems are larger than community-managed systems – some are of an entirely different order of magnitude, for example, discrete command areas on the Indus Plain (India/Pakistan) often exceed 500,000 ha and involve up to 150,000 farms (Oxby & Bottrall undated: 10) Bureaucratic–community-managed systems are generally found in the lower riverine plains where the scale, complexity and expense of infrastructure has led to the principal role in planning, design and construction being taken by governments or large private corporations.

The main problem with bureaucratic–community management is providing an effective link between users and the water supply/irrigation authority. Usually 'accountable leaders' are elected or chosen to perform this role and there may be subleaders linking individual users, or groups of users, managing the lateral channels with the accountable leaders.

Paying for water supplies

A few governments allocate water as a 'free input' making no charge, with the aim of improving the agriculture of a locality or region or particular group of rural people. In effect water is provided as a form of welfare payment. Most water supply managers, however, are faced with the need to recover installation costs and to generate income for administration, maintenance, repairs and possibly expansion or modernization. This means charging the water users or those who consume the products the users produce (Tiffen 1985).

Water supplies can be allocated to those who can pay for them; the recipient being charged for the *volume consumed* (this demands volumetric measuring devices, timekeeping and record-keeping). Alternatively, charges can be made according to the *flow rate* (which demands measurement of flows, record-keeping and particular care over the subdivision of flows to users). In India and Sri Lanka a notched weir is traditionally used to subdivide canal or streamflows for distribution to irrigators. This relatively simple means of allocating according to rate of flow is known as karahan-kota: the deeper and the wider the notch in a farmer's weir the greater the rate of flow to his/her land per unit time. A charge can simply be made by reference to notch size and duration of supply.

Rather than measure flows an authority can charge according to the *hectarage irrigated*, usually at a rate fixed according to the crop(s) being grown. Sometimes a charge is levied *per irrigation*, based on the area irrigated regardless of the quantity of water actually used. This can be very unfair if the user's supply is interrupted or if flows are feeble, it can also allow a user to apply excessive water if the chance arises. It is possible to charge a water user *according to what he harvests or sells* (in practice assessing just what is sold and what the crops actually fetch can be difficult).

There is much debate over the relative merits of the various charging systems (Table 4.2 lists the range of alternatives possible and notes the advantages, disadvantages and requirements of most). It is probably best if cultivators are charged according to the quantity of water delivered to them rather than be taxed/charged on the crops they sell or the area they irrigate, as this encourages efficient use of water rather than penalizes farmers for productive farming. Alternatively, a water user could be charged at a basic rate according to the hectarage farmed and at a higher rate for any water consumed in excess of the norm for such an area/crop in that locality (Oliver 1972: 143).

Education and farmer attitudes

Inefficient water distribution resulting from poorly designed channels, poor maintenance and most other technical reasons can, given the money, usually be cured. However, maldistribution frequently has an institutional cause or stems from the attitude of individual users and when these are the cause improvement may be difficult and slow (Pasternak 1968; Bromley 1982: 66). Lining a canal with plastic sheet or concrete or clearing weed from a choked channel system may be far less difficult than altering deeply entrenched attitudes or prejudices which lead to maldistribution of water and other inputs to irrigation. Providing adequate, reliable water supplies is no guarantee that farmers will apply it to their crops in an efficient and appropriate manner.

Theft of water supplies is common and most supply networks have to be policed, especially at time of peak demand and above all at night. Ideally, those caught stealing water should be swiftly and publicly punished by an authority above the local community level, because local officials are more likely to favour or be afraid of local individuals (Coward 1980: 46). Without clear rules, official support

Table 4.2 Water charges

Charging strategy	Advantages/disadvantages
Volume consumed	Requires measuring devices/timekeeping/ record-keeping
Rate of flow	Requires measurement of flow/record-keeping
Hectarage farmed – annual charge	May bear no relationship to water actually received – so can be unfair. Does not discourage over-irrigation
Hectarage farmed – per irrigation	Requires effective policing/assessment of true hectarage. Does not discourage over-irrigation if chance arises. The charge is fixed so it is unfair if the duration or volume delivered are low or interrupted
Hectarage of specific crop(s) – annual	Authority can encourage cultivation of a particular crop. Does not discourage over-irrigation. Unfair if delivery of water is inadequate
Hectarage of specific crop(s) – per irrigation	Marginally fairer than annual charge for specific crop(s). Does not discourage over-irrigation
Basic charge on area farmed plus surcharge for excess water above norm for locality/crop(s) for that area	In theory a good strategy to control water use, encourages good water management. Needs good monitoring and could for that reason be costly
Charge according to sales of crop(s)	Assessing what is sold, where and the profit it generates can be very difficult, even if there is a 'controlled market'
Charge according to size of harvest	May be difficult to assess and if there is a glut or storage problem rather unfair to farmer
No charge made	Over-use of water not discouraged. No revenue generated. May aid section of community

and effective sanctions against offenders, those charged with policing water supplies have an impossible task.

The behaviour of water users, especially small-scale farmers and above all those with land well down the supply network is conditioned by:

1. Their *interdependence* – water users are interdependent if

supplied from canals, river diversions or groundwater. If such sources cannot meet peak demand, those furthest away from the source down a lateral, or downstream where supplies are drawn from a stream or river, or with shallow wells where users share a common groundwater, obtain water at the sufferance of others.

2. *Institutional uncertainty* – the rules governing distribution may not be equitable, may be subject to alteration, may not be properly enforced.
3. *Cautious optimization* – if due to (1) and (2) water supplies are unreliable the user will steal or apply too much water when it is available. Over-watering commonly results from users being unsure of supplies. Understandably in such a situation they take as much as they can, when they can, in case the next release is delayed, reduced or fails to take place.

Water users who do not get reliable supplies are hesitant to use fertilizers or adopt improved crop varieties. They are likely to divide their labours between irrigated crops and rainfed crops which have some drought tolerance (intended as a security against failed irrigated crops) and may end up with low overall yields. Improved security of water supply could do much to reduce such inefficiencies. Improving the reliability of water supply demands alterations in water user/supply authority attitudes so as to:

(a) Improve discipline of users and of water supply managers who are in immediate contact with users;
(b) Arrive at means of settling disputes between individual users, groups of users or between users and supply authorities.

Once water supplies are reliable users' attitudes generally change. They are more willing to pay higher water charges, which can aid management in meeting the costs of maintenance and improvements. Above all, reliable water supplies are likely to reduce cautious optimization, users will apply only what they need at a given point in time and this will cut water wastage, help prevent waterlogging and salinization.

Over irrigation and the desire to steal water may arise because farmers see the application of a lot of water as a way of controlling weeds at minimal cost or because they have a belief (which is common in South and South East Asia) that a good throughflow of water somehow 'cools' the padi rice and raises yield. It seems likely that the latter belief is unfounded. Where there are clay soils which allow less infiltration as they become moist, and/or occasional heavy rainfalls, care should be taken to ensure that water users do not unnecessarily apply water. A rigid water supply schedule may be practised, but it should be possible to educate the users to withhold application if there has been rain or if the ground is sufficiently watered. Those in receipt of water and those managing the water supply must try to be vigilant and flexible (ASAE, 1981: 48). Educating farmers and improving management control of water application may well become easier in the future as microelectronics enable the application of water to be better monitored and as radio communications allow water users to hear regular bulletins on irrigation and educational programmes which might improve both their farming practice and their attitudes.

Chapter 5

Assessment of water resource plans

Introduction: general principles of evaluation

If problems and inefficiency are to be avoided or minimized water resources development plans should be assessed before implementation. *Pre-project evaluation* attempts to assess the likely economic, environmental and social costs and benefits of a proposed project, programme or policy. In practice in developing countries project assessment is more common than programme or policy assessment. Not only must these costs and benefits be forecast as accurately as possible, they must also be assessed to establish which are significant and what their likely magnitude will be. The assessment, then, has to be presented in a suitable manner to decision-makers, usually in the form of a report or an impact assessment document. Forecasting is seldom, if ever, completely accurate so that even with the best pre-project assessment, problems are likely to arise unexpectedly during or after implementation. Evaluation must not only forecast likely costs and benefits, it must also establish that the proposed development fits in with long-term regional and national development trends/policies or programmes. If it does not then support for it is unlikely to be easily found and, if found, is unlikely to be sustained.

Hindsight and common sense play an important part in the planning process; *post-project evaluation* (retrospective analysis) provides information which can help the planner avoid repeating past mistakes.

In-project evaluation (ongoing evaluation or monitoring) should be conducted for every scheme; without it management cannot be properly in touch with the performance of the development. Such evaluation can warn managers of the need to take remedial action or to modify their management strategy. In-project monitoring is often inadequate in developing countries and is a major cause of development problems. If a project cannot be monitored continuously, due to shortage of funds or skilled manpower, then it should be assessed at regular intervals, say once every year or every five years. Not only can such monitoring

/evaluation improve management it can also provide valuable information to help future developments (Commonwealth Secretariat 1978: 24).

Evaluations can be misleading. Assessment is as good as the assessors. Irrigation projects or other developments are too often judged 'successes' or 'failures' according to engineering or economic criteria. They may be technically or financially satisfactory but environmentally or socially disastrous. A common shortcoming of pre-project evaluation is that long-term costs and benefits are not adequately weighed against short-term gains or losses. Irrigated agriculture can run into a wide range of troubles and cause a diversity of problems; it would therefore make sense to use a multidisciplinary team to evaluate irrigation developments. In practice this assessment is often done by engineers and/or economists with inadequate consultation with social and environmental scientists.

The irrigation management literature is largely concerned with illustrating the poor water supply practices in effect around the world and with suggesting organizational solutions to these problems. The irrigated agriculture literature generally treats irrigation water as just another input along with labour and fertilizer. There is rarely any recognition of the special nature of irrigated agriculture nor any attempt to consider the water supply and cultivation practices together rather than in isolation (Bromley 1982: 3).

In order to make a meaningful assessment, whether pre-project or post-project, it is necessary for the assessor(s) to have some idea of the objectives of a development. In the case of irrigation development this may not be straightforward because it is initiated for a wide variety of reasons. Sometimes irrigation is seen as a quick way to produce crops for export to earn foreign exchange; sometimes to produce food to reduce imports, feed city populations or improve regional or national diets; for regional development, employment generation, sedentarization of nomadic or semi-nomadic peoples; the stabilization of degenerating rainfed cultivation may be the goal; there may be a desire to sustain otherwise transient agriculture, or a wish by a government to exercise greater control over farmers; to use irrigation and agricultural extension services to increase contact with rural people to aid modernization and, not infrequently, to raise taxation. Socio-political motives for irrigation development are often of equal or greater importance than maximum returns on capital invested or highest cost-benefit ratio (Kovda *et. al*. 1973; Bottrall 1981a).

Irrigation can attract powerful backing irrespective of its economic promise or other merits. Bureaucrats and politicians like to promote irrigation because it is visible, relatively apolitical, appeals to popular romanticism ('making the desert bloom') and can be finished in a time-span short enough to fit in with terms of office. The promotion of irrigation or other water resources developments may not only win a politician votes in the locality where the development takes place, it may also quell criticism of inactivity further afield. Aid donors sometimes back irrigation development for similar reasons. Many irrigation projects are show-pieces, not especially productive, but impressive to look at and somewhere that visiting dignitaries can be taken to show how 'progressive' the country is. Multinational construction companies with interests in providing the infrastructure or producing export crops are also very active in promoting projects. Irrigation can be used to establish a presence in disputed territory – some would say this has been the case in parts of Kenya and Israel (Carruthers & Clark 1981: 205). There are thus many reasons why water resources are developed and a number of them give little consideration to economic viability or the real needs of people.

Evaluation should not only try to reveal what might or what has gone wrong, it can sometimes reveal unplanned benefits which may then be better exploited. There is always a risk that assess-ments can be misused; 'rigged' to understate problems or hide them in jargon so that funding bodies, cautious governments or a suspicious public can be more easily won over. There have been cases where assessment has delayed or prevented proposed develop-ments at considerable cost, and in some developing countries this is a reason for reluctance to carry out assessments.

Evaluating environmental impacts

An increasing number of developing countries have adopted or are considering the adoption of legislation which requires at least larger development projects (and increasingly programmes and policies), including water resources development projects, to be subjected to pre-project *environmental impact assessment* (EIA). There is no concise, universally accepted definition of EIA, but most would accept that it is an activity designed to identify and predict the impact of an action on the biogeophysical environment and on man's health and well-being, and to interpret and communicate information about the impacts (Munn, 1979: xvi; Munn 1982: xvi). Many aid agencies, multinational corporations and the World Bank now use EIA to try to forecast the impacts of proposed develop-ment projects, legislative proposals, policies, programmes and aid donation on the environment and upon economies and societies (UNAPDC 1983; Ahmad & Sammy 1985).

A considerable range of impact prediction/evaluation techniques may be drawn upon by EIA. Whatever the methodology and techniques used it should be carried out sufficiently well in advance of implementation, ideally in the **earliest of planning stages** before any firm decisions have been made. If EIA is carried out early enough, and if it considers all the possible alternatives for achieving a desired planning goal (and the effects of no development), it can be a valuable aid to development planning and should help to ensure that resources are exploited to their full with the minimum of problems.

Development may have *direct impacts* (first-order impacts). These are relatively easily predicted, though assessment of their significance and magnitude may be less easy. There may also be *indirect impacts* (second or higher-order impacts). These are relatively difficult to predict because they occur at the end of a 'chain of causation' divorced in space and time (or both) from the development. For example, an irrigation project raises water-tables with the following consequences: (1) a particular tree species dies in the vicinity of the project because its roots are flooded; (2) a bird species nesting in that tree is deprived of nest sites; (3) as numbers of this bird decline an insect pest it preyed upon increases and damages crops; (4) the crop damage may be described as a fourth-order impact. Indirect impacts are often insidious and may be difficult to solve by the time they are apparent.

Impacts direct and indirect may be positive (beneficial to man or the environment in general) or negative (harmful to man or the environment); they may occur in the immediate vicinity of a development or at a distance from it; they may occur immediately or after some delay, possibly even years. Impacts may be widespread but of little significance, localized and insignificant, localized but very significant or widespread and very significant.

Techniques have been developed for EIA since the early 1970s; many of these originated in the USA or Europe and have only recently been applied in developing countries. In general EIA has not been used long enough in developing countries for its value to be fully assessed or for it to be more than partially adapted to developing country needs. Despite these weaknesses there is increasing use and increasing improvement of EIA in developing countries, and it has been valuable for assessing irrigation or water resources development impacts (OAS 1978; Interim Mekong Committee 1982).

Typically, developing countries seek EIA methods which are cheap, rapid and reliable. Simple matrices like the *Leopold matrix* and *overlay maps* are commonly used; they are easy to prepare and do not require especially skilled staff, but, they do not identify indirect impacts. Increasingly, systems analysis modelling and networking approaches are being used, although much more costly, these can identify indirect impacts – typical of such approaches is the *Sorensen network* (Munn 1979: 30).

Evaluating social costs and benefits

Some would argue that *social impact assessment* is merely one end of the same spectrum with EIA at the other, that the objectives, approach, methodologies and results obtained are sufficiently similar not to warrant separate treatment. Others argue that EIA is easier and more reliable because physical and, to a lesser extent, biological systems are relatively predictable, whereas human behaviour and hence social impacts are notoriously unpredictable. EIA tends to rely upon quantitative data, social impact assessment upon qualitative. The consequence of this is that social impact assessments tend to make less reliable, less precise predictions; nevertheless such predictions are valuable forecasting aids. And for post-project evaluation and monitoring social impact assessments predictive shortcomings are less important.

Social impact assessment techniques include: use of check lists, social surveys, questionnaire surveys. The main social variables assessed are: demographic impacts (size of population, composition, vital rates, mobility), human ecology (spatial arrangements, housing and neighbourhood studies, land-use patterns, accessibility), community/institutional impacts (amenities and services, community cohesion, community organization), cultural aspects (life styles, world views, beliefs). A problem with both environmental impact assessment and social impact assessment is that impacts that in themselves are harmless may act together to cause troubles – these *cumulative impacts* are difficult to recognize.

Evaluating economic costs and benefits

Evaluation of the economic impacts of water resources development usually depends upon *cost–benefit analysis* (CBA). This purports to describe and quantify social costs (disadvantages) and benefits (advantages) of a development (policy, programme or project); however, it is not always easy or desirable to consider social impacts in terms of monetary units. Benefits are considered against the capital outlays necessary to acquire them and they are then judged in the light of their *opportunity costs* i.e. the value of the utilities which have to be foregone to pay for them. Cost–benefit analysis attempts to allow for all the gains and lossess viewed from the standpoint of society, but what set of individuals constitutes society? Is it, say, the water resource supply administrators, or the water users or an entire nation? Also, traditional CBA views developments from the standpoint of present-day society, future developments are not adequately considered. Environmental and social impact assessment do claim to consider the possibility that factors affecting a development may alter in the future, and usually an attempt is made to establish whether identified impacts are irreversible or likely to have cumulative effects.

What should water resource assessment consider?

A balanced pre-project assessment of a water resources development proposal might include the following elements:

1. Whether the proposal is technically practical.
2. Assessment of short-term against long-term costs and benefits.
3. Assessment of who benefits or pays costs.
4. Environmental impact assessment.
5. Whether proposals are economically viable.
6. Whether proposals are legally and administratively pertinent – does it 'fit' existing laws and administrative procedures?
7. Assess if the proposals are socially acceptable, do the people want the development?
8. Whether the proposal is politically feasible and is compatible with local, regional and national interests.
9. Whether the project is flexible enough to adapt to unforeseen changes and altered circumstances.

There should be a fundamental shift from what generally prevails in developing countries today, namely *crisis management* (short-range preoccupation and 'technological fixes', which often do not work, to problems as and when they arise) to a more anticipatory *risk management* involving contingency planning and consideration of reasonably foreseeable futures (Laconte & Haimes 1982: 255–82).

Funding water resources development

Irrigated agriculture is presently the most productive kind of farming widely practised; it is also one of the most expensive.

Much of the irrigation development since the Second World War has been 'high tech', needing regular supplies of fuel, spare parts, fertilizers, seeds and so on. Many of these inputs have to be imported into developing countries. In the last few decades the costs of such inputs has climbed much more rapidly than the profit made on the sale of produce. Today the irrigator may have to sell say ten times the amount of produce compared with ten years ago just to buy inputs. This difficulty, often combined with poor management, means that many irrigation projects have become run down and need drastic rehabilitation of infrastructure, but unfortunately refurbishment of an existing project can often cost almost as much as construction of a completely new one. Also, when costly rehabilitation is carried out, it is no guarantee that the project will not immediately begin to deteriorate again. With the establishment of new irrigation and the rehabilitation of existing projects so costly, it makes little sense to fund them unless there is satisfactory management and an allocation of money for regular maintenance. A widespread problem in developing countries is that funding agencies give inadequate recurrent funds to ensure that there can be proper maintenance.

In the past, private companies have been important in developing irrigation, for example in India under the British Raj, the Madras Irrigation Company and the East (Orissa) Irrigation and Canal Company (Stone 1984: 19). During the last two decades there has been considerable private investment in irrigation. In southern India, for example, about $US15 million was invested in open wells and bamboo tubewells, which led to the opening up of roughly 30 million ha of irrigated farmland. This investment came largely from the farmers themselves – mainly the medium to large farmers who were able to take advantage of government credit schemes (World Bank 1982: 62). There has also been a marked increase in investment in irrigated agriculture in the developing countries by investors from developed countries, especially multinational corporations – in 1980, multinational corporations invested around $US15,000 million (World Bank 1982: 62). While that may seem a lot, it should be remembered that some multinational corporations have massive annual revenues, for example, according to Todaro (1981: 400) Exxon's in 1980 exceeded the entire gross national product of Sweden or Switzerland. Multinational corporations are adaptive; they can bring their own skilled management and 'know-how' in crop production and marketing to developing countries. They have the resources to carry out research, weather temporary set-backs such as pest attack or adverse market conditions, and may manage irrigation (and other ventures) with greater efficiency and continuity with less corruption than can some of the less stable developing countries.

Multinational corporations are not charitable institutions; they operate to make a profit. Some do not deserve the criticism often heaped upon them, but no matter how enlightened they almost invariably produce commercial crops for export and do little to raise food production in the host country. The *Brandt Report* (Independent Commission on International Development Issues 1980: 73) suggested that multinational corporations could do more to assist bilateral and multinational aid agencies in studies and surveys to assess the costs and benefits of proposed (not necessarily commercial) irrigated and rainfed agriculture projects, and could help in their execution when they are found to be viable: '. . . a very substantial mutual interest lies in harnessing the economic strength and experience of the multinationals for development.'

Various authors have commented on the widespread problem of non-existent or inadequate water charges, plus in many cases inefficient collection of charges. The rational reallocation of existing water supplies which would probably occur if charges were raised, is seldom considered as an alternative to uprating irrigation supply systems, usually at great expense (Clark 1970: vii). Usually it is the larger farmers pressuring the authorities who keep water prices low, although there are some countries, like Malaysia, where the government keeps charges down to help the rural poor.

There are alternatives to financial aid or loans. Governments or voluntary bodies may offer aid in the form of visiting experts; for example, Chinese extension workers have been seconded to a number of African nations to help them adopt or improve irrigated rice cultivation; the United States Peace Corps, British Voluntary Service Overseas and other bodies send personnel with appropriate training to help improve a developing country's agriculture and water supply systems.

Participation of affected people

In some countries irrigation infrastructure has been successfully built with local materials and manual labour with little cash investment. Sometimes such labour can be recruited in times of hardship in return for food wages, which costs the country little or none of its foreign exchange, and may offer a means of preventing dependency and a sense of shame which might be generated if food aid were simply distributed. A number of aid agencies now prefer to support such work-for-food schemes in the hope that they might improve agriculture to the point where future aid will become unnecessary as the infrastructure created boosts yields, halts soil erosion and/or improves moisture availability. In India, for example, the Drought Prone Areas Programme has provided employment for 21 million poor labourers on labour-intensive irrigation development, drainage, land reclamation and road-building. There are limitations on work-for-food development initiatives; mass unskilled labour is best suited to tasks like canal, road, ditch or terrace construction where workers are spread out and do not crowd one another. Also the use of such labour only makes sense if it offers more benefits than would the use of modern earth-moving or transportation machinery.

Sometimes people can be mobilized to improve irrigation or communications infrastructure or construct soil or moisture conservation measures without the distribution of food or monetary wages, if they are sufficiently inspired and motivated by aid agencies or governments. However, such efforts must be occasional (because people have their own farming duties to pursue), and limited in scale (because there are problems in controlling, housing and feeding large numbers of workers). Local cooperative organizations, youth groups and in some countries the armed services, have done much to improve agriculture, fight soil erosion and construct roads. For example, in Kenya the National Youth Service has done much irrigation and soil conservation work since the mid 1960s, and in Latin America, especially Brazil, the armed forces are active in road construction and the clearance of land for colonists.

The use of mass labour (in India, China, and elsewhere) has had mixed results. If a labour force fails to 'identify' with the work they are engaged upon (which is a risk if they are not local people likely to benefit from the work), then workmanship and productivity can be poor. If a government or aid agency is unable to generate enough enthusiasm with large groups of workers then it might find it more worth while to pursue smaller (community self-help) projects which have proved successful in a number of developing countries (Agarwal 1981).

Winners and losers

Because water resources can be used in a variety of ways there are often opportunity costs involved in developing sources for agricultural use. Even those who get to use water resources can sometimes end up as 'losers' because they may exchange adequate subsistence rainfed farming for risky market-orientated irrigated production of cash crops. There may be increased incidence of water-related disease after irrigation development or there may be greater economic polarization.

There is a real risk of jealousy and unrest if water resources development benefits one village or group of people but not others in the vicinity. The authorities managing the Geżira Scheme have tried to distribute the benefits of irrigation more widely by encouraging nomadic pastoralists to pasture their stock on sections of the Scheme that have just been harvested or which are being left fallow. The Gezira Scheme shows what can be done to widen the range of people who benefit from irrigation development, unfortunately it is not typical. Too often those who benefit from irrigation (the 'winners') are special-interest groups who wield power or who have money, while often those who most need the benefits irrigation can bestow tend to be losers.

Part II

WATER RESOURCES AND AGRICULTURAL DEVELOPMENT: TECHNOLOGY AND PRACTICE

Chapter 6

Using rainfall, runoff and floodwater, fog, mist and dew

Introduction

The FAO have predicted that by the year 2000 approximately 84 per cent of the world's total farmland will still rely on rainfall and will yield just over half of all crops (Biswas *et al.* 1983). (A recent overview of the extent of rainfed cultivation and the potential for expansion was provided by Higgins *et al.* 1984.) In most developing countries irrigation projects are costing more and more each year. Mexico, for example, made great advances with irrigation in the 1950s and 1960s, but since the early 1970s due to rising costs has increasingly turned to rainfed farming. One project, the Plan Puebla, a rainfed farming development scheme (begun in 1967), has shown that new crop varieties and cultivation methods can quadruple yields of maize and allow diversification into higher-value crops at lower start-up and operating costs than irrigation.

Over 600 million ha of the world's rainfed croplands receive less than 500 mm/y rainfall (Oregon State University 1979). In India alone, of the 138 million ha of arable farmland, 80 per cent is rainfed and 65 per cent of that suffers from inadequate or unpredictable rainfall (or both), (Unesco 1974: 78). Much needs to be done to improve the security of harvests for those farming such lands, and where possible to increase yields, diversify crops and extend the area under cultivation (Ambroggi 1980: 41).

In many developing countries where rainfed cultivation is practised, crop yields are decreasing. For example, the rainfed grain harvest in Niger fell from around 500 kg/ha in 1920 to around 350 kg/ha in 1978; the prediction for the year 2000 is that yields might well be in the region of 200 kg/ha if nothing is done to reverse the decline (Grainger 1982: 11). Over large parts of developing countries much could be done to remedy falling crop yields, sustain production and increase and/or diversify crops if rainfed cultivation were improved to make optimum use of available moisture (Postel 1985a). However, only recently has there been much attention paid to improving rainfed cultivation, and that pitifully small when compared with the efforts which have gone into installing expensive irrigation schemes, which not infrequently malfunction. Farmers throughout the tropics would benefit from improved rainfed farming because:

1. It is better to breed crops and develop cultivation techniques which 'fit' the environment, rather than attempt to modify the environment to suit the crops, which is what irrigation tries to do.
2. Environmentally appropriate crops and cultivation techniques are much less dependent on expensive, often finite inputs such as fertilizer or fuel.
3. Crop production which involves drastic modification of the environment can cause environmental problems such as pesticide pollution or contamination of streams with saline water or excess fertilizers.
4. Improving what is already widely practised by small farmers – traditional rainfed cultivation – will almost certainly prove less difficult than attempting drastically to change cultivators' ways, which is often the case if irrigation is introduced.

An important element in developing improved rainfed cultivation strategies is the production of 'environmentally appropriate' crops. Few international research centres have concerned themselves with developing improved, sustainable rainfed crops or cropping systems suitable for small farmers. The outstanding exceptions are the International Crops Research Institute for the Semi-Arid Tropics (ICRISAT), Hyderabad, India, and the International Centre for Agricultural Research in Dry Areas (ICARDA), the Lebanon, although interest has recently increased (Manassah & Briskey 1981: 342).

Although the area of rainfed rice grown is far greater than the area of irrigated rice, development efforts have largely ignored the former to concentrate on the latter. Typically an African, South Asian, South East Asian or Latin American peasant gets only 10 to 15 per cent of the yields with rainfed rice that irrigated rice could give. Improving rainfed rice production might not give as high yields as irrigated rice, but could still give great returns and would be especially valuable in areas where there is an unsuitable physical environment, lack of water and/or social resistance to adopting new techniques which preclude or make difficult the introduction of irrigation.

Improving rainfed rice production will not be without challenges; a large number of improved rice varieties will probably have to be bred to suit the varied local environmental conditions of the tropics (this is also the case with many other crops). Another problem faced by rainfed rice cultivators is that of weed control, and it may be that real improvements of yield await better tillage techniques and **safe** herbicides. Nevertheless, efforts are now being directed towards developing new rainfed rice varieties and the IRRI in the Philippines has already reportedly made some progress (Anon 1982a: 6).

Rainfed cultivation: strategies

Under *rainfed farming*, crops are dependent upon direct precipitation. Unlike irrigation which alters the natural environment considerably, rainfed cultivation only manipulates nature in an indirect way, for example through the timing of planting to coincide with optimal weather conditions, relatively sparse seeding to try to ensure crop plants do not compete with one another for water and the conservation of soil moisture (Hall *et al.* 1979: 4; Heathcote 1983: 169).

The goal of rainfed cultivation, whether a region's precipitation is well distributed but meagre or moderate but markedly seasonal, is to capture the maximum amount of moisture in the soil and to hold it there for as long as possible to support crops. Those practising rainfed cultivation do not face the same problems (of interrupted water supply) or costs (like paying a water supplier or pumping groundwater and maintaining on-farm water supply systems) faced by irrigators. Although not dependent on others for water supplies, rainfed cultivators are practising a high-risk enterprise, permanently facing the danger of moisture-deficient soils. Unreliable rains and droughts are frequent, for example, the Indian subcontinent has a drought of major proportions roughly once every twenty years, one or two Indian states being affected by drought approximately every five years.

Rainfed cultivation usually demands less constant attention than irrigated farming, so its practitioners are not so tied to the land and can more easily absent themselves for part of the year to engage in wage labour. The economies of a number of regions, for example the Zambian 'Copper Belt', depend on cheap labour provided by people who are able to delegate farming to their wives for a portion of the year. Any reduction in the cheap labour supply caused by agricultural changes which increase on-farm income, or lead to greater demands for farm labour, might be resented by such powerful interest groups as mining and industrial corporations or even governments.

Rainfed cultivation practices which ensure the optimum use

of water and the conservation of soil moisture are also useful to those practising irrigated agriculture because they offer means of reducing the amount of water needed, and they reduce the risks of crop loss should water supplies fail for some reason.

The character of the soil has considerable bearing on rainfed cultivation. For some soils which become hard during the dry season, rains are needed to soften them before farmers with only hoes or simple ploughs can cultivate. On the other hand, there are some heavy clay soils which become too sticky to work after prolonged rain. Some soils (often black earth 'regur' soils) change from too hard to too sticky so fast that farmers must choose the critical moment and work quickly if they are to obtain a crop. Sometimes the timing and amount of rainfall is such that soil conditions are so quickly unfavourable that the cultivator can only manage to prepare a portion of his land, or is delayed so that he plants late and his crops have insufficient time to mature before drought or the next rains strike them. The problem is most critical in regions with seasonal rainfall where the cultivators have only hand-tools. Where that is the case the introduction of animal- or tractor-drawn ploughs can greatly improve the security of harvest by ensuring that an adequate area is planted early enough.

Heavy clay soils hold water within reach of plant roots better than sandy soils, but may be poorly permeable and so shed water without storing much. Clay soils may also hold water so firmly that plants have difficulty extracting moisture. Lighter soils may thus offer rainfed cultivators more potential despite their theoretically poor water retention (Arnon 1972).

Strategies for improving soil moisture availability and use

The aims of the rainfed cultivator should be to maximize the intake of moisture into the soil, prevent it escaping then make the best possible use of it. There is a wide range of strategies available to the rainfed cultivator, some tried, tested and widespread, some effective but less widespread and some still largely experimental (Table 6.1).

Optimizing the use of soil moisture

Dry seeding and *sparse seeding* are widespread means of making the best use of available soil moisture. Large areas of India (at least 18 million ha), Sri Lanka, Africa and the New World depend on dry seeding in advance of seasonal, and often unreliable, rains (Domroes 1979). *Fallowing* is a strategy often adopted to accumu-

Table 6.1 Strategies for improving soil moisture availability and use

Technique	Advantages	Disadvantages
Improving soil moisture intake		
Plant cover crop(s) – provides ground cover and slows runoff (list of suitable plants given by Wrigley 1981: 88–90)	Cheap/easy/reduces soil erosion, can add organic matter to soil, may fix nitrogen, might provide a useful crop/be grazed	
Leave crop stubble/ debris – slows runoff	Cheap/easy reduces soil erosion, can add organic matter and nitrogen to soil, may be grazing opportunities	May harbour pests and diseases
Strip cropping – strips of arable land and pasture, stubble or perennial crops arranged roughly along contour	Cheap/easy/reduces erosion, may be symbiotic benefits between crops, i.e. one may benefit another by shading, supporting, deterring pests, fixing nitrogen and so on	May limit mechanization, grazing may be difficult without damage to cropped strips
Mulches – organic or inorganic material spread to slow runoff and hinder evaporation/weed growth	Can probably reduce evaporation, can deter insects, can reduce mud-splashing of crop, can reduce weed growth	Debate over real value
Tillage – to improve infiltration into poorly permeable soils	May speed planting when rains come	Can cause erosion, may cause oxidation of soil organic matter, requires labour or animal power
Chemical treatment – application of 'wetting agents' which speed infiltration	Cheap/easy	Not much tested, possible pollution risks, costs
Runoff retention bunds and terraces – structures designed to delay runoff and thereby improve infiltration	Cheap, demands little other than local labour, reduces soil erosion, may help prevent streams flooding, can help recharge groundwater	May be difficult to persuade farmers that effort of construction is worth-while

Table 6.1 (continued)

Technique	Advantages	Disadvantages
Reduction of evaporation losses		
Mulches – organic or inorganic, e.g. peat, plant debris, straw, vermiculite, ash, sand, dust, gravel, sawdust, etc.	Reduces evaporation?/ rain-splashing/ may deter pests/ reduces soil erosion/ discourages weed growth	Most presently used have to be applied after moisture has infiltrated and are lost by ploughing/ tillage or have to be removed before next cropping cycle
Anti-evaporation compounds: sprayed-on emulsions of water and oil or wax, hexadecanol, bitumen, asphalt, latex	Easy to apply/binds friable soils/reduces wind erosion and runoff erosion. Ideal for establishing shelter-belts or woodland in dry localities	Costly/pollution risks?/may degrade
Tillage of topsoil	Might reduce evaporation by hindering capillary movement in soil	Considerable debate over value
Removal of weeds – use of herbicide, pulling or burning	Can reduce interception losses	Useful only if vegetation cover is sparse and does not slow runoff yet intercepts precipitation on leaves, pollution risks
Planting hedges/ shelter-belts – should not compete, with nearby crop plants (which are usually shallow-rooting)	Reduction of evaporation/wind damage to crops/may increase dew and precipitation downwind. May provide fodder, fuel-wood, fruit, may fix nitrogen in soil, may provide compost for soil improvement	Takes time to establish/may need protection from animals/people until established
Structures to act as wind-breaks – brushwood, palm frond or stone barriers	Reduction of evaporation/shelter for crops from wind damage	Labour and materials needed/may hinder cultivation/mechanization

Table 6.1 (continued)

Technique	Advantages	Disadvantages
Reduction of evapotranspiration losses		
Cultivate crops with low evapotranspiration		Finding such crops
Removal of weeds from around crops, especially deep-rooting species	Reduction of evapotranspiration	May decrease infiltration/increase soil erosion because runoff not impeded, will probably need to establish low evapotranspiration cover crop in place of weeds
Spray crops with compound to reduce albedo – e.g. kaolinite and water	Cuts solar heating and air turbulence around crop. May reduce pest damage	Could reduce photosynthesis or respiration of crop and thereby cut yields, must be non-toxic and not distasteful if it lingers on the crop
Spray crop with compound(s) to hinder transpiration – e.g. wax emulsion, silicone, latex emulsion	As above	As above
Apply compound to alter crop physiology – e.g. chemicals which cause crop stomata to close	Quick-acting	As above
Reduction of percolation losses		
Add hydrophytic material to soil – organic matter or new 'Agrosoak' compounds	Easy and fast	Costly/must be imported
Insert plastic, bitumen or sheet rubber/plastic below crop roots – insertion with special ploughs or by hand	Also retains nutrients and thus reduces need for fertilizer application	Costly/may need a lot of labour and/or mechanical equipment/can deteriorate fast/may cause soil drainage problems or salinization

Table 6.1 (continued)

Technique	Advantages	Disadvantages
Optimizing use of soil moisture		
Dry seeding – seed sown before main rains break	Widespread/easy/may be necessary on some heavy soils that become difficult to work when wet	Risk of premature germination before enough rain falls
Sparse seeding – density of seeding is in inverse proportion to aridity	Optimizes moisture use by giving each plant room to spread its roots out and collect moisture	Poor yield per unit area, thus best if labour costs are low/mechanization used
Fallowing	Widespread	Debate over real value of fallowing
Sequence cropping – one crop planted to mature after another	No tillage between plantings needed (tillage allows more moisture to escape)	
Relay cropping – crops planted to grow up through previous crop	May be symbiotic advantages; one crop provides nitrogen, shade, support or discourages pests	Care needed to select appropriate combinations/more research needed
Intercropping – mixture of two or more crops or varieties of one crop, one a high yielder the other a lower yielder, but robust to withstand drought or disease or pests	Secure but decreases overall yields	More research needed

late soil moisture where annual precipitation is insufficient to safely support cropping (fallowing may also be used to allow land to regain its fertility or to allow salts to leach out after cultivation or to combat pests and diseases). For example, in the Maghreb (North Africa) fallowing is used to obtain cereal crops by rainfed cultivation every second or third year (i.e. a one-year or two-year fallow). There has been debate over the real value of fallowing in regions in receipt of more than 500 mm/y rainfall, the argument being that few soils are capable of holding over 250 mm of rainfall in their root zone, and most store much less (Grigg 1970: 170; Manassah & Briskey 1981: 351). Rainfed cultivation depending on fallowing can be a risky practice with the chance of crop loss being as high as one in five. In some parts of Africa acacia trees which produce gum arabic are traditionally grown during long fallows (often of over ten years' duration) with periodic crops of millet, sorghum or pulses. Where fallows have to be shorter, soil fertility might be improved with little loss of accumulating moisture if nitrogen-fixing leguminous shrubs were grown, for example *Stylosanthes*.

Intercropping, although widely practised by peasant farmers and shifting cultivators, has only recently attracted the attention of agronomists, but may have great potential (Beets 1982). In Mozambique, groundnut/maize/pumpkin rainfed intercropping by peasants can produce more food from a given area than the best modern farming methods in general practice, and seem to make good use of scarce soil moisture. Traditionally, some grain cultivators increase their security of harvest where there is a risk of drought by mixing a short growing season crop variety (which usually has mediocre yields) with a slower growing but better yielding variety. The former ensures some sort of harvest if rains are poor, the latter maximizes crop yield if there are good rains (Hall *et al*. 1979).

Reduction of evaporation losses

The ideal anti-evaporation treatment is one which is cheap, easy to apply and which does not hinder (ideally increases) infiltration, allows air to reach plant roots and soil micro-organisms, and if possible is of a higher albedo (reflects more sunlight) than the natural soil surface so that there is less convective air movement to carry away moisture. Gravel, ash, cinder, dust or sand *mulches* (see Fig. 6.1) go part way to meeting such requirements and are in widespread use, mainly in the belief that they cut evaporation losses (Hall *et al*. 1979: 33; UNEP 1983: 192). Jackson (1977: 79) is, however, of the opinion that mulches are of most value in increasing infiltration by retarding surface flow, and he is by no means alone in being sceptical of the value of mulching for moisture conservation (Wrigley 1981: 88, 91).

Except where there are suitable, convenient, low-cost materials mulching is really only suitable for crops which give high returns. It can help improve the quality of soft fruit and vegetables by reducing mud splashing and insect damage. Plastic sheets, popular with fruit and vegetable growers in developed countries,

Fig. 6.1 Ash mulch and cinder block wind-break used in Lanzarote (Canary Is.) to reduce evapotranspiration losses.

Technique reputedly discovered by accident after a volcanic eruption in the eighteenth century.

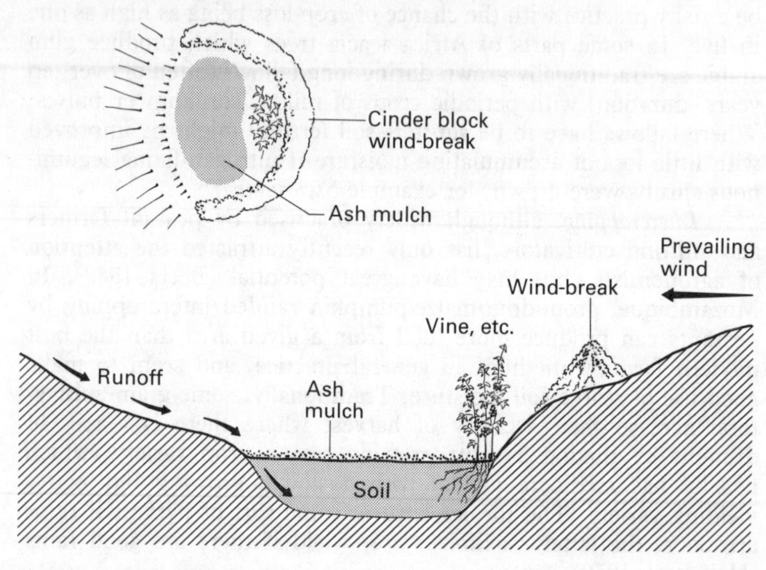

Source: Hall *et al*. 1979: 205.

are presently of limited use in developing countries because they are too expensive, too vulnerable to animal and sunlight damage and to theft. Latex, oil-emulsions or sprayed asphalt have been successfully used to bind friable soil or sand and conserve moisture, indeed in some arid regions such treatments have enabled the establishment of tree or shrub shelter-belts to check stabilize the soil and provide firewood. Oil-films have been used in parts of Libya and Algeria to create a 'North African green belt' mainly of eucalyptus trees. This green belt begun in Algeria in 1970 will, when completed, stretch for over 1,200 km from Morocco to Tunisia and will, it is hoped, help halt the deterioration of lands fringing the Sahara. Similar attempts to establish green belts have been made or are in progress in the Rajasthan Desert (India) and in Australia.

Opinions also vary on whether tillage of the topsoil reduces evaporation (Arnon 1981: 136). Deep tillage which disturbs the ground below about 10 cm certainly seems to waste soil moisture and thus *zero-tillage* is being increasingly advocated, even though it involves the laborious removal of weeds by hand or the use of herbicide to prevent unwanted transpiration losses (Wrigley 1984: 107).

Planting tree or shrub shelter-belts can cut soil erosion and evaporation losses. *Wind-break* is a term generally applied to non-living structures such as brushwood or woven palm-frond fences or to smaller-scale plantings. *Shelter-belts* are larger-scale plantings which give protection from the wind for at least twenty times their height downwind.

Care is needed in siting wind-breaks or shelter-belts to ensure that they are at 90 degrees to the most damaging winds (not necessarily the prevailing wind). Cross-sectional shape and the permeability of the barrier to airflow are also important; a shelter-belt or wind-break should not be too windproof or damaging eddies may form in its lee. Also, it should not be overlooked that shelter-belts take time to establish (Webster & Wilson 1966: 244; Cooke & Dornkamp 1974: 62). Care must be taken that local people understand the need for and support the establishment of shelter-belts or the erection of wind-breaks. If they do not, damage is almost certain. It is also important that the shrubs or trees used for shelter-belts or wind-breaks are deep-rooted so that they do not compete with nearby (more shallow-rooted) crops for moisture. Use of species with high evapotranspiration should be avoided, in Portugal eucalyptus planting has been blamed for lowered water-tables (*Ambio*, 1986: XV, pp. 6–13). Some plants which meet these requirements also have the advantage of being useful for supplying firewood, forage or because they fix nitrogen, their leaves can then be used for compost to enrich the soil – the 'ipil ipil' tree (*Leucaena leucocephala*) has attracted much attention for possessing such properties. There have been some impressive claims for the ability of shelter-belts or wind-breaks to cut evaporation losses; for example, in semi-arid parts of the USSR it has been reported that grain yields in the lee of tree shelter-belts increased by 20 to 30 per cent and forage yields by up to 200 per cent (Adams *et al.* 1978: 103). In Niger, according to Grainger (1982: 13, 79), tree shelter-belts in the Maji Valley, established since 1975 by the aid agency Cooperative for American Relief Everywhere (CARE), have increased millet yields by up to 15 per cent. Some of the rise in yields may be due not only to reduced evaporation but to an increase in dewfall (Oliver 1972: 119).

Reduction of transpiration losses

The removal of high-transpiration-rate herbs, trees and shrubs may sufficiently improve soil moisture availability to make rainfed cropping more secure and more productive. In South West Africa for example, the trees *Tamarix austro-Africana* and *Acacia albida* were found in one study to have depleted 68 per cent of the moisture from sandy soils (Balek 1977: 73). *Brachystegia*, a tree typical of great expanses of African savannas can reputedly transpire 90 per cent of the precipitation received by the soil (Balek 1977: 73). In some regions replacing deep-rooted, high-transpiration-rate trees and scrub with selected grasses allows recharge of groundwater sufficient to maintain shallow wells, stream and spring flow year-round.

Many tropical plants exude resins, waxes or have varnish-like coverings or hairs on their leaves and stems to reduce water loss. Nature might be profitably copied by physically sealing crops to reduce transpiration with sprayed-on films of low-viscosity wax, silicone, latex or plastic: these must be permeable to oxygen and carbon dioxide, but not water vapour. Alternatively, a substance might be applied to a crop which causes partial or total closure of stomata, for example, abscissic acid, alkenyl succinic acid, atrazine, phenylmercuric acetate. Some of these chemical *anti-transpirants* can reputedly cut transpiration by 40 per cent with little hindrance to plant metabolism (and therefore little depression of crop yields). At present anti-transpirants are largely experimental and costly, but already there are ultra-low-volume application sprayers which could reduce the expense of application. Probably these treatments have greatest promise as a last-resort measure when drought threatens a crop or an irrigation supply fails (Sutcliffe 1979: 83). There are some hopes that biochemists will develop compounds that can be sprayed on to crops to counter the persistent growth-inhibiting hormones often produced when plants are subjected to moisture stress. Crops often survive short droughts, but have their yields depressed by these hormones (Carruthers & Clark 1981: 23).

There have been a number of experiments with spraying solutions of diatomaceous earth or kaolinite (celite) on to crops and soil, these being light-coloured, inert compounds that are supposed to reflect sunlight and thereby cut evapotranspiration losses. Soya-bean crops treated with the latter solution reportedly had their evapotranspiration losses cut by roughly 15 per cent (Slater & Levin 1981: 157); Carruthers & Clark (1981: 31) reported reductions of 22 to 28 per cent in similar trials. In neither case do the reductions in evapotranspiration seem to have cut crop yields to any marked degree.

Reduction of percolation losses

The addition of hydrophilic material to sandy soils can greatly increase their water retention, although organic matter like compost may not be as effective in doing this as some have claimed. A wide range of materials have been tried, including seaweed, peat, crushed brown coal, waste paper and many other compounds (National Academy of Sciences 1974). Spectacular claims have recently been made for the effectiveness of starch copolymers or granular (plastic) polymers, for example polyacrylamide – 'Agrosoak', which if added to a soil are supposed to be capable of holding very large quantities of moisture (Anon 1982b). Compounds like Agrosoak seem especially useful for establishing tree seedlings (*Unasylva*, 1986, No. 152, p. 38).

Percolation in free-draining soils can be physically hindered by films of plastic, rubber or bitumen, either spread out and then buried beneath the planting bed, or injected/introduced as a ribbon a few metres wide by a special plough at a suitable depth. Such techniques can cut percolation losses considerably, and can reduce the need for costly fertilizers, but they are presently too costly for most developing countries (for details of costs see Gischler 1979: 112).

Rainfed cultivation: crops and crop improvement

Until recently the efforts of plant breeders were mainly directed towards improving yields through better crop response to fertilizer applications under optimum, not suboptimum or marginal farming conditions. Some crop breeders are now developing crops more suited to the real needs of small farmers, many of whom cultivate in harsh environments. Plant breeders can improve drought resistance/avoidance in a number of ways:

1. By selecting plants which have short growing seasons.
2. By developing crops which become dormant and then recover after drought.
3. By reducing transpiration. It might be possible to develop crops which can reduce their transpiration without their yield suffering too much by closing their stomata during the hottest part of the day and fixing carbon dioxide at night or in wet or dull weather (Anderson 1981: 103; Heathcote 1983: 78).
4. By developing deep-rooting crops which can reach more soil moisture; some sorghum varieties have been developed which are capable of this.
5. By developing crops with extensive shallow roots to take advantage of light precipitation which infiltrates only the surface layers of the soil.
6. By altering plant morphology to reduce transpiration losses, but not yields. For example, short-stemmed varieties of wheat and barley have been bred which give about the same or better yields as long-stemmed, but which transpire much less (there being less leaf and stem area bearing stomata). Another option is to develop crops which shed their leaves, especially the older,

less photosynthetically active ones, when there is moisture stress; there are varieties of wheat and barley with this ability.

Over large parts of Africa, maize has replaced traditional rainfed cereals. A hectare planted with maize can give 1,000 to 5,000 kg of grain per harvest, traditional cereals as little as 100 and seldom over 800 kg/ha – the change-over is hardly suprising. However, if rains are unsatisfactory maize is likely to fail completely, the traditional cereals may well survive and give a harvest. There is, in Africa at least, a real need to concentrate on improving the yield of hardy traditional crops. The most promising is probably sorghum. Plant breeders, notably at ICRISAT, are presently trying to improve its disease resistance, raise yields and, most important of all, shorten the growing season. Where there is a highly seasonal rainfall, a short growing season crop may be able to mature before the worst drought season. Already varieties have been developed which mature in 70 rather than 100 days (Cross 1985). It may also prove possible to raise sorghum yields from the present rainfed average of 200 to 300 kg/ha to 900 kg/ha, and with irrigation possibly to 5,000 kg/ha (Anon 1985a).

Millet yields might also be raised and the growing season shortened. Recent work suggests an increase from the present rainfed farming maximum of around 800 kg/ha with traditional varieties (and a 200 mm/y rainfall regime) to 2,000 kg/ha (already obtained in trials by ICRISAT), possibly even to 3,000 kg/ha is possible (Anon 1985b). One variety of pearl millet (*Pennisetum* spp.) needs only sixty days from sowing to harvest (Hall *et al.* 1979: 200).

There have also been considerable successes in producing barley able to thrive on only 100 to 125 mm/y rainfall, and short-stemmed, water-efficient wheats (Ritchie 1979: 2). While crops with such short growing seasons can reduce the risk of crop loss due to drought, and may in areas that receive higher rainfall permit two or even more crops a year, they are not without risks, an early-maturing field may well attract the undivided attention of pests. There is also a risk that converting from single to multiple crops in a year will give rise to soil fertility problems and even erosion.

Drought resistance varies between species and varieties and from one individual plant of one variety to another of the same variety. The vulnerability of most plants to drought increases at certain stages of growth: germination, early seedling stage and when flowering or setting fruit are usually the most critical periods. If moisture is scarce at just one of these periods, even though it be adequate for the rest of the year it may prevent a crop – for this reason very limited irrigation, possibly for only a few days a year may have considerable potential. The salinity of water available to crops greatly affects drought resistance; both saline soils or irrigation water with high concentrations of salt being likely to cause water stress in crops.

There are other means by which plant breeders might help rainfed cultivators, apart from increasing drought resistance:

1. Develop crops which give small farmers very secure food supplies with minimum inputs.
2. Develop crops with a high value to unit weight, and if possible, a capacity to withstand rough treatment during storage and transport. These would include so-called *industrial* and *hydrocarbon* crops which are rich in alkaloids, essential oils and hydrocarbons (waxes, xylose and latexes) or could be used to produce alcohol by fermentation, or useful fibres (Belmares *et al.* 1979). Such crops may well increase in importance if petrochemicals become scarce (Table 6.2).
3. Develop crops with *inhibited germination*. In the wild, some plants have seeds which will not germinate unless there has been enough rainfall to ensure there is adequate soil moisture for growth to maturity. Few rainfed crops have this valuable trait. If the plant breeder cannot develop crops with inhibited germination it may be possible to treat seeds chemically to achieve the same result. Inhibited germination could increase the security of harvest in many regions where rainfed cultivators sometimes make 'false starts' and lose their crops when adequate rain fails to fall (UNESCO 1962: 187).

Some pessimists claim that getting small farmers in remote rural areas to adopt improved varieties or new crops is difficult (Smith 1982: 99), yet there have already been many successes in disseminating new crops to dryland cultivators. A cross between wheat and rye (triticale) which can withstand drought and saline soils better than either parent species, and which also provides a more nutritious grain, has been quite widely adopted, particularly in Mexico (Smith 1982). In the 1960s cultivation of a (non-cereal) grain-producing plant, the amaranth (*Amaranthus* spp., see Table 6.2), spread relatively suddenly from northern India and Nepal to the Deccan Plains (India); it is now widely grown by small farmers from northern India down into upland southern India. Amaranth was a major crop in Latin America before the 1520s, but was suppressed by Cortés and his followers and is today little grown. The grain of amaranth is rich in lysine, an essential amino acid which is virtually absent in other cereals; if poor farmers who presently grow sorghum or wheat were to adopt the crop they could considerably improve their diet. Amaranth is an ideal quick-maturing crop for seasonal rainfall regions which have more than 600 to 800 mm/y rainfall, and it will also grow at high altitude (Vietmeyer 1982). It makes excellent poultry feed and has a further quality which may hasten its adoption in South and South East

Table 6.2 Plants with potential as dryland crops

Plant	Useful quality
Food and forage	
Sorghum (e.g. *Sorghum bicolor* race dura.)	Grain, drought resistant
Pearl millet (*Pennisetum americanum*)	" " "
Finger millet (*Eleusine coracana*)	" " "
Acha or hungry rice (*Digitaria exilis*)	" " "
Australian channel millet (*Echinochloa turnerana*)[a]	and forage
Triticale (*Triticum* × *Secale*)[b]	" " "
Palmer saltgrass (*Distichlis palmeri*)[c]	" " "
Amaranth (*Amaranthus hypochondriacus, A. cruentus, A. edulis A. caudatus*)[d]	and forage " " "
Tepary bean (*Phaseolus acutifolius*)	Pulse, nutritious, grows under poor moisture, high yields.
Chick-pea (*Cicer arietinum*)	Pulse, low moisture needs
Pigeon pea (*Cajanus cajan*)	" " "
Cow-pea (*Vigna unguiculata*)[e]	Pulse, good yield under arid conditions
Soya bean (*Glycine max*)	Pulse, already important
Marama bean (*Tylosema esculentum*)	Pulse
Safflower (*Cathamus tinctorius*)	Oil-rich seed, withstands drought
Bambarra groundnut (*Voandzeia subterranea*)	Oil-rich 'nut'
Cassava (*Manihot ultissima*)	Carbohydrate-rich food or alcohol production feedstock, resists drought
Yeheb (or Ye-eb) nut (*Cordeauxia edulis*)	Edible seed and forage, resists drought well
Cashew nut (*Anacardium occidentale*)	Valuable nuts, fruit and useful wood
Prickly pear (*Opuntia ficus-indica*)	Fruit or fodder
Almond (*Amygdalus communis*)	High-value nuts
Olive (*Olea europea*)	Oil-rich fruit
Date (*Phoenix dactylifera*)	Fruit
Buffalo gourd (*Cucurbita foetidissima*)[f]	Oil and starch
Mesquite ('tamarugo') (*Prosopis* spp.)[g]	Protein-rich seed, forage, fuelwood, wind-break, fixes nitrogen
Saltbush (*Atriplex* spp.)	Forage in dry and/or saline environments
Cassia sturtii	Salt-tolerant, drought-resistant forage?
Cassia (*Cinnamonum cassia*)	Essential oil (cinnamon substitute)

Table 6.2 (continued)

Plant	Useful quality
Food and forage (continued)	
Acacia spp. (e.g. *Acacia aneura*, *A. senegal*, *A. nilotica*)	Gum arabic (where there is 200 to 400 mm/y rainfall), wind-break
Guar (*Cyamopsis tetragonoloba*)	Gum and oil, tolerates drought and saline soil
Ipil Ipil (*Leucaena leucocephala*)	Fodder, fuel, wind-break, shade and nitrogen fixation. Fast-growing, will flourish on poor soil with 400 mm/y rainfall
Stylosanthes	Fixes nitrogen, stabilizes degraded soils in drylands
Neem (*Azadirachta indica*)	Leaves discourage insect pests – can use as mulch to control soil pests or steep in water to make 'insecticide'. Useful wood
Hydrocarbon and Industrial	
Jojoba (*Simmondsia chinensis*)[h]	Wax, oil-cake and forage
Euphorbia (e.g. *Euphorbia lathyris*)[i]	Candelilla wax and hydrocarbons
Guayule (*Parthenium argentatum*)	Hydrocarbon-rich latex, can be used for rubber production
'Russian dandelion' (*Taraxacum kok-saghys* and *Scorzonera tau-saghys*)	Used for rubber production in USSR
Canaigre (*Rumex hymenosepalus*)	Tanning liquor
Agave (e.g. *Agave lecheguilla*)	Fibre
Creosote bush (*Larrea tridentata* and *L. divaricata*)	Hydrocarbons, will survive on poor soils and 50 mm/y rainfall

[a] Needs only one rainfall to germinate and then little further moisture to mature.
[b] Yield compares with that of wheat, but it withstands drought and saline soil better. Already adopted by smallholders in some countries.
[c] Used by North American Indians, yield could probably be much improved.
[d] The food value of the grain is especially good, and the plant resists drought well. It seems suitable for either large-scale or small-scale farming.
[e] IITA are reported to have produced a variety with a 40-day growing season able to yield 1,000 kg/ha under Nigerian smallholder cultivation.
[f] This seems likely to become a much more important crop in the future, yields of oil and starch are heavy even with poor soil and low rainfall.

[g] Mature trees will reportedly yield 10,000 kg/ha per year in areas with only 250 to 500 mm/y rainfall. The seed needs no cooking, a useful quality in fuel-short regions.

[h] Can reach groundwater 30 m deep with a long tap root. The oil has begun to replace sperm whale oil. Useful production probably needs over 200 mm/y of rainfall.

[i] Yields of 20 barrels/ha/y reported. *Euphorbia* latex is of lower molecular weight than *Hevea* latex or man-made rubber which may give it marketable qualities.

Note: An extensive list of dryland plants used by rural peoples may be found in Adams *et al*. (1979: 136–9), Arnon (1972: 84–17) and in the *Journal of Arid Environments*, 11 (1986): 17–59; a list of savanna tree species with potential is given by Kowal & Kassam (1978: 237).

Sources: National Academy of Sciences 1975; Adams *et al*. 1978: 55–8; Ritchie 1979; Hall 1981; Hutchinson *et al*. 1981: 11–12; Kazarian 1981; Madelay 1982; Mannasah & Briskey 1981; Vietmeyer 1982; Myers 1983.

Asia – being regarded as neither pulse nor grain by Hindus it can be consumed during times when people are observing fasts (according to Kipling – see *The Portable Kipling*, Penguin, Harmondsworth, 1982: 132).

What are the limits of arable cultivation?

Although it is frequently carried out, simply mapping rainfall isohyets does not give a reliable indication of the limits of cultivation. A complex of environmental, socio-economic and technical factors determine the productivity of land and the utility of crops, a change in any one of which may turn unproductive marginal land into productive or *vice versa* (Furon 1963: 138; Biswas 1979a: 257; Andreae 1981: 86; Kaduma 1982). Prevalent crop or human diseases, or lack of law and order may also make an area marginal. Some of the world's poorest people live on marginal land, and in some developing countries comprise a high proportion of the total population. The farming of marginal land has to be improved, but due to the risks of failure and the slow and meagre returns generally made, even if development efforts are 'successful', funds for agricultural improvement tend to go to better-watered regions (Brady 1981).

Runoff cultivation

Runoff cultivation is a generic term applied to cultivation methods which rely upon tillage and/or planting patterns, banks or terraces to delay and retain runoff (overland flows), to increase moisture infiltration and reduce soil erosion. Although a few tropical crops do quickly cover the ground surface with enough luxuriance to retard runoff and thereby reduce soil erosion and increase moisture intake, most do not. Cultivation with most crops on all but level ground and gentle slopes is likely to require ground cover crops, terraces, ploughed ridges and so on if it is to be sustained.

Contour farming, strip cropping and terraces

The simplest way of retarding runoff is for the cultivator to plant or plough along the contour (*contour farming*). Another possibility is *strip cropping* which involves alternating strips of crops with strips of grass, forage or cover crops. The method can be used on quite steep slopes (between 3.5 and 8.5 degrees), the strip width being varied according to the gradient (Cooke & Dornkamp 1974: 43). The cropped strips and the grass, forage or cover crop strips can be rotated to maintain fertility and combat pests and weeds. A problem with both contour farming and strip cropping (and with some forms of terracing) is that if care is not taken to construct suitable *spillways* (drop structures) to safely shed excess water, downslope gullies can be eroded at roughly 90 degrees to the strips or terraces. If slopes are of less than 11 degrees (26 per cent) simple grassed channels usually suffice, but where there are steeper gradients more sophisticated structures are required (Fig. 6.2a illustrates a common drop structure: the *lock-and-spill drain*).

One of the oldest and most widespread means of retarding runoff (and retaining a level strip of soil for planting) where ground is sloping is *terracing*. There is considerable diversity in form and construction of terraces, but they may be divided into two broad groupings: *bench terraces* constructed to enable farmers to crop steep slopes, which considerably modify the natural terrain; and *contour terraces* which are designed to slow runoff but which do not radically alter the general slope profile (Webster & Wilson 1966: 108–14; Sheng 1981). In a review of terracing suitable for the humid tropics, Sheng (1977) recognized four broad types of *bench terraces* (Fig. 6.2b, c). Reverse-sloped bench terraces (Fig. 6.2b iv) are especially suited to steep slopes where draining excess runoff may be a problem, most of the types of terrace illustrated in Fig. 6.3b, c, d are reverse sloped, and will function satisfactorily where gradients are between 7 and 30 degrees (Fig. 6.3a depicts simple terraces without bench construction). Bench terraces and some other terraces can be built and cultivated by manual labour or by animal- or tractor-drawn machinery. *Hillside ditches* are disconti-

Fig. 6.2 Farming sloping land (a) *Lock-and-spill drain*, vital for removing excess runoff from terraces. Construction removes much of the silt from the channelled flow. (b) *Types of bench terraces*: (i) level bench terraces; (ii) outward sloped terraces; (iii) conservation bench terraces; (iv) reverse-sloped terraces. (c) *Detail of simple bench terrace*.

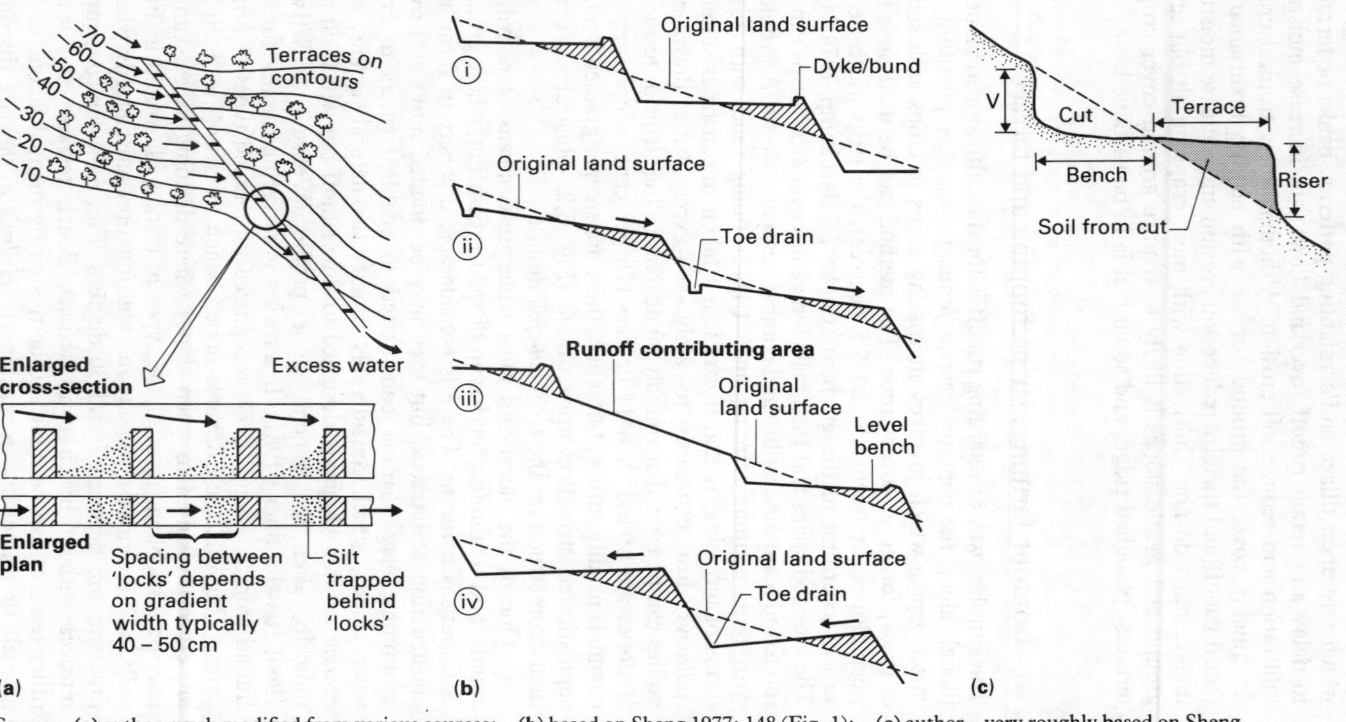

Sources: (a) author much modified from various sources; (b) based on Sheng 1977: 148 (Fig. 1); (c) author – very roughly based on Sheng 1977: 151.

nuous, narrow, reverse-sloped terraces constructed to break up long slopes into a series of shorter slopes. The hillside ditches carry excess runoff to suitable drop-structures, like lock-and-spill drains, and cultivation is practised on the slope **between** the along-contour hillside ditches. Usually some form of contour ploughing or strip cropping will also be required to control runoff. *Individual basins* can be dug into steeper slopes (25 to 30 degrees), and semi-permanent or permanent crops can be planted in these basins. Fertilizers tend to remain longer in such basins and they can be mulched to conserve moisture and more easily kept free of weeds. Basins are especially useful where the soil depth and/or quality is very variable (Fig. 6.3b). *Orchard terraces* as the name suggests are planted with fruit or other tree crops, either directly along the terrace or in basins dug into the terrace. *Mini-convertible terraces* are terraces planted with field crops, with fruit trees or other permanent crops in between (Fig. 6.3c). If there is a need to intensify production the slopes in between the terraces can be converted into further terraces, or if there is less demand for food crops or a labour shortage, then all terraces can be planted with tree crops. *Hexagons* (Fig. 6.3d) are arrangements of farm roads which envelop a plot of land which can be accessible to tractors. The method is used for mechanized orchard farming on level ground, but can be applied on slopes of up to 20 degrees. If the roads are planted with food crops the technique can be used by small farmers (Sheng 1977: 149).

The role of terracing in increasing moisture infiltration and storage is greater than is usually acknowledged – there is a tendency to view terracing more as a means of stabilizing soil and reducing erosion. Denevan (1980) pointed out this misconception with respect to terracing in the New World, 85 per cent of which he claimed was in drylands (where there was less than 900 mm/y rainfall and a dry season of more than five months' duration) and had been constructed more to conserve moisture than control soil erosion.

Spectacular examples of terrain-modifying terracing are to be found in Latin America (the Andes got their name because of the widespread terracing found by the Conquistadores) and parts of Southeast Asia settled by *émigré* Chinese. In some cases terracing enables cultivators to work land close to city markets which is not built upon because it is too steep. In Indonesia, notably Bali and Java, and in the Philippines, notably Luzon, sophisticated contour terraces retain enough rainwater or the flow of small streams so that padi rice can be grown on steeply sloping land.

Contour terracing was widespread throughout much of Central and South America (on 12- to 25-degree slopes) before the Spanish Conquest, nowadays it survives and is still important

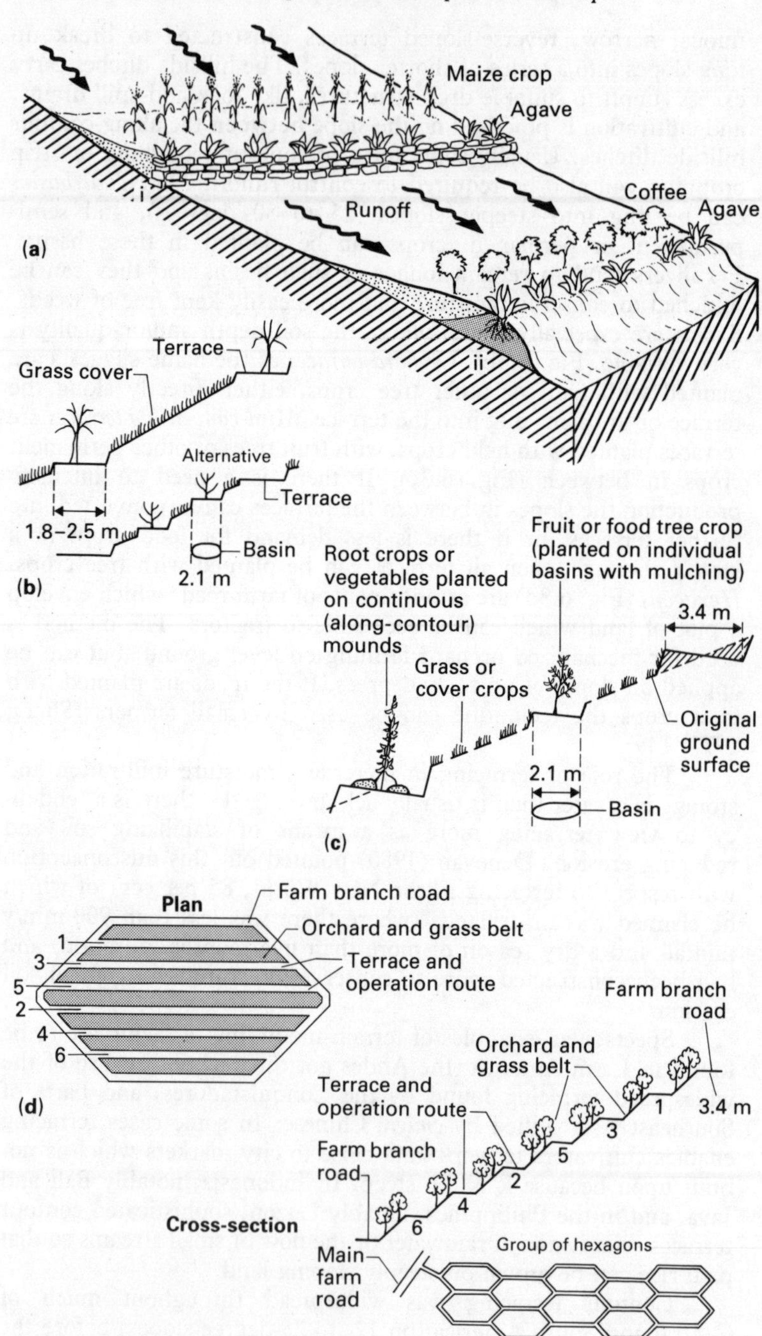
Maize crop
Agave
i
Runoff
Coffee
Agave
(a)
ii
Terrace
Grass cover
Alternative
Terrace
1.8–2.5 m
Basin
2.1 m
(b)
Root crops or vegetables planted on continuous (along-contour) mounds
Fruit or food tree crop (planted on individual basins with mulching)
3.4 m
Grass or cover crops
Original ground surface
2.1 m
Basin
(c)
Plan
Farm branch road
Orchard and grass belt
Terrace and operation route
1
3
5
2
4
6
(d)
Farm branch road
Orchard and grass belt
Terrace and operation route
Farm branch road
3.4 m
Cross-section
Main farm road
Group of hexagons

Sources: (**a**) to (**d**) from Sheng 1977: 150.

mainly in Peru and Mexico in regions which receive upward of 400 mm/y rainfall. *Conservation bench terraces* (see Fig. 6.3a) which channel scarce rainfall on to the cultivated plot, are cheap to construct, typically costing in the late 1970s around $US70 per (planted) hectare on moderate slopes with soft soils, and requiring little labour to farm (UNEP 1983: 117). In Arizona (USA) researchers found that similar terraces could yield an average of 2,600 kg/ha of sorghum where only 246 mm/y of rain fell, and up to 4,400 kg/ha where rainfall was better (the sorghum was grown in rotation with wheat and then grass fallow to maintain fertility, this would appear to offer a promising dryland rainfed farming strategy.

Considerable areas of terracing similar to that in use in Mexico, plus large areas of more sophisticated irrigated terraces (watered from hillside streams diverted along stone-lined channels which follow the contours but with just enough gradient to keep them flowing) lie abandoned in the highlands of Central and South America. In Peru alone Gischler and Jauregui (1984) estimate 80 per cent of pre-Hispanic terracing lies abandoned. In some locations such systems are still in use, but the users often seem unwilling or unable to maintain and repair the systems they have inherited. Archaeological evidence and recent experiments (notably those of UNESCO) in highland Peru suggests that present-day farmers could renovate and extend many ancient channel and terrace systems to obtain good crops, and in some regions to do this would be more cost effective than developing new lands (Donkin 1979; Klee 1980: 22; Hutchinson *et al.* 1981:3; Gischler and Jauregui 1984: 11). However, some caution should be exercised. In parts of Latin America there is archaeological and geological evidence of tectonic movement and flash flooding. In Peru, where there was roughly 40 per cent more cultivated land in pre-Conquest times than now, well-laid-out irrigation systems were destroyed (e.g. along the Moche River of northern Peru) (Saunders 1986). This *should* serve as a warning to those presently developing large-

---

Fig. 6.3 Farming sloping land (**a**) *Simple terraces*: (i) use of stone retaining wall stabilized with agave plants (which can be harvested for fodder, fibre and other uses); (ii) use of earth-bund also planted with agave. These types of terrace are common in Mexico where they are called bordos or metlepantli.

Terraces can collect and concentrate rainfall to support crops in low-rainfall regions, the less rain the greater the space between terraces. (**b**) *Orchard terraces:* (i) terrace proper planted with tree crop; (ii) basins dug on contour or into terrace and planted. Latter useful where ground is uneven and soil variable. (**c**) *Mini-convertible terraces.* (**d**) *Hexagons.*

scale gravity irrigation – infrastructure *should* be able to withstand earth movement and floods.

Terrace cultivation is not especially demanding of labour, but the construction of terraces can be. In Ethiopia, for example, 1 km of simple field terraces required 150 man-days in the early 1980s (Cross 1983). In some regions where terracing may appear to have promise, the farmers may be so involved in growing crops to subsist that they have little time and no one to organize their construction (Rambo 1982: 89). It is not uncommon for peasants who are unfamiliar with terracing to be sceptical that the efforts involved in constructing them will yield worthwhile rewards. The establishment of demonstration terracing may help overcome this problem and would give the development agency wishing to promote them, a chance to check that the techniques really do work in an area (Carpenter 1983: 129).

A cheap, rapidly installed and effective method of stabilizing and cropping steep slopes is *wattling and staking* (*clayonnage*); this has been widely used in Jamaica (Fig. 6.4; FAO 1977: 289–93).

Ridging and broad beds

Wherever gradients are less than about 7 degrees, runoff is still a problem but terracing is impractical. *Contour ploughing*, *contour bunds*, *strip cropping* or *tied ridging* can be adopted (Fig. 6.5a). The latter, a widely used technique, can have marked effect: when an area of traditional rainfed cultivation in northern Nigeria adopted the method, cotton yields rose from 224 kg/ha to 2,242 kg/ha (fertilizers were also applied to the land; Jackson 1977: 177). The 'ties' trap moisture and reduce soil erosion, but unfortunately intense rainstorms can seriously damage them and this has led ICRISAT to explore alternative ways of cultivating gently sloping land (gradients of about 0.5 per cent) in regions with intense rainfall – conditions not uncommon in the semi-arid tropics.

A system of *broad-beds* and *grassed drains* which channel runoff into storage tanks to be used for supplementary irrigation of rainfed crops has been developed by ICRISAT (Fig. 6.5b, c). Trials suggest that such a system providing roughly 50 mm/y supplementary irrigation could increase yields of millet, sorghum, chickpea or groundnut between two and fivefold, compared with those of simple rainfed cultivation under the same environmental conditions. It would also greatly decrease the risks of crop loss if rainfall is poor. Bullock-drawn water carts fitted with sprinklers offer a cheap and effective means of applying the supplementary irrigation and can operate even if the terrain is undulating (Lal &

Fig. 6.4 Farming sloping ground: wattling and staking.
Stakes set in rows on contour. Angle of stakes indicated in (i). A trench is cut just above stakes (ii). Brushwood wattles are packed into trench (iii). Soil from next trench used to bury wattles of first (iv). Cereals sown at g (at g′ in very loose soil or if wattles/trenches are wide spaced, say 1.25 m apart); other crops may be grown at s–p.

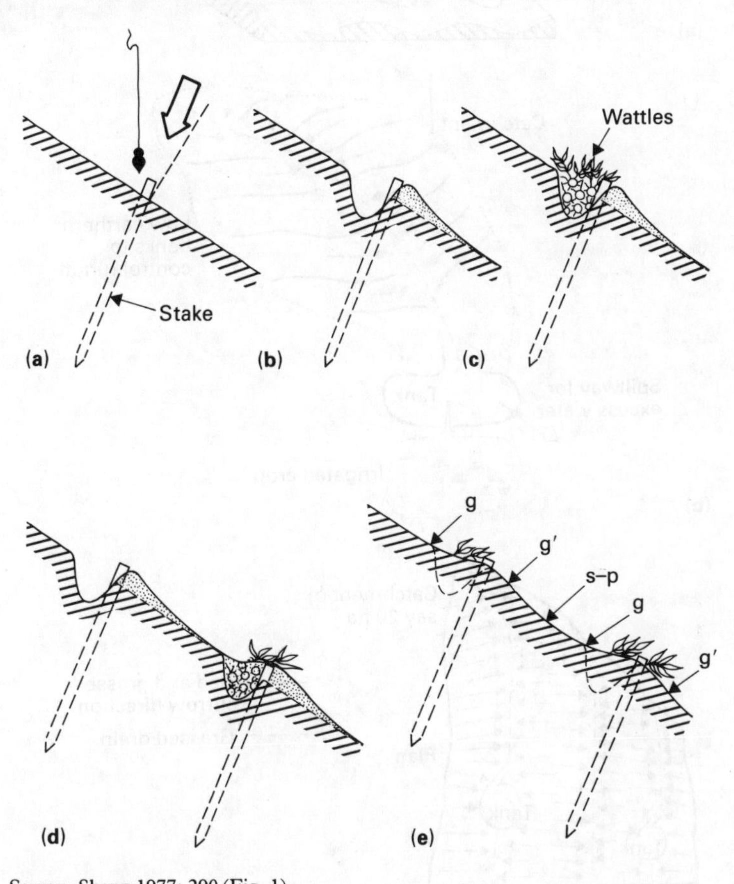

Source: Sheng 1977: 290 (Fig. 1).

Russell 1981: 259; Manassah & Briskey 1981: 364, 391; Anon 1984).

Broad-bed and grassed drain systems have other advantages – they need less tillage than many other cultivation methods, just an occasional harrowing to reduce weeds and periodical reforming of the beds with a bullock-drawn moulder. Because there are fewer

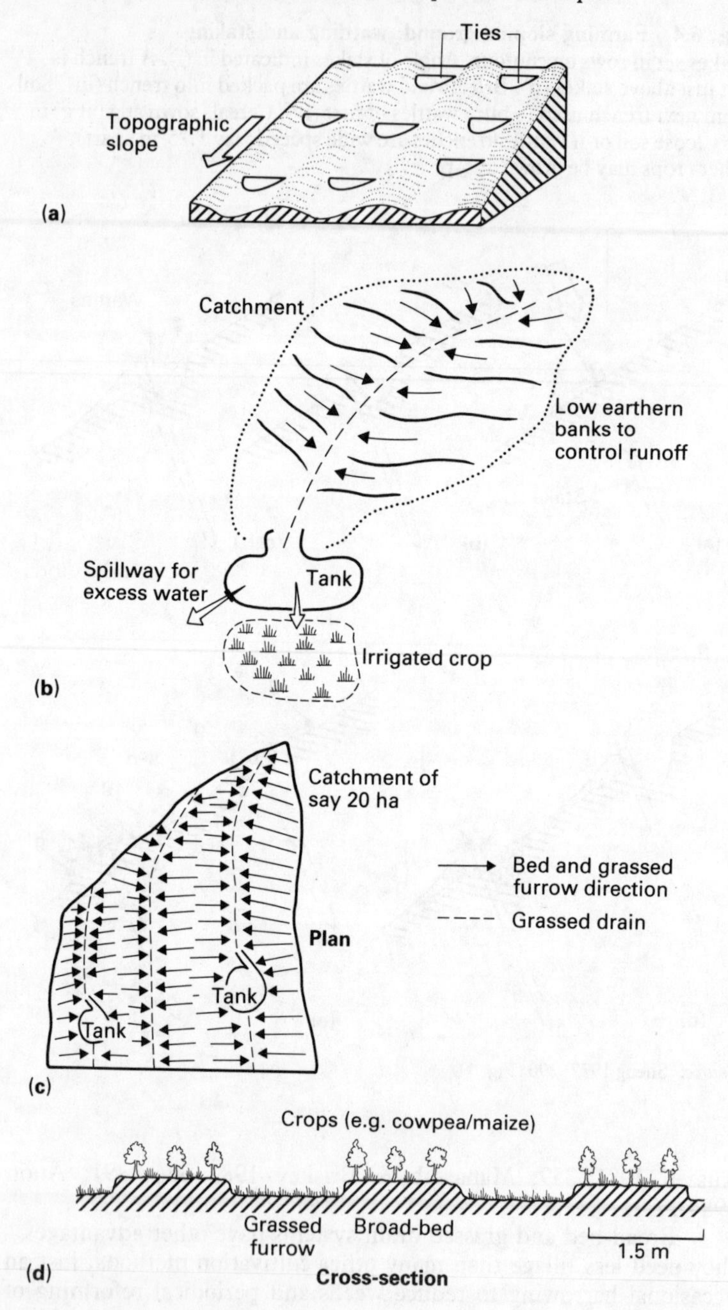

Sources: (a) author; (b) Manassah & Briskey 1981: 361; (c), (d) Manassah & Briskey 1981: 366.

furrows per hectare and less tillage, soil erosion should be reduced to a level below that usual with traditional rainfed farming. The broad-bed and grassed drain system is ideal 'intermediate technology' for those farmers who have access to draught animals. Broad-beds are also suitable for mixed cropping – the planting of crop combinations so that they might benefit one another by fixing nitrogen, providing shade and support. The grass drains can be grazed provided the livestock are tethered or carefully supervised, but above all the grass traps silt in the runoff and reduces the siltation of the storage tanks.

Collection and concentration of precipitation

Even if a region gets very meagre rainfall, lacks suitable supplies of groundwater and is too remote or separated by land too rugged for pipeline or canal to bring in water economically, farming may still be possible if precipitation can be collected, concentrated and supplied to cultivated plots. *Rainfall collection, runoff concentration* and *rainwater harvesting* are all broadly synonymous terms used to describe the collection and concentration of rain falling on natural slopes, cleared and compacted slopes or sealed catchments, to irrigate crops or supply water to livestock or meet domestic needs (Medina 1976; Lawton & Wilkie 1979; UNEP 1983; Bruins *et al* 1986; Pacey & Cullis, 1986).

If infiltration and evaporation losses could be prevented, a mere 10 mm of rain falling on 1 ha of catchment could yield around 100,000 litres. The problem is to find a suitable catchment and then to cut infiltration and evaporation losses as much as possible, both cheaply and simply (National Academy of Sciences 1974).

Fig. 6.5 Methods of using runoff and avoiding soil erosion (**a**) *Simple tied ridge and furrow (box ridging)*. The ties are made with hoe or disc-plough. (**b**) *Natural catchment and ephemeral stream with contour bunding to reduce rate of runoff and to guide water to storage tank(s)*. The catchment produces rainfed crops (which may fail in poor-rainfall seasons) and the tank supports irrigated cultivation (the area of which varies according to quantity of water collected). (**c**) *ICRISAT system of broad-beds and grassed furrows/drains*. An improvement on (b). In the dry season water from tank(s) can be taken by bullock-drawn water carts equipped with sprinklers to sustain rainfed crops in catchment and boost yields in normal-rainfall seasons (i.e. protective/supplemental irrigation). (**d**) *Cross-section of broad-beds/grassed furrows/drains of* (**c**).

Excavations by archaeologists have shown that in the past some rural peoples managed to crop quite arid regions using little more than manual labour and local materials to create catchments, supply channels, silt traps, storage reservoirs and fields (Shanan *et al.* 1969; Schecter 1977; Evenari 1981; Nabhan 1986). Interest in rainfall collection and concentration to support crops has also been stimulated by recent developments in earth-moving equipment and sealant materials, which have reduced the costs and difficulty of preparing catchments and can improve collection efficiency. Increasing competition in many developing countries between industry, domestic and agricultural water-users for limited stream and groundwater supplies is also likely to make rainfall collection more attractive.

Rainfall collection and concentration could have considerable potential for small farmers in remote regions where the rainfall is low. It could allow reasonably secure, reasonably good crops to be obtained where presently there is little chance, and where rainfalls are better it could increase the diversity of crops beyond those presently grown by rainfed cultivators (Hutchinson *et al.* 1981: 1; Bruins *et al.* 1986).

Rainfall collection and concentration agriculture seems to have been practised in the Negev Desert (Israel) between 1000 and 600 BC. During the Nabataean – Roman – Byzantine period (roughly 250 BC to AD 630) techniques were developed to quite a high level in the Negev, but after the Arab Conquest in the seventh century AD the systems declined, probably as a consequence of the inmigration of pastoralists who were unsympathetic to arable cultivators. Excavations show that Nabataean farms typically consisted of several cultivated, terraced fields of 0.5 to 2 ha each on the lower slopes of arid hills or in valleys – where the soil was deeper. Rainwater was collected from the hillsides by 10 to 15 ha catchments, prepared simply by removing loose rock and stones. As these slopes had little or no plant cover they were of little value for grazing so water collection interfered with no other land use. The catchments were subdivided into plots, each of roughly 3 ha, by earthen bunds to ensure that runoff was slowed enough to prevent erosion and to keep supplies running at a manageable rate even during heavy storms so that simple channels and storage cisterns were not damaged. The ratio of catchment area to cultivated plot area ranged from 20 : 1 to 30 : 1, depending on the rainfall and the local soil conditions. It would appear that cultivation was worth while even where annual rainfall was as low as 50 to 80 mm/y (in one case production seems to have been possible where there was only about 24 mm/y).

The reconstruction of some of the ancient Nabataean systems

proved that they worked and led the Israeli Ministry of Agriculture to support commercial ventures which adopted similar techniques. The indications are that yields of 3.5 to 5.0 tonnes/ha cropped of winter wheat is possible where there are only 100 to 240 mm/y of rain. Other crops which have been successfully tried include: olives, carob bean, pomegranate, almonds, apricots, vines, vegetables and cut flowers (National Academy of Sciences 1974: 26; De Ploey & Yair 1985).

Because the soils of the Negev are loess or loess-like and become more or less impervious after only very light rain (3 to 10 mm will seal them) mere clearance of loose debris is enough to prepare catchments for rainfall collection. In regions with less favourable soils, light rainfall may not yield runoff (it is worth noting, however, that Cox & Atkins 1979: 676 suggest that loess-type soils are quite widespread in developing countries).

Experience in the Negev has helped establish the size ratio of planting plot to catchment, what are suitable crops and what yields can be expected with a given rainfall regime. A training farm established at Wadi Mashash (Israel) has in recent years helped to extend rainfall collection/concentration cultivation to Tunisia, Afghanistan, India, Botswana, North-East Brazil, Niger, Australia and other countries (Evenari *et al.* 1968, 1971 and 1975; Goudie & Wilkinson 1977: 74; Pereira 1977: 192; Cox & Atkins 1979: 131; Gishchler 1979; Gischler & Jauregui 1984; Bruins *et al.* 1986). While much valuable experience has been provided by Israeli experiments in the Negev, it should not be overlooked that other countries, notably India, have long traditions of runoff agriculture and, as Pacey & Cullis (1986: 130) pointed out, these may be more relevant to hydrological conditions in large parts of the South. The Negev has much of its annual rainfall during a cool winter growing season, other drylands may have significant amounts of rain during hot seasons when evaporation losses are much greater.

Improved (sealed or artificial) catchments

Not all soils shed rainwater as readily as loess-type soils. The quantity of precipitation required to yield runoff is known as the *threshold precipitation* and this varies with terrain and vegetation cover as well as soil characteristics. It is quite possible for a locality to have frequent light showers which fail to exceed the threshold and simply evaporate or soak away. Special treatments may reduce infiltration and/or render the surface hydrophobic (water repellent) and so make rainfall collection possible in localities with unsuitable soils; treatment can also improve the collection efficiency of existing, untreated catchments. The problem is to find and apply an effective, sufficiently cheap, durable and non-harmful compound (Hutchinson *et al*. 1981: 9–16; UNEP 1983: 49–61) (Tables 6.3 and 6.4).

Table 6.3 Treatment, 'threshold' precipitation (i.e. the minimum rainfall needed to produce useful runoff from a given experimental slope) and runoff efficiency (i.e. yield as a percentage of the total precipitation on the catchment)

Treatment	'Threshold' (mm)	Runoff efficiency (%)
Sprayed silicone compounds	1.8	81
Concrete	1.1	84
Aluminium foil	0.4	88
Butyl rubber sheet	0.3	90
Gravel-covered plastic sheet	1.2	92
Paraffin wax	0.5	95
Asphalt (two-phase)	0.5	96
Polyvinylfluoride	0.0	100

Source: Frasier 1975: 42.

Table 6.4 Costs of improving catchments, based on a 500 mm average annual rainfall, 1975 prices, the expected life of each treatment and 6 per cent interest on the costs of treatment

Treatment	Expected life (years)	Costs ($US/m²)	Cost per 1,000 litres yield ($US)
Sprayed silicone compounds[a]	5–8	0.10–0.15	0.60–0.19
Paraffin wax	5–8	0.25–0.33	0.13–0.39
Butyl rubber sheet[b]	10–15	1.67–2.50	0.49–1.05
Sheet metal[b]	20	1.67–2.50	0.40–0.68
Soil grading and compaction	5–10	0.04–0.06	0.07–0.19

[a] Myers *et al*. 1967 quote sprayed asphalt as costing $US 0.73 to 2.70 per m².
[b] In rural areas the risks of theft may make metal foil, metal sheet, plastic and rubber sheet less attractive. Plastics are particularly vulnerable to wind, sunlight and livestock or pest damage.

Sources: Frasier 1975: 43; UNEP 1983: 52–3.

In addition to the treatments listed in Table 6.3, the application of sodium salts (common salt or sodium methyl silanclate) on certain clay soils can seal catchments by causing the soil colloids to break down, resulting in the choking of pores and fissures with greatly reduced infiltration. A 45 kg/ha application of common salt has been found to increase runoff from clay – loam soil by 70 per cent (Arnon 1981: 135). Sodium salts are cheap, often easily available and have the advantage of reducing weed growth on the catchment. However, care is needed to ensure that the runoff is not too saline or there will be salinization problems and crops might suffer.

In the West Indies, Gibraltar, southern USA, Australia, Hawaii, Botswana and the Sudan, catchments have been sealed with a range of materials. Sprayed asphalt has proved to be one of the most cost effective, and since a number of oil-producing developing countries actually find it difficult to dispose of such heavy by-products of petroleum refining, supplies are likely to be cheap (Myers *et al*. 1967).

In many developing countries there may already be potentially useful rainfall-collecting surfaces which are unused. In the USA there have been proposals to use the interstate highways in drier states to collect water to support pasture and other crops grown in roadside strips (UNEP 1983: 58). In Botswana and parts of Nigeria peasant farmers have already recognized such potential for themselves and it is not uncommon to find plots watered by runoff from surfaced roads, school yards, threshing floors or other hard surfaces (Frasier 1975).

If cultivated plots, watered by concentrating rainfall, are not too free-draining and are kept weeded and well hoed or mulched, enough moisture can often be stored in the soil for crops in regions where rainfall is seasonal or erratic and evaporation rates are high. However, where soils drain freely and/or if rainfall is erratic it makes sense to store collected rainwater wherever this is possible. Because relatively small volumes of water are involved, an underground cistern or roofed tank should be used to reduce evaporation losses; in all cases the tank or cistern should be filled via a system of silt traps to prevent it becoming choked up. There is an increasing number of designs for low-cost, easily constructed cisterns and tanks. Bateman (1974: 15–18) gives a brief bibliography.

Microcatchments

In some regions the soil may be suitable for rainfall collection, but the topography may not slope enough for collected water to flow to farm plots. Richer farmers may be able to overcome this problem by using mechanical graders to form *roaded catchments* – large plots cleared of vegetation and compacted (and sometimes sealed) to shed water on to cropped plots (Fig. 6.6a). The small farmer without access to graders or tractors must find some alternative, and *microcatchments* offer considerable potential.

Microcatchments (microwatersheds) of various forms can be constructed on level or sloping ground with little more than a hoe, or by tractor and disc plough (Fig. 6.6) (Bateman 1974: 11, 17).

Perennial crops like pomegranate, vine, pistachio, apricot, olive, carob, almond and to a lesser extent citrus fruits are ideally suited to microcatchment planting, although barley, wheat and forage crops have often been grown. Microcatchment systems were used by the Carthaginians and Romans, indeed much of Rome's food supplies were provided by microcatchment farming in the mountains of Tunisia, Cyrenaica, Tripolitania, the Mureotis region of Egypt (between present-day Alexandria and the Libyan border) and parts of Libya. Some of these ancient systems still function, but most have long been abandoned. Archaeological studies show the Romans and Carthaginians grew olives, vines, figs, carob, dates and barley in *composite microcatchments* (impluvium or meskats; Fig. 6.7a).

It seems the ancient impluvium gave crops mainly where rainfall today averages 150 to 160 mm/y, but in some cases they were operated where there is presently as little as 50 mm/y. The systems probably declined in the eleventh century AD because produce was more easily won from irrigated lands of the Nile Delta, and because labour became difficult to find after the loss of slave labour. The influx of semi-nomadic herders may also have led to the decline of impluvia farming, but climatic change does not seem to have been to blame; so given cheap labour the technique should function in low-rainfall regions today (El Kassas 1979). In some regions composite microcatchments are in use and according to Ruddle & Manshard (1981), in the late 1970s there were over 70,000 ha of meskat-type systems growing fruit in the Khost Province of Afghanistan alone. Composite microcatchments may be found in Australia, Botswana, the Canary Islands, Israel, Mexico and Pakistan. However, their potential is far from fully exploited.

Microcatchments, whether of the single or composite type, are more efficient than the hillside catchment and separate fields means of collecting/concentrating rainwater because there is no need for construction, management and maintenance of channels,

---

Fig. 6.6 Microcatchments (**a**) *Roaded-type microcatchment.* (**b**) *Microcatchment easily constructed on sloping terrain with hoe.* (**c**) *Variant of* (**b**) *suitable for level terrain.*

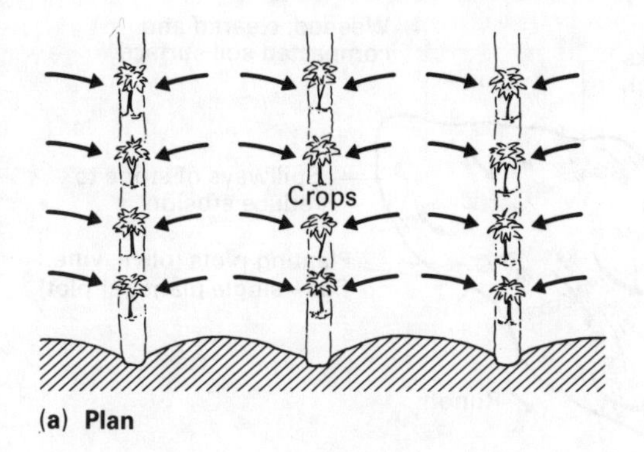

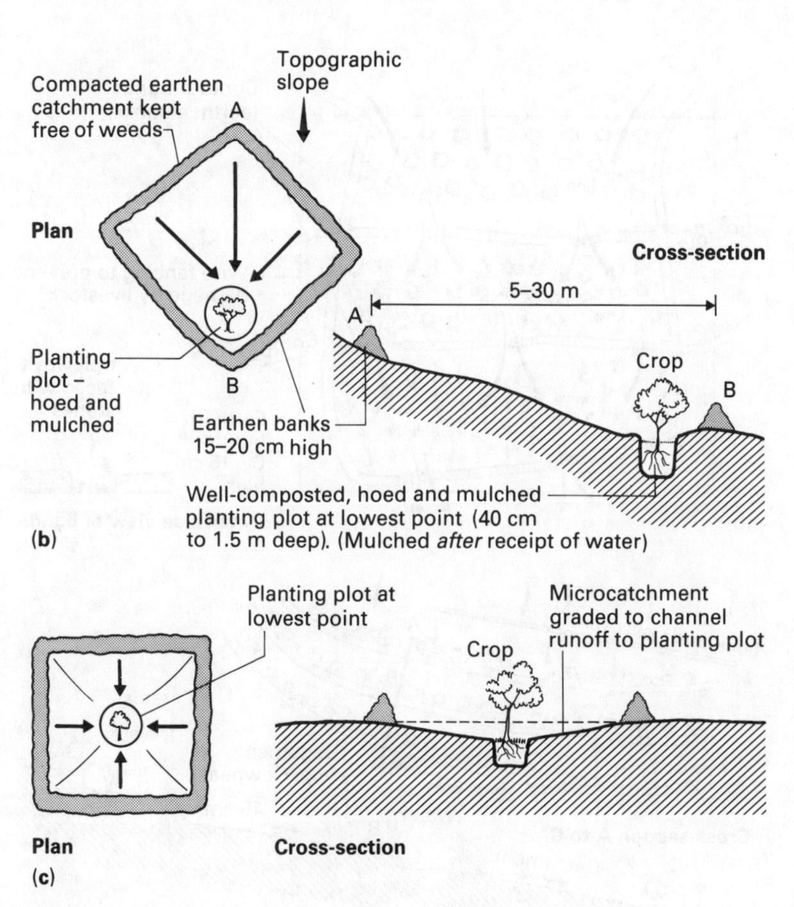

Sources: **(a)** National Academy of Sciences 1974: 13 (Fig. 5) (modified); **(b)** National Academy of Sciences 1974: 30 (Fig. 20) (modified); **(c)** author.

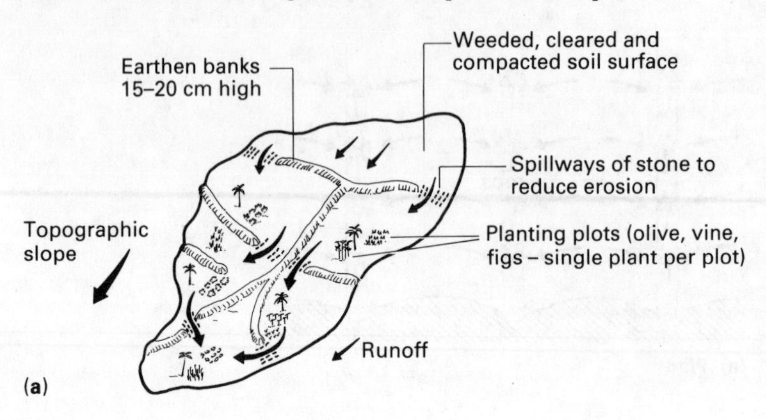

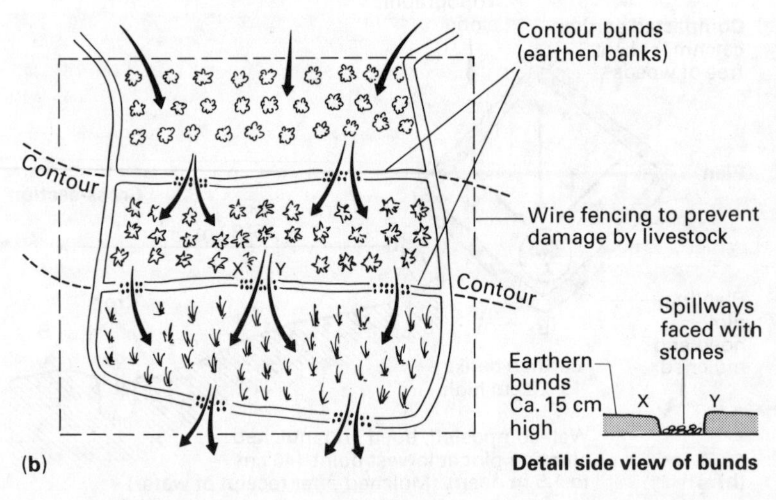

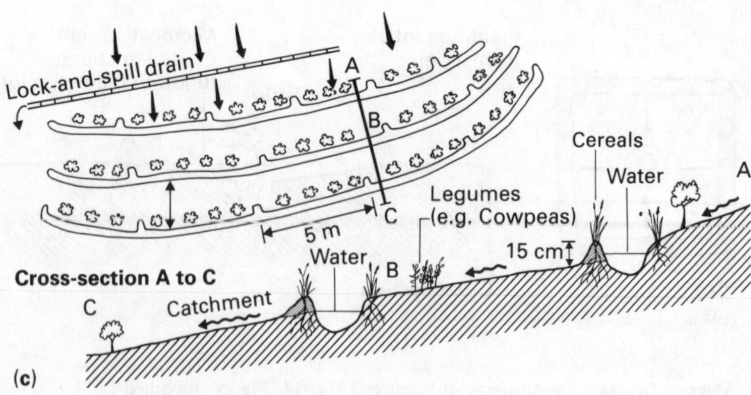

Sources: (a) redrawn with changes from Gischler 1979; (b) and (c) drawn with some change from Lewis 1984.

cisterns and so on, and there is less loss of water between precipitation and it reaching the cultivated plot. Topography is less important with microcatchments, they can be used on slopes of up to 5 degrees and where terrain is uneven.

Studies in the USA (National Academy of Sciences 1974: 30) suggested that microcatchments suitable for developing countries could be constructed for around $US20 per hectare of catchment. More recently, experiments in the Negev indicated construction costs of $US10 to 40 per hectare cultivated (Postel 1985b: 4). Most of this cost is labour and it would probably be much less in most developing countries with abundant manpower.

Studies in Kenya at the Baringo Pilot Semi-Arid Area Project (BPSAAP) suggest that typically microcatchments require 300 man-hours per 0.5 ha cultivated to construct, and 160 man-hours per 0.5 ha cultivated in subsequent years to maintain (Fig. 6.7b, c) (Pacey & Cullis 1986: 156). Traditional, rainfed cultivation required about 100 man-hours per 0.5 ha cultivated per annum. However, microcatchment yields were significant improvements over rainfed cultivation yields: crops of 2,000 kg/ha of sorghum have been obtained, compared with the average for rainfed cultivation in the locality of 45 kg/ha (Lewis 1984).

Given suitable extension services and cautious pilot studies to check for problems the establishment of microcatchments should be reasonably straightforward since sophisticated materials are not needed. Often the main problem is control of livestock to prevent them damaging catchments or crops. The potential for microcatchments to help small farmers in low-rainfall areas is considerable. Stern (1979) suggested that a 'typical' dryland rainfed cultivator requires 10 ha to feed a family on a rather poor diet of sorghum and millet, even if only 1,000 m^2 of that rainfed 10 ha were set aside to form a 700 m^2 catchment supplying a 300 m^2 vegetable plot, the farmer and his/her family would obtain a much-improved diet.

Utilizing seasonal and ephemeral streams and flows

Check-dams

In the drier tropics rain does occur, there may even be 400 or 500 mm/y, but this can fall in one or a few short-duration storms

---

Fig. 6.7 Composite microcatchments (**a**) *Meskat-type composite microcatchment.* (**b**) *Composite microcatchment of type used at BPSAAP, Kenya.* (**c**) *Contour-ridge (Negarim) microcatchment of type used at BPSAAP, Kenya.*

causing sediment-laden 'flash floods' to course through creeks (wadis), or down slopes as overland flow. Those wishing to retard and make use of such waters must contend with considerable difficulties: the flows are sudden and strong (which puts structures under great stress) and the waters are very silty (which can easily choke up channels, tanks and fields). Nevertheless, some of the world's oldest irrigation techniques, techniques which are still important (especially in the Middle East, the Sudan, Ethiopia and Southern Asia), successfully utilize such waters (Lawton & Wilkie 1979; Heathcote 1983: 203).

The terminology is a little confused: *wadi farming, flood-channel irrigation, check-dam farming, bund irrigation, water spreading* and *spate irrigation* are all terms applied to the same or similar systems of earthen or stone dams and banks designed to spread water over the ground to moisten it and/or to trap wet silt which can then be planted with crops (UNEP 1983: 127–36). As far as possible in the following section an attempt is made to distinguish between the different forms of check-dam or water-spreading systems; however, there is a considerable degree of overlap between some techniques (Nabhan 1986: 5–11).

Check-dams are simply built barrages which trap water and sometimes water-borne sediments to create a moist planting plot; hopefully the plot of soil behind the check-dam is moist enough to support crops through to harvest. At their simplest, check-dams require only labour and local materials, they are therefore suitable for poor farmers in remote areas (Dennel 1982b: 171). An example of a traditional check-dam farming technique is the Mexican bolsa. Bolsas are earthen-walled basins or bordered gardens which catch water diverted from seasonal creeks (arroyos). After the flows from the arroyo have wetted the bolsa and have ceased, the cultivator plants seeds and covers the moist bolsa with a mulch of dry dust or sand to reduce evaporation. As little as the equivalent of 300 mm/y rainfall fed into a bolsa can sustain a crop of wheat, maize, beans, cotton or vegetables (Dobyns 1967).

Similar to the bolsas are the mahafurs constructed in North-west Arabia. These are shallow excavations usually between 20 and 100 m diameter surrounded on three sides by earthen bunds 1 to 4 m high, rather like 'ring doughnuts with a bite taken out' in the direction of the expected flow. They are sited at points in wadi beds where they will escape destructive flows yet collect and retain enough water for livestock or moist soil for crops (Gonzales 1978: 114–18). Wadi farming was widely used in Roman Tripolitania, North Africa.

In Ethiopia, the Sudan and Botswana small check-dams often associated with shallow excavated tanks catch 'creeping flow' (gentle overland flow) passing down slight slopes. These are known as haffirs and support crops in the moistened plots formed upslope of the earthen-bank check-dam. Provided it falls in a relatively cool season as little as 80 mm of rain can be sufficient to support crops if haffirs are constructed (UNEP 1983: 158–64, 219). A system of earth-banks called khadin is used in western Rajasthan (India) to retain monsoon runoff and soak plots so that crops can be raised (Pacey & Cullis 1986: 135). Khadin trap silt as well as moisture and this helps to maintain the fertility of the cropped plots; additionally the constant seepage from the plots prevents excessive salt accumulation, which can be a real problem with some of the other traditional irrigation methods in the region.

Water-spreading

Water-spreading techniques aim to spread the water from seasonally flowing creeks or overland flow from slopes on to land which can be cropped or grazed. The water being spread out and retarded has a much greater opportunity to infiltrate and recharge soil moisture or even groundwater than it would if allowed to flow unhindered. Water spreading can also reduce the damage caused by flash floods suddenly flowing down a channel. It is commonly practised in China's loess soil region (UNEP 1983: 174), Tunisia, the Arabian Peninsula, and can be found in many other drylands or regions of seasonal rainfall (Fig. 6.8a, b).

Water spreading does have drawbacks. The banks and barrages cannot always withstand spates, and break so that land below them is eroded or covered with infertile deposits of sand or gravel. Water spreading is often used to improve pasture, with the risk that livestock will trample and damage the embankments; seeding them with unpalatable or prickly shrubs both deters grazing and trampling and the plant roots bind and strengthen the embankments. Where farmers can afford the materials and time, embankments can be strengthened with stone, brushwood, concrete or gabions (wire-mesh baskets which are easily transportable and can be filled with locally available stones to give a robust, cheap structure which allows some flow to pass through and is therefore ideal for constructing check-dams).

Little can be done to increase the security of harvest of farmers who use water spreading, except in those situations where a considerable depth of moist soil can be accumulated. However, provided the farmer or rancher can accept the risk that there may sometimes be little flow to spread, the gains from water-spreading can be attractive. In Australia (New South Wales) water-spreading

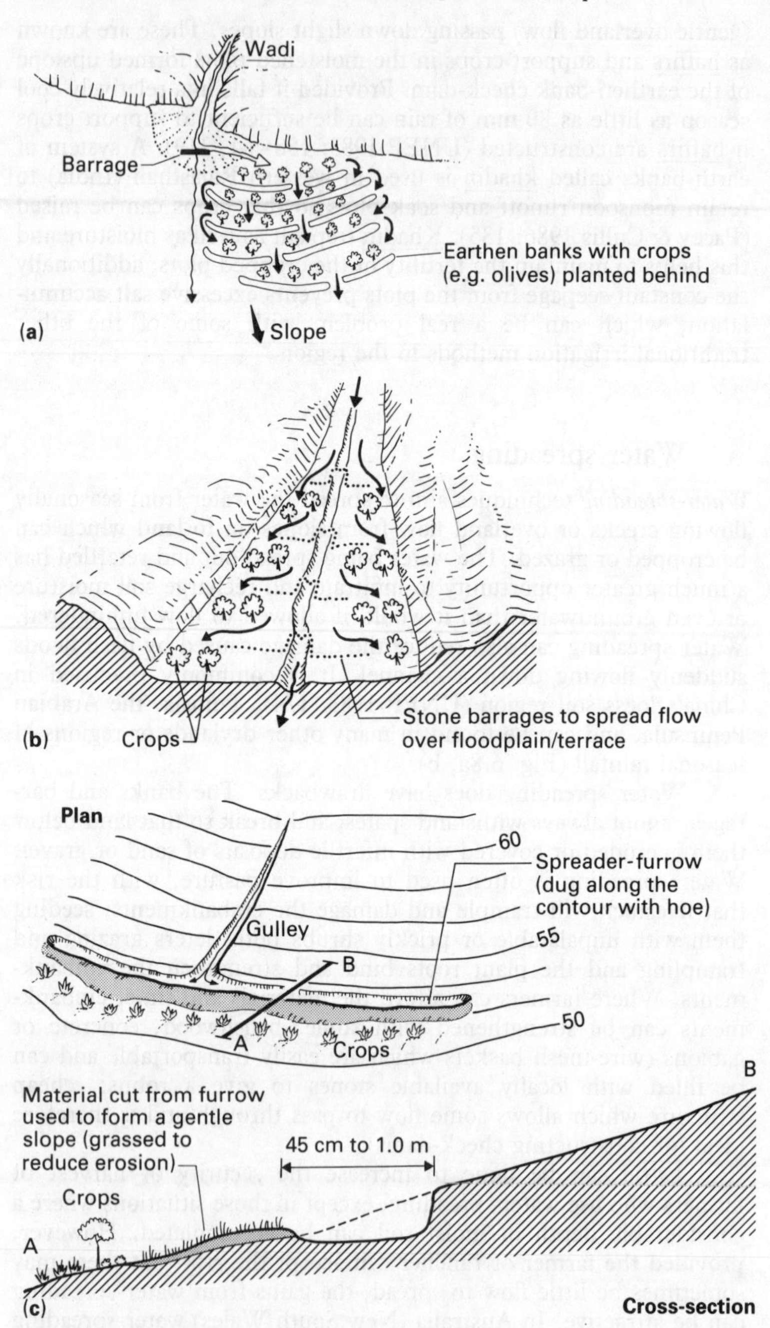

Sources: (a) based on National Academy of Sciences 1974: 27 (Fig. 17); (b) based on Cox & Atkins 1979: 131; (c) based on Stern 1979: 34–5 and Gischler 1979: 56

methods like those in Fig. 6.8a improved the carrying capacity of dryland pastures from 0.18 to 2.66 sheep/ha (National Academy of Sciences 1974: 34).

Where storm water runs to waste, eroding gullies or stripping away the soil, one solution is to construct *spreader-seepage furrows* (Fig. 6.8c). These are a sort of hybrid between the contour bench terrace and earthen embankment water-spreading. The farmer or community work gang can construct spreader-seepage furrows with simple hand-tools. A furrow is dug across a gully or point where runoff is concentrating, along the contour for at least 20 m either side (one or two days' work for one man with a hoe). The excavated soil is used to form a low earthen embankment which is stabilized by seeding it with grass. Provided slopes do not exceed a gradient of 25 per cent and the soil is not too stony the system works well, preventing erosion and improving infiltration. At present a United Nations University research project is working to improve contour spreader-seepage furrows and contour bunds traditionally used in Ethiopia. Hopefully this work will improve techniques and give useful insight into the problems involved in getting rural peoples to construct such structures. Where there is no tradition of contour bunds or similar structures there may be difficulty persuading farmers that work on them is not better spent on their existing rainfed cropland (Anon. 1983b).

Storage of seasonal and ephemeral flows

Small reservoirs

If small reservoirs are constructed where vegetation cover is sparse and seasonal rains or storms deliver brief heavy runoff, they face the threats of sudden silt-charged flows or, during dry periods, evaporation losses which seriously deplete their storage or cause the stored water to become saline. There is also a greater need to consider the risk of percolation losses when constructing smaller

---

Fig. 6.8 Using ephemeral flows (a) *Simple earthen bunds for spreading floodwater* (known as jessours in Tunisia). The bunds are about 1 m high and each have 'weeps' (low-efficiency drains to prevent excess puddle formation behind the bund) at roughly 25 m intervals. **(b)** *Stone or brushwood dams for diverting and spreading floods over floodplain and low terrace fields.* **(c)** *Contour spreader-seepage furrow using a low earth bank and furrow to achieve gentle infiltration along a wide front to combat gullying and to support crops where rainfall is poor.*

reservoirs. For example, in Jamaica in the 1970s the European Development Fund financed the construction of 230 'microdams' for irrigation supply and small-scale rural hydroelectricity generation; many of these microdams failed to fill or remain filled because inadequate soil surveys had neglected to map porous limestone in their beds (Johnson 1979). Careful survey and treatment with sodium bentonite or lining with puddled clay, sheet rubber or concrete may overcome problems of percolation losses, but it is usually quite costly. Siltation and evaporation losses remain a threat to small reservoirs even if their beds do not leak.

Tanks

For millenia in tropical Asia, especially Southern India and Sri Lanka, simple earthen bunds and/or excavated hollows have been used to retain runoff for use during the dry season. In some regions there is little alternative to using *tanks* because clay soils so hamper infiltration that most rainfall just runs away to waste. Tank storage permits millions of Indian farmers to grow a second, dry season, irrigated crop in addition to their rainfed crop planted during or just after the wet season rains.

Tanks are generally sited so as to reduce the risk of flood damage and siltation; ideally they collect only gentle overland flows or slower-moving channel flows, both of which are relatively free of silt (Fig. 6.9a). Tanks are also sited so that any overflow from one is caught by others downslope, minimizing the wastage of water. However, even the best-sited tanks do in time silt up, but because they are shallow it is not too difficult to clear them out at the end of the dry season when water levels are low. The detritus removed is usually quite fertile, and in the past it was cost effective to hire labourers to remove it and spread it on farmland. The UNEP (1983: 85) however, suggested that rising transport and labour charges have made such practices much less economically viable.

The importance of tank irrigation can be gauged from a recent estimate that out of 34.5 million ha irrigated in India, at least 4 million ha were watered with supplies from tanks. On average, in India a single tank will irrigate 9 ha (UNEP 1983: 154). The costs are quite low – an estimate in the 1960s put it at $US1,000 per hectare cultivated which was much less than canal- or tubewell-supplied irrigation at that time (UNEP 1983: 157). Against this must be weighed the fact that evaporation, and to a lesser extent seepage, wastes a lot of the water stored. Another problem where rainfall is especially unpredictable (as is the case over much of South Asia) tanks fill to quite different levels each year. Farmers therefore cannot really do much more than guess how much land they will plant or how intensively they can irrigate

Fig. 6.9 Storage tanks (a) *South Asian tank.* (b) *Chinese canal-linked accumulator reservoirs ('melons-on-a-vine' system).*

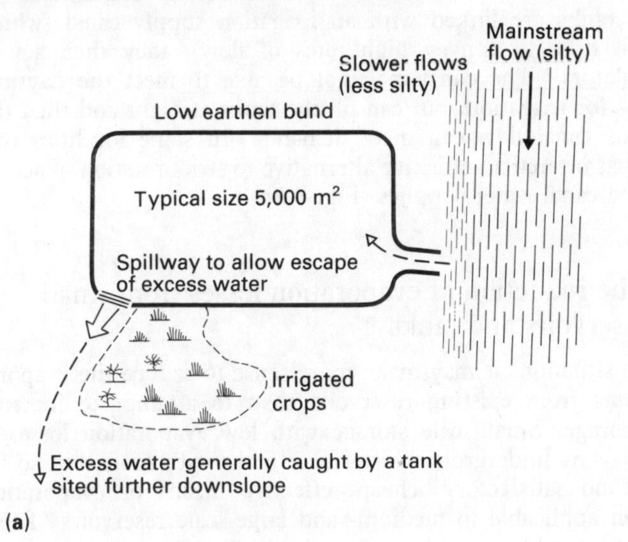

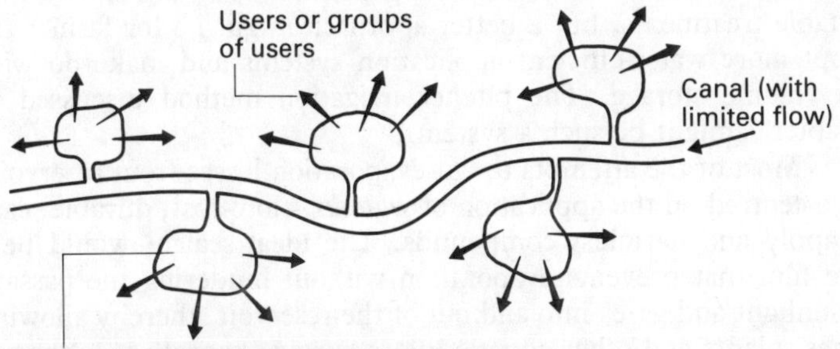

Sources: (a) author; (b) based on Bottrall 1981c: 227.

before the rains have actually filled their tanks.

Tank construction requires few inputs other than manual labour; it is therefore possible to use public relief/aid (such as work-for-food schemes) to facilitate their adoption. Tanks also offer considerable opportunities for pisciculture and the cultivation of aquatic vegetables such as water-chestnut (*Trapa* spp.). As with water spreading, control of livestock is important if tanks are not to be damaged and thereby easily breached while filling in the wet season.

An interesting development, which in China has shown promise, is the *melons-on-a-vine* system. Newly constructed or existing tanks are linked with an irrigation supply canal (which does not have to convey high rates of flow), they then act as 'accumulators'. The canal may not be able to meet the daytime demands for irrigation, but can fill the tanks at night and then the tank plus canal flow can meet demands. In some localities this system has offered an effective alternative to strict rotation of access to limited canal water supplies (Fig. 6.9b).

The reduction of evaporation losses from small reservoirs and tanks

In some situations it may make more sense to reduce the evaporation losses from existing reservoirs than to attempt to increase water storage. Small-scale storage with low evaporation losses is possible using underground cisterns or tanks with roofs, but so far there is no satisfactory, cheap, effective means of evaporation reduction applicable to medium- and large-scale reservoirs (Table 6.5). Considerable effort has been expended in an attempt to find suitable treatments, but a better approach might be for farmers to adopt more water-efficient application systems and make do with low-volume storage. The pitcher irrigation method discussed in Chapter 7 might be such a system.

Most of the attempts to cut evaporation losses from reservoirs have centred on the application of *sealants* – low-cost, durable, easy to apply and harmless compounds. The ideal sealant would be a thin film that prevents evaporation without hindering the passage of sunlight and gases into and out of the reservoir, thereby allowing fishes, plants and other aquatic life to survive beneath it (although it would be beneficial if the treatment also discouraged mosquito larvae). Some thin-film treatments can reduce evaporation losses by up to 40 per cent, and because so little of the compound need be applied they are fairly cheap. However, on reservoirs of over 40 ha size normal wave action soon breaks up the films so far tried (Furon 1963; Bateman 1974: 15; National Academy of Sciences 1974; Pereira 1977).

Sand-filled reservoirs for subsurface water storage: venetian cisterns and percolation dams

Where ephemeral flows carry much detritus it may make sense to allow a reservoir to silt up, indeed it may be difficult to avoid such an occurence.

Table 6.5 Sealing small reservoirs to reduce evaporation losses

Treatment	Advantages	Disadvantages
Floating plastic beads or balls	Easy to apply, sturdy, oxygen can get through, discourages mosquito larvae	Suitable only for small reservoirs, excludes light
Floating rafts of expanded polystyrene	As above	Costly, excludes some light, suitable only for small reservoirs
Floating rafts of foamed cement	As above, durable, might be produced locally	Suitable only for smaller reservoirs, excludes a lot of light
Sealant film of low-melting-point wax	Relatively cheap, easy to apply, allows oxygen and light to pass	Relatively expensive, excludes oxygen, suitable only for quite small reservoirs
Ultra-thin (mono- or multi-molecular layer) films, e.g. hexadecanol, aliphatic alcohols or commercial compounds like Esso Standard EAP 2000	Easy to apply, very effective at reducing evaporation, does not prevent all sunlight from reaching water, discourages mosquitoes	Not durable – very vulnerable to wind and wave damage on reservoirs over 40 ha size

Although the storage capacity of a sand- or gravel-filled reservoir is only 25 to 35 per cent of that of an equivalent unsilted reservoir, it has the advantage that as soon as the water level falls more than about 60 cm from the surface, evaporation losses virtually cease (Furon 1963; Balek 1977: 125). Sand- or gravel-filled reservoirs (*venetian cisterns*) can withstand flash floods better than normal reservoirs and offer a secure, clean flow suitable for livestock, domestic supply or small-scale irrigation (a venetian cistern and pitcher irrigation can be a useful combination for a small farmer in a remote poor-rainfall region). A further advantage of sand- or gravel-filled reservoirs is that where malaria is endemic, unlike normal tanks or reservoirs, they offer mosquitoes no breeding sites (UNESCO 1962: 374).

The main problem in constructing sand- and gravel-filled reservoirs is to ensure that only coarser sediment, not fine silts, accumulate so that there is adequate interstitial space for water storage. A way of ensuring collection of suitable sediments is to construct a check-dam of boulders, brushwood or stone-filled gabions which only partially obstructs streamflow, allowing fine sediments to escape but trapping the coarser sediments (Dixey 1950). Such techniques have been used in Botswana, Namibia, Ethiopia, Mexico, southern USA and many other countries. Pereira (1977: 193) describes the successful use of check-dams to create venetian cisterns in Ethiopia for watering high-value export vegetable crops. It is possible without too great an expenditure to improve the most basic sand-filled reservoir to increase its storage and reduce water losses between fillings. Figure 6.10a, b illustrates an intermediate technology approach which could achieve such objectives.

Where there are springs with inadequate flow to meet daily needs of livestock or cultivators, or if they run dry for part of the year, sand-filled reservoirs may offer a suitable means of uprating them (sand or gravel may have to be transported to the spring to form the storage layer).

Percolation dams are check-dams designed to retain water and wet sediments from spates long enough for it to infiltrate and recharge groundwater. Between 1967 and 1970 in a drought-stricken part of Maharashtra State (India), USAID grain was used to pay labourers on a work-for-food basis to build forty percolation dams to recharge floodplain aquifers which could then be tapped by shallow wells in the dry season. The programme was a success. Wells dug before infiltration was improved ran dry in most dry seasons; after the percolation dams were built, wells continued to provide water even during droughts (Pacey *et al.* 1977).

Although there is often abundant labour in rural regions which could theoretically be used on construction to improve water supplies, it may be very difficult to make use of it in practice. The Maharashtra experiences and also similar successes in Karhataka State (India) helped, at least in those regions, to overcome rural peoples' scepticism of the effectiveness of soil or water conservation measures.

Before any attempt is made to build reservoirs, planners should check whether catchment land-use could be altered to improve streamflow and groundwater recharge. Watershed managers can often use contour bunds, terraces, check-dams and percolation dams to improve the runoff and groundwater recharge regime.

Inundation canals

In North-west India and Pakistan there is a long history of constructing *inundation canals* or pynes – simple channels cut to meet a

Fig. 6.10 Intermediate technology check-dam and sand-filled reservoir (**a**) Flow partially obstructed by gabions on concrete base (1) and coarser sediment deposited. (**b**) Gabions replaced or sealed with concrete (2) and gabions placed on top to repeat sediment trapping, the gabions are then replaced with concrete (3).

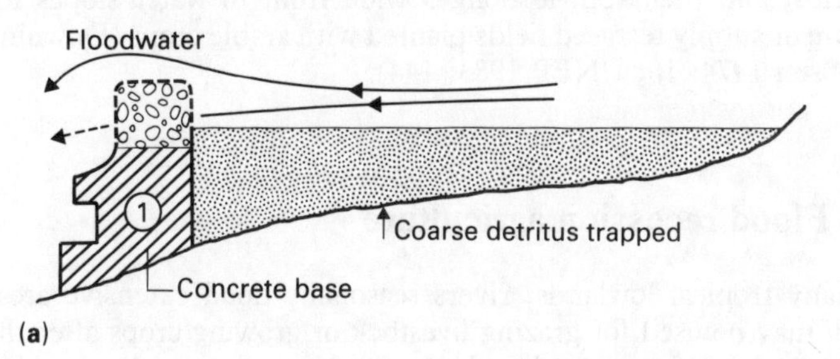

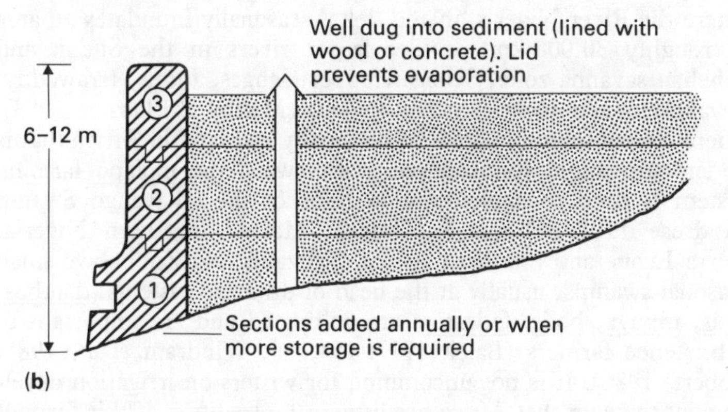

Source: Pacey *et al.* 1977 (redrawn).

stream or river which is subject to seasonal spates. A pyne has no headworks or barrage, it is simply led off at roughly right angles, its bed well above normal river level so that it is only filled when floods rise on to the floodplain or terrace surface into which it is cut. The pyne is led in such a way that from its head to the land it irrigates it slopes less than the river or stream course (i.e. the pyne almost follows contours), consequently flows along it are kept gentle. The gentle rate of flow and the high level of the pyne intake allows it to escape flood damage and ensures that most of the floodwater's silt is dropped in its uppermost section where it cannot choke up too much of the channel or irrigated fields. The technique is common in the Sind (Pakistan/India) where floodwaters from the

Indus or its tributaries are led to farmland or the floodplains. Inundation canals are also quite common in Mesopotamia, Kashmir and throughout the foothills of the Himalayas as far east as West Bengal. The Kashmiri – Bengali inundation canals are known as kuhls and lead water from highland streams or springs along the contour, like the irrigated terrace systems of Central and South America, and then spill it along a wide front to water slopes for pasture or supply terraced fields planted with arable crops (Downing & Gibson 1974: 10; UNEP 1983: 144).

Flood recession agriculture

In many tropical lowlands, rivers seasonally flood extensive areas which may be used for grazing livestock or growing crops after the waters recede, for example, along the Niger (especially in Mali where the River Niger's 'inland delta' seasonally inundates an area of roughly 30,000 km^2), Nile, larger rivers in the Sudan and Sahelian savanna zone, Senegal, upper Ganges, Indus, Irrawaddy, Hwang-Ho and Rokel (Sierra Leone) (Maltby & Turner 1983). There are considerable areas of seasonally wet land in Africa that is under cultivation, for example in Botswana the molapo farming system is used to exploit the margins of the Okavango Swamp (Andreae 1981: 65) and in Guinea, Malawi, northern Nigeria, Sierra Leone and Southern Africa, Tanzania and Zimbabwe small seasonal swamps, usually at the head of drainage systems (dambos, vleis, mbuga, bolis, fadamas) are cultivated and are important to subsistence farmers (Balek 1977: 154, 16; Windram *et al.* 1985; Roberts 1986). It is not uncommon for writers on irrigation development to claim that Africa has little or no tradition of this form of cultivation. Richards (1985: 72), however, pointed out that West Africans do make effective use of wetland environments, indeed West African smallholders have displayed special enterprise and inventiveness. In this field he provides (1985: 74) a list of types of smallholder irrigation in West Africa – indigenous African irrigation is also described by Goldsmith and Hildyard (1984: 291–301).

Early floods or severe flooding may damage crops, late flooding may delay planting and increase the risk of them being damaged in the next flood season. Despite the risks of crop loss, the cultivation of wetlands and floodlands tends to be more sustainable than most tropical rainfed farming because salts are washed away, and fertile silt is often deposited by floodwaters. Furthermore the cultivators are likely to be better nourished than their drier-environment counterparts because they are able to fish, hunt waterfowl and other aquatic game. There may also be opportunities (presently under-exploited) to collect aquatic weeds to mulch crops and/or make compost (Junk 1980).

There would seem to be considerable scope for improving *flood recession agriculture* and for extending the area under such use. Floodplains and swamps can be made to drain faster by cutting drainage channels, thereby extending the growing season and improving security of harvest, and if the channels remain full, providing a source of irrigation water during dry periods. In some regions simple earthen banks could be thrown up to form 'polders' (shallow lagoons) in which the depth of water could be controlled for growing rice. This has been done in the south of Guyana, near the coast and along the Demerara River where they are used for sugar rather than rice cultivation. *Raised fields* or *floating beds* (layers of soil on reed or brushwood rafts) could be used in many floodlands, or even on lakes to open up new areas for cultivation, to extend the growing season and to avoid flood damage. This is done around some Andean lakes and in parts of Central America (Welcomme 1979: 236–44; Klee 1980; Ruthenberg 1980; Arnon 1981: 55; Lambert *et al.* 1984).

Despite some developments, vast untapped potential for flood recession agriculture remains. In Brazilian Amazonia alone there are somewhere between 64,000 and 128,000 km^2 of floodplains and levees (várzeas) which are periodically inundated. Where the flooding is by 'white-water' rivers, fertile sediment may be deposited, giving the várzeas considerable agricultural potential, unfortunately not all of the Amazonian rivers are white-water rivers, some are acidic or nutrient-poor and form little in the way of useful várzeas. Even if only a fraction of the floodlands are suitable there is vast potential for rice, beans, jute and other crops – already there is one large-scale rice production project in the várzeas of the Rio Jarí and there is increasing small-scale production in lower (eastern) Amazonia (Lima 1956; Petrick 1978).

In parts of Bangladesh, Brazilian Amazonia, Burma, Ecuador, Indonesia, Sri Lanka, Thailand, Vietnam and West Bengal (India) there are vast areas of tidal swamps which are flooded either twice daily or monthly during spring tides by fresh or brackish river water backed-up by the tide. Although acid sulphate soils can be a problem, tidal swamps yield a lot of rice in the aforementioned countries. There is also scope for greatly expanding and improving production from coastal swamps (IRRI 1984).

In many developing countries there are, or soon will be, large hydropower projects which can greatly alter river flow regimes. This alteration may either greatly benefit wetland and flood recession agriculture, or seriously disrupt such practices. When planning such projects, all benefits and disadvantages for all agricultural sectors must be considered.

Flood-resistant agriculture

Some of the rice grown in Thailand, Burma, Vietnam, Bangladesh and West Africa is 'flood rice' (floating or deep-water rice), that is, varieties capable of growing fast enough to keep abreast of rising floodwaters or able to withstand brief submergence; some of these rice varieties can grow as much as 6 m in length compared with the usual 0.3 to 1.0 m (Clay 1978; Pearse 1980: 231; Wrigley 1981: 103–4, 203; Richards 1985). According to Farmer (1979: 311–14), roughly 18 million farmers in South Asia face considerable risk of flooding and have responded by growing flood rices, but in many parts of the world flooding is a serious threat to, or constraint upon, rice production and flood rice is not grown. In such regions the adoption of flood rice varieties could do much to improve yields and security of production. With these goals in mind IRRI and research stations in India, Bangladesh and Thailand are trying to develop high-yielding rice varieties which can withstand about 1 m of flooding (Bangladesh Rice Research Institute 1975; Clay *et al.* 1978; IRRI 1984).

Flood-resistant, raised-bed cultivation systems were used in pre-Conquest Central and South America (Denevan 1970; Darch 1985; Farrington 1985). One, the chinampas system, protected crops during high water and allowed irrigation during the dry season. A type of floating bed cultivation system survives near Mexico City (Harris 1980: 73–315; Farrington 1985). In South and Central America at least, in the past wetland agriculture, especially drained field cultivation, was highly successful and gave sustained yields for a long time over a large area. Such success warrants the attention of present-day agricultural planners, not least because many tropical wetlands are under-used.

Pump irrigation

Traditional water-lifting devices have long been used to raise water from rivers or lakes, either to irrigate land which remains above water all year round or to extend the growing season and boost the yields of crops planted on floodlands after floodwaters recede. Modern pumps mounted on pontoons, or with their intakes mounted on pontoons, have greatly increased the opportunities for such *pump irrigation*. Considerable areas of pump irrigation have been established along the Senegal River (Senegal, Mauritania and Mali), on the Nile below the Aswan High Dam and on many other tropical rivers.

Fog, mist and dew collection

There is some disagreement in the literature over the quantities of dew formed in the tropics, its actual value to natural vegetation and its potential for supporting crops (Stone 1957; UNESCO 1962; Monteith 1963; UNEP 1983: 188). Much of this confusion arises because of the difficulties in measuring dewfall and in distinguishing between dew formed from condensation of atmospheric moisture, dew formed by distillation of soil moisture and *guttation* (the moisture exuded by plants from their leaves). Little is known of dew formation in or upon the soil, as opposed to that condensed above ground level.

Evenari *et al.* (1971) measured nightly dewfall in the Negev at ground level and at 1 m height and recorded an average of 35 mm/night – which they felt was of little value to agriculture as it would soon evaporate after sunrise. Furon (1963: 58) pointed out that were deserts like the Negev better vegetated, a six or sevenfold increase in dew might be expected because the leaves would increase the surface area on which it could condense (for dew to form, atmospheric moisture must come in contact with a cold surface). Studies in the UK suggest that a light wind (less than 5 m/sec) helps dewfall, but it also aids the collection of mist, which makes the precise measurement of dew difficult under such conditions.

Dew may provide some useful moisture for livestock grazing early in the morning, and may marginally reduce evapotranspiration. Plants do absorb dew from their foliage, but whether or not some are able to take up dew from the soil is uncertain; some xerophytic plants certainly have extensive shallow roots which might be able to draw dew from the surface layers of the soil (Pereira 1977: 102).

In the past on some of the UK's uplands, simple structures known as *dew ponds* were fairly common. These were shallow depressions 20 to 40 m in diameter lined with straw and clay. On clear nights the ponds were supposed to have cooled more than the surrounding land thanks to their insulating layer of straw, and to have acted as condensing surfaces (Hubbard & Hubbard 1907; Dixey 1950: 66). However, it has been argued that most of the moisture they collected was derived from rain rather than dew or mist. Nevertheless, they might be improved upon by using modern insulating materials and could be useful for watering stock in tropical drylands.

Fog and mist are presently insignificant sources of water, but in some localities they might offer promise. There are natural plant communities in the tropics which are supported by such precipitation, for example, the lomas vegetation association of the coastal deserts of Peru and Chile seems, at least in the past, to have intercepted enough sea mists not only to sustain growth but also to maintain groundwater so that springs flowed. Such sea mists, known as camanchacas in Chile/Peru occur between southern Ecuador and Chile (as far as 30 deg. south). Records show that many springs ran dry after loss of the lomas vegetation through overgrazing and woodcutting (Goudie & Wilkinson 1977: 66). Near Lima (Peru) reafforestation with *Casurina* has proved that trees can support themselves by trapping passing mists and fog. A study by Walter (1971: 375) suggests that lomas vegetation might intercept as much as 1.8 litres/h when the windspeed is 1 m/sec. As the mists in the regions covered, or once covered by lomas vegetation, often last all day and occur on around 120 days of the year, the amount of water that can be collected is considerable. Pereira (1977: 102) cited Norfolk Island pines (*Araucaria heterophylla*) growing in Hawaii which collected an average of 760 mm/y from mist and low cloud. Mist and fog support vegetation in Cape Province (South Africa), in western Muscat and Oman and in Western Australia (Israelsen & Hansen 1962: 6; Furon 1963; Gischler 1979: 76; UNEP 1983). There are also a number of *cloud forests* ('elfin woodland') in tropical highlands where cloud and mist are supposed to play a major role in supporting the plants.

There have been a number of experiments with structures designed to collect mist/fog. In Peru netting-surfaced traps (each of 6,000 m^2) were found to collect 50 m^3/day, and in the Negev simple polythene sheet traps reportedly yielded monthly up to 3,631 mm/m^2 – enough to support trees (Gindel 1965). In Chile 90 m^2 polythelene net screens yielded a daily total of between 1 and 15 litres/m^2, enough to supply pitcher irrigation systems and grow crops (Gischler & Jauregui 1984: 12–15). Doubtless, in the right situations (for example: Cape Verde, Namibia, the Sultanate of Oman, Baja California, Ecuador, Peru and Chile), carefully designed trapping and collecting devices might provide enough water for livestock or domestic use, possibly even to supply crops or help establish trees.

However, there are claims for *aerial wells* which though fascinating must be viewed with some scepticism. For example, in the nineteenth century a Russian engineer in the Crimea discovered a system of stone conduits running from the remains of fountains in the town of Theodosia into the mountains to piles of crushed limestone. These piles it has been suggested, trapped passing mist, dew or low cloud (Furon 1963; UNEP 1983: 191). Balek (1977) described similar structures in Southern Africa, and Knappen (Dixey 1950: 69) mentions towers with loose stone infill sited on some European mountains supposedly with the ability to trap mists.

Controlled-environment cultivation

Systems of *controlled-environment cultivation* have been developed which require minimal inputs of fresh water. Some are already in commercial operation producing high-value crops in very inhospitable environments, but they are very expensive to install and run. At present it is only the rich and arid oil-exporting states and developed countries which can afford such systems. If integrated with some form of aquaculture, controlled-environment cultivation may be a little more cost effective, and it is possible that the cost may fall with future innovations. However, at present the costs of plastic enclosure, motors, fuel and the skilled management needed to run most of the systems is too high for widespread adoption (Gale 1981a, b; Slater & Levin 1981; DeCachard & Balligand 1982).

Hydroponics – techniques for growing crops in sand or other suitable inert medium by supplying a carefully – controlled supply of nutrient-enriched water – like controlled-environment cultivation, are presently too expensive for widespread application in developing countries, but they should not be entirely dismissed. Cooper (1983) described a hydroponics system to enable small farmers to grow fodder (hybrid Napier grass) which used very little water (550 m^3 to grow 30 tonnes of fodder) and which gave a crop worth at least twice the cost of production. It should not be overlooked that there are tropical lands with growing populations, with little hope of developing groundwater, with few streams, little rain and soils which do not favour the collection/concentration of rainfall. Small islands in particular may be forced to consider controlled environment cultivation and/or hydroponics, possibly supplied by desalination of seawater.

The importance of improved rainfed cultivation that is accessible to small farmers

At the start of the next century roughly four-fifths of the world's arable farmland will be non-irrigated. By that time much of the world's easily developed irrigated land will have been opened up. The problem will be how to intensify, diversify, extend and sustain the production of numerous, remote, small cultivators who will probably have little or no money, no access to groundwater or supplies from canals or streams. The answer must largely be the development and promotion of improved rainfed cultivation, runoff collection and concentration and the use of ephemeral flows and floods, together with the cultivation of new moisture-efficient crops and improved varieties of existing rainfed crops. Innovations must help farmers make optimum use of available moisture, and the improvement of security and sustainability of harvest must be given equal, if not higher, priority than obtaining increased and higher-value crops. In some regions larger farmers might be encouraged and aided to convert to improved cultivation methods which yield more than enough to compensate them for the redistribution of some of their land to help the very poor, i.e. the landless and smallholders with inadequate holdings (Bottrall 1981a: 28).

The routes to improved rainfed cultivation, the use of stream-flow, runoff or floodwaters will have to be cheap and accessible, yet today's research and investment is often directed at developing high-technology, costly cultivation systems, frequently unsuited to the environmental conditions in which most developing country farmers operate. Much agricultural development, especially of irrigation, has sought to modify the environment to suit particular crops, crops which economists or administrators have decided should be grown. A more sensible approach would be to select crops and cultivation practices to suit particular developing country environments and human needs.

In the past the Romans, Nabataeans, Incas and many other peoples developed agricultural strategies which fitted their often harsh environments. Two encouraging developments are (1) that researchers have begun to study ancient farming systems in the hope that this may help provide cultivation strategies suitable for poor, remote smallholders; and (2) that a number of research institutions are seeking strategies and crops for small farmers in marginal lands. Doubtless some of these strategies and crops will have potential for large-scale commercial farming as well as small-holders.

Chapter 7

Irrigation

The present role of irrigation in developing countries' agriculture

Irrigation methods that are available today offer the potential, *if all goes well*, for doubling or even quadrupling crop yields and for considerably reducing risks of crop failure. They can also offer considerable indirect benefits, like improving the sustainability of production. Unfortunately, irrigation commonly fails to yield its promise, fails to repay the investment made in it and, worse, may directly or indirectly generate serious environmental or socio-economic difficulties.

Irrigation is much discussed but seldom clearly defined. To some it may mean frequent and regular application of water, to others as little as one annual watering may merit being called 'irrigation'. In addition to the vagueness of terminology there is also a tendency for bureaucrats to publish the *planned* rather than the often less impressive *actual* irrigated areas their country has developed. Consequently, estimates of irrigated areas vary considerably. A wide definition such as the following is therefore more useful: *irrigation is the practice of applying water to the soil to supplement the natural rainfall and provide moisture for plant growth* (Wiesner 1970: 23). A way of solving the problem of defining irrigation so that the documenting of irrigated areas becomes more precise would be for authorities to record, honestly, the net returns per cubic metre of irrigation water applied to the land and the amounts of water used (Carruthers & Clark 1981: 82).

Moisture conservation plays an important part in many tropical cultivation strategies, both irrigated and non-irrigated. Irrigation development has had greater attention from researchers and much more funding, but in the future money and attention must be directed towards optimizing the use of natural precipitation, and where irrigation is practised to reduce the demand for water so that supplies are less likely to fail, enabling greater areas of crops to be grown (Wiener 1972).

There are basically three irrigation strategies: where rainfall is low and the bulk of crop needs must be met, *complete irrigation* is practised; where rainfall is usually adequate, but improved quantity, quality and intensity, i.e. more than one crop a year or crop diversification are desired, *supplemental irrigation* may be used; where there is a risk of inadequate rainfall *protective irrigation*, i.e. the application of water when drought threatens to damage a crop may be used to improve the security and yield of harvest. In view of the sheer numbers of small-scale farmers in lands subject to unpredictable rainfall fluctuation, security of harvest is as great, if not a greater need than increased crop yields.

Between 1965 and 1977 there was an established 31.7 per cent increase in irrigated land in developing countries (Arnon 1981: 68). One recent source suggests that there are over 160 million ha of irrigated land in developing countries (World Bank 1982); another more cautiously suggested that by AD 1990 there will be 119 million ha (Ruddle & Manshard 1981: 135) and another put it a little higher at over 145 million ha (Jurriens & Bos 1980). A conservative 'guesstimate' would be that about one-fifth of the total harvested land in developing countries is presently irrigated. This uses about 60 per cent of the supplies of fertilizer available to those countries and produces over 40 per cent of their annual crops, particularly foodstuffs.

In terms of agricultural improvement about 50 to 60 per cent of the increase in agricultural output achieved over the last twenty years in developing countries has been from new or rehabilitated irrigated land. It should not be forgotten, however, that for much of the last twenty years development effort has been directed at improving irrigation, so these figures are hardly surprising and may not be too much to be proud of. Table 7.1 indicates the expansion of irrigation between AD 1900 and 1978.

Irrigation may be said to be the dominant form of agriculture in South and South East Asia, China with probably about 49 million ha (estimates vary between 30.8 and 99.0 million ha) and India with about 39 million ha account for more than half of the developing countries' irrigated area (Carruthers & Clark 1981: 81; World Bank 1982: 62). Egypt's agriculture is virtually wholly irrigated, Peru's 75 per cent irrigated, Japan's 60 per cent, India's roughly 50 per cent, Iraq's is about 45 per cent and Pakistan's about 38 per cent (Arnon 1981: 56). Morocco, Bangladesh, Cuba, the Korean Republic (South Korea), the Philippines, Thailand and the Sudan have considerably extended their irrigated lands in recent years (a summary of eighteen of the most important Commonwealth irrigation projects was published by the Commonwealth Secretariat 1978: 65–75).

Sub-Saharan Africa has lagged behind in irrigation development compared with South or South East Asia. A large part of Africa's irrigation development has been in Egypt or the Sudan;

Table 7.1 Expansion of global irrigated areas 1900–1978

	1900	1930	1955		1965	1968	1978
Area (million ha)	44	80	120		173	163	201
Annual rate (million ha)		1.2		1.6	3.2		3.8
Dates		(1900–30)		(1930–55)	(1955–68)		(1968–78)

Source: Heathcote 1983: 188 (Table 13.2a) (figures rounded).

recently there have been more signs of activity elsewhere in the continent, particularly in Kenya. Between 1961 and 1971 African irrigation growth was slower than in any other continental grouping of developing countries – a mere 13.4 per cent. Various authorities have speculated as to the cause of this sluggish expansion, but no conclusive answers seem to have emerged (Chambers & Moris 1973: 19; Eicher & Baker 1982: 133–9). While African irrigation growth has in general been slow there has been an increasing interest in small-scale irrigation, some countries like Mauritius or the Malagasy Republic and some West African nations already have a history of smallholder irrigation and other states are making efforts to promote it without such a tradition. An important point made in a recent review of African smallholder irrigation development is that smallholder schemes are not miniaturized large schemes; they pose different problems (Bolton & Pearce 1984: 2; Blackie 1984).

The future role of irrigation in developing countries' agriculture

The *Brandt Report* (Independent Commission on International Development Issues, 1980: 94) identified irrigation and water management as the biggest single category of investment required in developing countries. Certainly irrigation development is important, but, even if all unused, potentially irrigable land – estimated by Jurriens & Bos (1980: 100) to be 450 to 500 million ha – were brought into production the majority of the South's farmers would still depend on rainfed cultivation. Carruthers & Clark (1981: 15) warned: '. . . it is extremely misleading to assert that "in agriculture, water holds the key to increased production through irrigation".' Presently many, if not most, of the better sites for irrigation development are being exploited, it is likely therefore that in the future only the less than ideal sites will be available. This means the costs of establishing irrigation will be more, and the risks of complications and failure significantly greater. There may be advantages in developing less densely settled, poorer-quality land with the aid of irrigation; it can reduce conflicts that may arise over ownership and the need for land redistribution which tend to plague more easily farmable areas, but it also means that yields are likely to be lower and running costs higher and so the returns on investment are likely to be sluggish. Not only is good irrigable farmland becoming scarce, demands on water supplies are increasing and it seems likely that by the end of the century water shortage will be a major constraint on agriculture, unless use becomes far more efficient than at present (Brown 1974).

It is relatively rare for much more than 20 per cent of the water diverted for irrigation to actually reach crops; sometimes it is even less than 10 per cent (Arnon 1981: 58). Inefficiency may be due to design or management faults, or both, but whatever the causes, performance of existing and planned schemes must be improved. It seems likely that more land has been irrigated, has become degraded and is now uncropped than was actually under irrigated crops in 1980 (Schulze & Van Staveren 1980: 15).

Low irrigation efficiency and poor sustainability reflect goals other than the maximization of yields. In India in 1969–70 only about 7 million out of roughly 30 million irrigated hectares were cropped more than once a year, reflecting that the goal is most often to gain security against unreliable rainfall (protective irrigation; Sigurdson 1977). Countries like Taiwan or the People's Republic of China, on the other hand, have put greater emphasis on boosting crop yields. The two goals may not be divergent, because once farmers feel better protected from the vagaries of the weather they will tend to adopt improved seeds and cultivation techniques which lead to more intensive cropping.

There are regions which still offer considerable potential for irrigation development, notably the Niger Basin, along the Brahmaputra–Ganges (especially in Bangladesh) and the Mekong River (Independent Commission on International Development Issues 1980: 81). There are also possibilities for developing the floodlands (várzeas) of Amazonia (Barrow 1985), for roughly doubling the area irrigated in Indonesia by draining swamps and using coastal lowlands (Carruthers & Clark 1981: 2), and by rehabilitating abandoned padi Burma might gain 250,000 ha of irrigated land. There are opportunities for irrigation development, the problem is that they may not come as cheaply as in the past. Developing countries will have to decide whether funds might be better spent in other ways, for example, on improving rainfed cultivation.

In the future one of the greatest problems facing those developing irrigation is going to be where to find water supplies. Basically there are five sources:

1. *New supplies* – from river diversion (with or without reservoir storage). If it is considered cost effective such supplies may be conveyed considerable distances from places with surplus to places in need of water. Increasingly, new supplies will be obtained by exploiting groundwater, which in many developing countries has been under-utilized.
2. *Conserving precipitation* – more efficient rainfed farming, the collection and concentration of precipitation and the cultivation of crops with low moisture demands. Compared with irrigation these strategies have been neglected.
3. *Flood agriculture* – making better use of river floodwaters, ephemeral streams, seasonal swamplands or the drawdown areas of large storage reservoirs.
4. *Use of saline and waste waters* – the development of water application and crop cultivation methods which with present crops, or better still crops with improved salt tolerance, will permit the use of irrigation return flows, sewage and industrial effluent, saline or brackish water or even seawater.
5. *Reduction of wastage* – one of the major routes of improving water supply availability is by cutting the waste of water on existing and future irrigation schemes and in the conveyance systems which supply them.

Demands for domestic water supplies and for sewage disposal will increase considerably by the end of the decade. The period 1981 to 1990 has been declared the International Drinking Water Supply and Sanitation Decade by the UN General Assembly (Agarwal 1980; Anon 1977b). If, as it is hoped, this initiative results in a reduction of water-related diseases and a lowering of death-rates in developing countries, then the UN should be thinking of declaring an International Water for Agriculture Decade as soon as possible to try to ensure that the extra people can be adequately fed.

Several nations will be approaching their maximum exploitable water supply by AD 2000 (Council on Environmental Quality and Department of State 1982: 26); moreover, some fear that many developing countries will suffer from destabilization of water supplies as a consequence of more intensive poorly managed land use. Another problem which may arise is climatic change. If the present-day environmental conditions continue more or less as they are the water resources situation in some developing countries seems likely to become difficult (Ambroggi 1980). It may well be, however, that the environment does not remain as it is, this might mean that for some countries water supplies become easier as rains fall more frequently or cloudiness increases, but some countries which presently seem in control of their water resources might suffer conditions which upset or hinder traditional and possibly even improved agriculture.

Irrigation practice

Irrigation systems can be divided into three subsystems which must all function satisfactorily if crops are to be satisfactory and sustained:

1. *The water delivery subsystem*;
2. *The farm subsystem*;
3. *The water removal subsystem*.

The water delivery subsystem encompasses the source, conveyance, allocation and distribution of water as far as the user's turnout. The farm subsystem consists of a network of channels, pipes or more often unlined ditches which distribute water from the turnout, or where the farmer(s) have access to water (this may be a tubewell, well, spring or stream) to the point(s) where it is actually applied to the crops. Sometimes water drains naturally from the irrigated land so well that there is no need for drains; however, often it is necessary to install a drainage system to control groundwater levels and/or to guide surface runoff to where it can be safely disposed of. When return flows are badly contaminated with salts or agrochemicals, a collecting pond or sump may have to be constructed to hold them until they are rendered safe or can be harmlessly released. The water disposal subsystem is important, without its satisfactory function there may be waterlogging and/or salinization of the irrigated land and contamination of surrounding land, streams, lakes or groundwater.

A major reason for the continuing inefficiency of many irrigation developments is that those who have planned and financed the projects ask few searching questions about their needs beyond the supply of water (i.e. they concentrate on item (1) just described). The provision of adequate drainage, efficient on-farm distribution networks, satisfactory land grading, the provision of agricultural extension services, supply of inputs, marketing of produce and so on tend to get left to different agencies, each charting its own course, determining its own schedule with no common objectives and poor overall coordination – or, worse, some or all of these vital aspects of irrigation are forgotten (ODI 1980: 5). Even if the engineering and management of water delivery is perfect, getting water to a user is no guarantee that it will not then be wasted. Yet many planners seem to imagine that delivery of water miraculously leads to better yields and improved socio-economic conditions (Nir 1974: 3).

Factors to be considered when selecting irrigation methods

Broadly speaking there are two irrigation development situations: when irrigation is introduced to a locality for the first time; and when existing irrigated agriculture is altered or improved. Whatever the situation, before any attempt at irrigation development the following considerations should be made:

1. Would proposed expenditure on irrigation be better spent on schools, roads, health care, rainfed farming improvement or something else?
2. Is irrigation the most economical and appropriate way of producing/increasing crops compared with say the use of fertilizers, alternative cultivation practices, better seeds or improved crop storage?
3. Will domestic terms of trade give enough incentive to irrigators to help development efforts work? Is security of tenure adequate?
4. Are there existing irrigation schemes in need of improvement, or is it better to start new ones?
5. Are the innovations proposed within the grasp of the participants? For example, the Helmand Valley Project, Ethiopia, encountered difficulties because planners overestimated the capacity of semi-nomadic people to adopt sedentary, irrigated farming (White 1962; Van Raay 1975).
6. Is the technical design adequate for supply, application and disposal of water?
7. To assess the suitability of the technical design it is necessary to consider:
   (a) *Soil factors* – these include the structure, texture, depth, infiltration rate, internal drainage rate, aeration, moisture-retention characteristics, level of salinity, presence of toxic substances, plant pests and diseases, fertility, risk of laterite formation.
   (b) *Topographic factors* – the degree of slope, size and shape of existing fields (if any).
   (c) *Climatic factors* – temperature, solar radiation, wind

exposure, air humidity, length of the growing season, variation of daylength.

(d) *Plant factors* – crops to be grown (their rooting characteristics, drought and salinity tolerance, length of growing season), rotation system, method of cultivation (close planting or row crops?).

Most irrigation strategies involve the management of a complex mix of physical and socio-economic factors which might be divided into four categories. Two of these are central to all irrigation schemes: *water supply activities* (supply, maintenance of supply, improvement of supply, and where appropriate collection of water charges); and *agricultural production and advisory service activities* (agricultural extension, agricultural research and advice on water application).

The remaining two less central types of activity: *commercial service activities* the supply of credit, agricultural inputs like fertilizer and aid with marketing and crop storage; and *basic infrastructure and social activities* – the provision of housing, roads, schools and health-care facilities and, if need be, land reform. The latter may be important where landholdings are too small or if cultivators are tenants with inadequate security of holding, where rents are exploitative or where sharecropping is practised (Bottrall 1981a: 75).

How much water will irrigation require, when should it be applied and what crops should be grown?

Without irrigation, crop yields are determined by the amount to precipitation received beyond the basic minimum required to enable the plants to reach maturity. For instance, a grain crop may need a minimum of 250 mm of rain, a reduction of only 10 mm from that may cause total crop failure, a reduction of less than 10 mm may cut yields markedly, yet, an increase of only 25 mm over the 250 mm might double crop yields. If rainfall is close to the minimum for satisfactory crop growth the planner must decide whether to use available water resources to enable a larger number of cultivators to obtain secure harvests and slight improvement of yield through developing supplementary or protective irrigation, or to supply more water to fewer cultivators practising complete irrigation.

Careful timing and regulation of flow to crops is vital; under-irrigation at best depresses yields, over-irrigation can also damage crops and cause waterlogging and/or salinization (Doorenbos &

Pruitt 1967; Kramer 1969). Wastage of water through over-irrigation deprives other potential users of supplies and wastes resources on supply system, maintenance and management; it is not efficient to convey the wrong quantity of water efficiently. An irrigation supply system should be able to provide enough water to meet crop needs during the most demanding combination of crop growth and environmental stress; this situation is termed *peak crop requirement*.

Basically, there are two ways of assessing when a crop needs water and when that need has been satisfied:

1. *Direct measurement* – for example, the cultivator or irrigation supply officer may feel the soil or weigh it, or check the soil suction pressure with a soil tensiometer, study a *lysimeter* (a lysimeter is basically a tank, filled with soil and planted with crops or other vegetation to match its surroundings, sunk into the soil and supported on a weighing device, or with some other means of measuring moisture changes in the soil), check whether plant growth is steady or watch indicator plants to see when they wilt and recover.
2. *Empirical methods* – use is often made of what is termed the *reference crop evapotranspiration* to decide what crops are most appropriate given the capacity of the water supply system. Reference crop evapotranspiration is estimated from evaporation pans or lysimeter measurements corrected for local environmental conditions, or a reference plot of well-watered rich soil covered with dense grass ('standard lawns'). The *Blaney-Criddle formula* is commonly used to calculate potential evapotranspiration, or use may be made of water budgets drawn up with reference to assumed crop moisture requirements, crop ability to extract moisture from the local soil and the rate at which the soil can be recharged with moisture. For a review of use of the formula see: Allen and Pruitt (1986) *ASCE Jnl. of Irrigation and Drainage Engineering*, 112(2), 139–155.

In practice, working assumptions are often made that plants cannot readily use more than half the available moisture in a soil, that irrigation is needed when soil moisture falls to between half and two-thirds of the available moisture and that the amount of water which should be applied would be that required to restore the soil moisture to field capacity (Fig. 7.1).

In practice the exact point on the curve in Fig. 7.1 where moisture falls low enough to check crop growth may not be easy to recognize in the field. The irrigator, therefore, is forced to err on the side of over-irrigation rather than risk crops suffering water stress which might reduce yields or even kill them. Deep-rooted crops are less vulnerable to checks in growth if the irrigator gets things wrong and delays application or applies too little. Table 7.2

Fig. 7.1 Rate of growth to moisture content of the soil.
Optimum growth varies according to aeration, water-holding capacity of soil and crop grown.

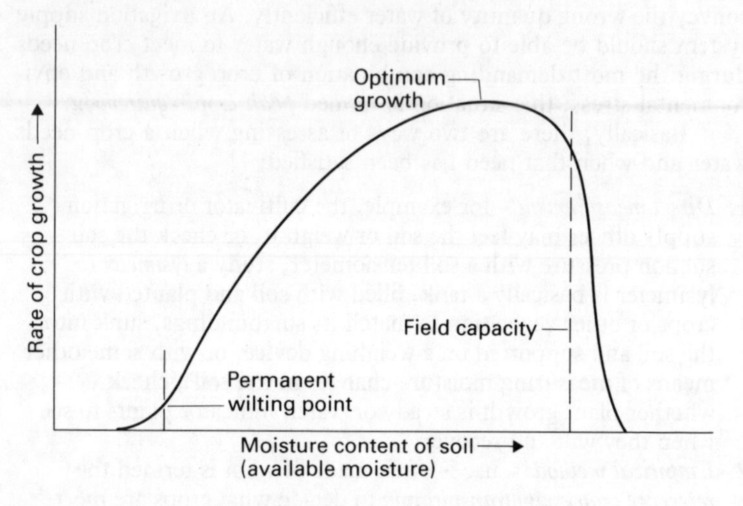

Source: Israelsen & Hansen 1962: 267 (Fig. 12.1).

lists a range of factors which determine the frequency with which irrigation should take place.

Commonly, irrigators assume enough water should be supplied to maintain potential evapotranspiration. There is, however, a level of watering well below that needed to maintain potential evapotranspiration which will support economically sound cropping (Carruthers & Clarke 1981: 60). Unless water is virtually free the most economic application will be less than that required for potential evapotranspiration and less than that for maximum yield per hectare – getting maximum yields is largely a matter of diminishing returns, little is gained for inputs of water, fertilizer or labour above a certain point. The problem is, how much less water than that required to support potential evapotranspiration should an irrigator apply? In recent years there has been considerable interest in *deficit irrigation*. This involves decreasing the amount of water or increasing the interval between applications – in effect subjecting the crop to planned moisture stress in the hope that it will save water without reducing yields too much; some crops may actually improve in quality if subjected to such treatment (e.g. cotton or sugar). The technique is little tested in practice and judgement of its value should await further analysis (Bottrall 1981a: 205).

Table 7.2 Frequency of irrigation

	Relatively frequent irrigation	Less frequent irrigation
Plant	Shallow, sparse, slow-growing roots	Deep, dense, fast-growing roots
	Fresh weight yield of vegetative organ desired	Dry weight yields of reproductive organ desired
	Quality dependent upon size of vegetative organ	Harvest for content of sugar, oil, etc
Soil:	Shallow soil – poor structure impeding root growth	Deep soil, good structure
	Slow infiltration and internal drainage, poor aeration	Good infiltration, internal drainage and aeration
	Small fraction of available water held at low soil moisture stress level	Large fraction of available water held at low soil moisture stress level
	Saline soils or water	Non-saline
	Fertility level high – nutrients concentrated in topsoil	Fertility level low – nutrients distributed in profile
	Root disease nematodes present	Constant water-table in reach of roots
Weather:	Planted at beginning of hot dry season	Planted ahead of hot dry season
	Major growth period – hot dry season	Major growth period before hot dry season
	High evaporation rates	Low evaporation rates.

Note: Factors in the left-hand column above conspire to increase the frequency of irrigation needed; factors in the right-hand column act to reduce the frequency of irrigation.

Source: Wiesner 1970: 180 (Table 33).

Water use efficiency

On-farm water losses are high. The amount of water which reaches a plot of land is usually less than 70 per cent of that received at the turnout, commonly it is a mere 20 per cent or less. The ratio of the amount of water applied to the crop to the amount of water delivered at the turnout, well or other point of supply is termed the *farm conduit efficiency*. Still more water is lost after conveyance to the cropped plot; evaporation takes place from the ground surface or leaves and water may run away without infiltrating. There is therefore very considerable wastage of water, yet there has been little standardization of measurement of water losses. Only the common methods are outlined in the following paragraph, but further details may be found in Yaron *et al*. (1973: 414); Stern (1979: 111); Carruthers & Clark (1981: 86).

The ratio of volume of water actually consumed by the crops to the volume of water applied to the field is the *field irrigation efficiency* (usually expressed as a fraction or percentage) and is probably the most commonly cited and useful measure. There is variation, at least in detail, in defining this measure. Finkel (1982: 73) for example, determined it by dividing the amount of water required by a crop in the field by the amount of water applied to the field. Another frequently used determination is to take the ratio of the average depth of irrigation water which is beneficially used to the average depth of irrigation water applied.

Another measure sometimes used is the *field application efficiency* – the ratio of the average depth of water reaching and remaining in the root zone to the average depth of water applied to the field. Under normal conditions this shows the amount of water potentially available to the crop, but gives no indication of the adequacy or uniformity of water application. Wade (1982) uses *water use efficiency* to indicate a project's technical efficiency – the ratio of the volume of water used for crop production to the volume of water diverted from or stored at source.

Water productivity is often used, and is the ratio of crop yield to quantity of water used, but unfortunately it is sometimes not clear at what point (source, turnout or crop application) measurements are taken. Measures of irrigation or water use efficiency should therefore be treated with caution, especially if any attempt is made to compare figures quoted by different authorities. Typical irrigation efficiencies for four irrigation methods are listed in Table 7.3.

Measures of irrigation efficiency are used to determine what crops can be grown in a locality with a given supply system. For example, assuming in a given locality maize requires 60 mm/month rainfall, irrigation or an equivalent combination of both. If the field irrigation efficiency is 0.5 then the maize plot will need 60/0.5 = 120 mm/month; were the efficiency 0.7 then only 60/0.7 = 86 mm/month would be needed.

Removal of excess water

Drainage is one of the most critical aspects of irrigation. Sustained cropping often depends almost as much on getting rid of excess

Table 7.3 Field irrigation efficiencies for different methods of irrigation

Irrigation method	Efficiency (%)
Surface – basin (except rice), border, furrow, corrugation	40–60
Flooded rice	30
Overhead sprinkler	60–80

Source: Stern 1979: 74 (Table 16).

water as on getting enough to irrigate with. All but the most freely draining soils require some sort of artificial drainage to cope with excess irrigation water and intense downpours. *Surface drains* will do this, but where soils are waterlogged or at risk of becoming waterlogged deep drainage, i.e. *subsurface field drains* (terracotta or plastic tubes with perforations, or mole-plough drains), deep dikes or pumped tubewells will be required.

When drainage is installed the rate at which pollutants like pesticides, herbicides or excess fertilizer reach streams or other bodies of water may drastically increase. Water disposal therefore requires careful attention.

The relationship between fertilizer application and crop water requirements

It has sometimes been claimed that fertilizers are wasted without increased application of water. Obviously, agrochemicals have to be dissolved in soil moisture for plants to utilize them, but provided fertilizers do not result in foliar growth that raises evapotranspiration, increased crop yields may be obtained without increased applications of water (or in the case of rainfed farming with natural rainfall; Wiesner 1970: 87; Hagin & Tucker, 1982). Nitrogenous fertilizers tend to increase leaf growth and may delay the maturation of crops, all of which may increase moisture consumption; phosphatic fertilizers may improve moisture utilization by accelerating crop maturation (Carruthers & Clark 1981: 60).

Where crops are planted in a fairly open pattern, fertilizer application may stimulate leaf growth enough to enclose a beneficial microclimate between the ground and the leaf canopy, while not promoting so much leaf that evapotranspiration is raised significantly – the net result can be a considerable reduction in evapotranspiration losses. It may be that a vigorous crop may lose less water in proportion to yield than a weakly growing one with a poor root system and sparse leaf cover.

It has been suggested (before the OPEC petroleum price rises) that in some circumstances fertilizer application might be more cost effective than irrigation (Wiesner 1970: 83); whether this is still likely is uncertain, but merits study. High-yielding crop varieties, it is often stated, use more water than traditional varieties. Certainly some HYVs may be less tolerant of moisture stress, but those with short growing seasons often use much less water or use the water at times of the year when it is available. To fulfil their potential many HYVs require artificial fertilizer applications and this should be applied when it can be carried into the root zone, but not when it will be rapidly leached away by heavy or persistent rainfall.

Salinization and irrigation

As good-quality water becomes scarce, those wishing to irrigate crops will be forced to consider poor-quality supplies and this will increase the risks of salinization. Soils may contain *residual salts* or salts may accumulate in a soil as a consequence of natural or human causes (*primary salinization* and *secondary salinization* respectively). Insufficient downward movement of water through the soil, the application of poor-quality irrigation supplies or groundwater too close to the ground surface may lead to salinization, individually or in various combinations. Most research has focused on the first two cases; however, Kovda (1980: 184), was critical of this and argued that the role of groundwater deserved far more attention. In loamy soil Kovda estimated the *critical depth* of the water-table was between 1.5 and 2.5 m below the soil surface. Were the water-table to rise above the critical depth for a given soil, Kovda argued, salinization was highly likely *even if good irrigation water was used*. In some clay soils, he suggested, there may be a significant risk of salinization even when the water-table lies deeper than 3 m below the ground surface. The critical depth may be found by the formula:

$L = 17 \times 8t + 15$ (Kovda 1980: 205)
where L = the critical depth (cm);
t = the average annual temperature (°C).

Water-tables often rise when canals and reservoirs leak, when farmers over-irrigate their crops and when clearance or alteration of vegetation cover increases groundwater recharge (Holmes &

Talsma 1981: 313). In view of the role of groundwater in causing salinization, drainage of irrigation schemes assumes an even greater importance, yet it is still an aspect of irrigation development that is often neglected, for example, Worthington (1977: 102) cited a large modern irrigation scheme where drainage was not installed until several years after 'completion', by which time considerable damage had already been done. It is by no means a unique example.

Salts are found in the soil solution (the moisture in the soil's interstices) and can be removed by drainage or suction, or linked with clay particles (*exchangeable salts*). A soil has a limit to the number of exchangeable salts (*cations*) which it can hold: this is the *cation exchange capacity* (expressed in mg equivalents per 100 g of soil; Withers & Vipond 1980: 110).

Socio-economic factors play an important role in salinization/sodification. For example, where land is held on short-term tenure, cultivators will tend to have little regard for long-term management of the soil to avoid salinization. Peasants sometimes prefer to exploit shallow salty groundwater rather than dig deeper wells or make use of government tubewells supplying better water (Farvar & Milton 1972; El Gabaly 1979).

Provided groundwater is not within 'critical distance' of the soil surface so that salts rise through capillarity, those being added in irrigation water can usually be leached down the soil profile (the removal of exchangeable salts, however, may require the addition of chemicals to the soil). The amount of leaching required to sustain viable irrigated agriculture depends upon the drainage, the irrigation water and soil salt content, climatic conditions, crop salt tolerance, the water management strategy and the cultivation strategy. Sometimes natural rainfall is enough to leach salts away; if it is not the cultivator must apply enough irrigation water to meet the *leaching requirement* which may be defined as *the minimum fraction of applied water that must pass through the root zone to prevent a reduction in crop yields from excessive accumulation of salts* (Holmes & Talsma 1981: 166). To meet the leaching requirement, irrigation applications must sufficiently exceed the amount consumed by the crop and evaporated from the soil surface to ensure a net downward movement of water to carry salts away – a balance should be reached whereby a certain water application holds the soil salt content steady year after year, i.e. a *salt balance* is maintained. Leaching can be done at each watering, after several waterings, annually or at even longer intervals.

Maintaining a salt balance may be quite difficult in practice, one problem is that the leaching requirement is determined from salt/water balance calculations which do not allow for the dynamics of salt/water movement in the soil. It is possible for the leaching requirement to be theoretically satisfied, but in practice more water need be applied to reach it. This is because salts in solution in groundwater rise through the soil by capillarity through fine pores and fissures, water moving down through the soil, hopefully carrying away salts, tends to move through the larger fissures so salt removal is less than 100 per cent effective (Kovda *et al.* 1973: 19).

Where risks of salinization are high, irrigation method and cultivation techniques must be carefully selected to slow salt build-up and make removal as easy as possible. This is especially important if the irrigation water itself is saline (Hoffman *et.al.* 1980; Johl 1980). If the risks of salinization are high and water for irrigation is scarce during the growing season, or where salts continue to rise, it may be necessary to *pre-irrigate* to wash salts out of the ground before planting. Salt levels in the soil are lowest after irrigation and highest just before the next application. Similarly, in seasonal rainfall regions, salts in the soil are highest at the end of the dry season(s).

It may be possible to adopt a planting pattern which takes advantage of points of relatively low salinity, for example at the sides of wide ridges. An irrigator who is unaware of these 'low' salinity points is at a grave disadvantage (Fig. 7.2).

When high levels of soluble salts like sodium chloride or sodium sulphate accumulate in the soil, provided there is less than 15 per cent exchangeable sodium a *saline soil* (solonchak or 'white alkali soil') may form – the sodium content of a soil is generally expressed as *percentage exchangeable sodium*, i.e. the percentage of the total cation exchange capacity occupied by sodium. Typically, saline soils have a pH of less than 8.5 (a pH of less than 7 is acid, pH 7 is neutral, above pH 7 is alkaline).

Abundant salts and more than 15 per cent exchangeable sodium leads to the formation of *alkali soils* the process being called *alkalinization* (sodification or soda salinization – as opposed to the production of saline soils which is *salinization*). Alkali or saline–sodic soils generally have a pH above 8.5. If they have little salt content, abundant sodium (exchangeable sodium levels above 15 per cent) and little calcium, clay particles in the soil matrix adsorb sodium and magnesium salts, swell, become impermeable and easily compacted (the process is known as *hydrophilization*). When this happens rainfall penetration and irrigation are hindered and plant roots may be starved of oxygen. Sodium hydroxides may also form and these dissolve organic matter in the soil – this may either be deposited on the soil surface as a dark crust, or leach away. When a dark crust forms the soil is often called *black alkali soil* (solonetz or non-saline–sodic soil). Typically, these black alkali soils have a pH between 8.5 and 10.0.

Alkaline soils may be formed from concentration of sodium

Fig. 7.2 Salt accumulation using high-, moderate- and low-salinity water and with different bed types.
Germination/growth is delayed or prevented where salt accumulation is high (shaded areas).

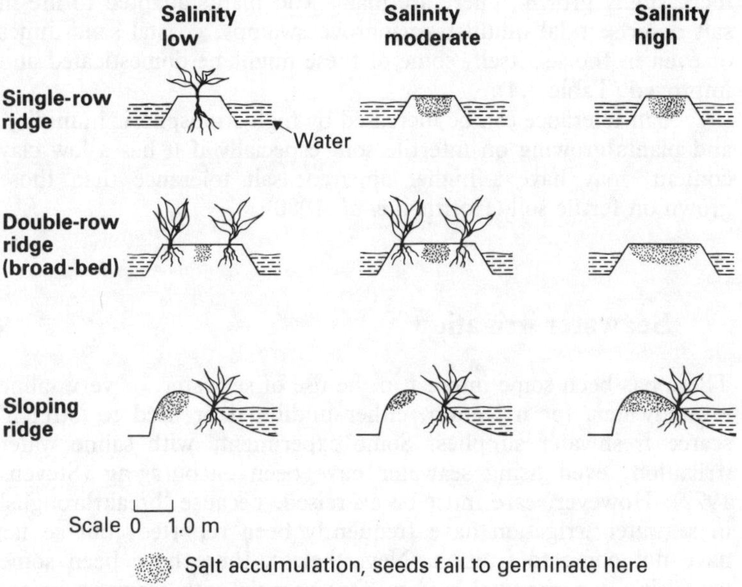

Source: Kovda *et al*. 1973: 327 (Fig. 10.5).

compounds already in the soil through the capillary rise of alkaline groundwater, or may result from attempts to rehabilitate salinized soil (rich in chlorides, sulphates and gypsum) by drainage and leaching. If there is inadequate calcium in soil being rehabilitated, sodium hydroxide can build up and in turn can lead to sodium carbonate formation (*solotization*).

Where sulphate ions are reduced to sulphides due to anaerobic conditions, problems may occur if the soil is drained. A very acid soil results and if there is not enough calcium present to neutralize the reformed sulphate ions the result may be an *acid sulphate soil*. These are common in tropical wetlands, especially coastal lowlands, and are one of the few soils which are unsuitable for rice cultivation. The effects of high salinity levels may not all be damaging; for example, there have been claims that plants grown under saline conditions better resist viral diseases and urban air pollution (particularly damage due to high levels of atmospheric ozone; Shainberg & Oster 1978; Gupta 1979).

Salt-tolerant crops

Cereals, notably triticale, rye, barley and wheat, and forage crops with improved salt tolerance have already been developed and are increasingly grown. There are many wild plants adapted to life in salt deserts, tidal mudflats mangrove swamps, coastal sand-dunes or even in the sea itself; some of these might be domesticated and improved (Table 7.4).

Salt tolerance can be increased by high atmospheric humidity, and plants growing on infertile soil, especially if it has a low clay content, may have a higher apparent salt tolerance than those grown on fertile soil (Hoffman *et al.* 1980).

Seawater irrigation

There has been some interest in the use of seawater, or very saline groundwater, for irrigation, either undiluted or used to 'stretch' scarce freshwater supplies. Some experiments with saline water irrigation, even using seawater have been encouraging (Stevens 1972). However, care must be exercised, because 'breakthroughs' in seawater irrigation have frequently been reported, but so far have not come to fruition. Nevertheless, there have been some promising experimental results, for example with seawater irrigation of barley. In one set of trials freshwater-irrigated control plots gave 2,390 kg/ha, seawater-irrigated plots gave 458 kg/ha, and in later stages of the experiment this was raised to 1,580 kg/ha which is impressive considering that the world average barley yield was about 2,070 kg/ha in 1976 (Epstein & Norlyn 1977; Epstein *et al.* 1979). Experiments with barley suggest that carefully selected salt-tolerant varieties might be grown using seawater, success is probably more likely if the soil is very free-draining. There should be no shortage of suitable sites: according to Hollaender (1979: 92) sandy soils and stable sand-dunes cover over 13,000 million ha of the world and are presently virtually unused.

A seawater irrigation system operated by the Monastery of St Nicolas de Ugarte at Sestão (near Bilbao, Spain) was described by Boyko (1968: 105, 195), and another seawater irrigation system was described by Stevens (1972). Both systems rely on planting normal crops in beds of sandy soil around which seawater is channelled in ditches. The beds are raised sufficiently to ensure that the crop roots remain above the level saturated by the seawater but close enough to it to benefit from non-saline moisture formed (presumably by the evaporation of the seawater and its condensation within the open-textured soil). Brushwood wind-breaks around the beds help

Table 7.4 Plants with some potential as salt-tolerant crops

Plant (use)	Maximum salinity tolerated (ppm)	Source
Forage		
Suaeda spp.	35,000	
Palmer saltgrass (*Distichlis palmeri*)	35,000	
Saltbush (*Atriplex* spp.)	25,000–35,000	San Pietro 1982
Rhodes grass (*Chloris gayana*)	26,000	
Spartina spp.	35,000	Hollaender 1979
Suwanee Bermuda grass (*Cynodon dactylon*)	12,000	
Salt-grass (*Paspalum vaginatum*)	10,000	National Academy of Sciences 1975
Wheat-grass (*Agropyron elegatum*)	—	
Polypogon sp.	7,000	
Salicornia spp.	35,000	
Suadia spp.	35,000	
Grain		
Eelgrass (*Zostera* spp.)	35,000	
Spartina alternifolia	35,000	San Pietro 1982
Palmer saltgrass (*Distichlis palmeri*)	35,000	
Surfgrass (*Phylospadix*)	35,000	Hollaender 1979
Zizania aquatica	—	
Barley (*Hordeum* spp.)	35,000	Various experimental trials
Wheat (*Triticum* spp.)	15,000	Various experimental trials
Miscellaneous[a]		
Nipa palm (*Nipa fruticans*)	35,000 S, T	
Mangroves (*Rhizophora* spp., *Avicenna* spp. and others)	35,000 W, t, F	
Mulberry (*Morus nigra*)	10,000 f	
Date (*Phoenix dactylifera*)	8,000 f	
Coconut (*Cocos nucifera*)	— f, x	
Carthamus tinctorius	3,500 o	
Sarsason	3,500 a	
Sagebrush (*Artemisia* spp.)	—	
Salt cedar (*Tamarix pentandra*)	— W	
Creosote (*Larrea tridentata*)	— H	
Beet (*Beta vulgaris* and others)	12,000 S	
Guar (*Cyanopsis tetragonoloba*)	— o, g	
Jojoba (*Simmondsia chinensis*)	7,000 o	

Table 7.4 continued:

Plant	Maximum salinity tolerated (ppm)	Source
Algae[b]		
Red algae (*Porphyra* spp.)	35,000	f, c, a, e
Brown algae (*Macrocystis* spp.)	35,000	a, c
(*Laminaria* spp.)	35,000	a, c
(*Undaria* spp.)	35,000	a, c
Green algae (*Monostroma* spp.)	35,000	e
(*Enteromorpha* spp.)	35,000	e
(*Ulva* spp.)	35,000	e
(*Spirulina* spp.)	35,000	e
(*Chlorella* spp.)	35,000	e
(*Tolypothrix* spp.)	35,000	e
(*Dunaliella* spp.)	35,000	e

[a] Key to miscellaneous plants uses: S = Sugar, T = thatch, W = wood, t = tannin, F = fodder, f = fruit, x = fibre, o = vegetable oil, a = alcohol, H = hydrocarbons, g = gum.
[b] Key to algae uses: f = fucoidon, c = carageenan, a = alginates, e = edible.

Note: Some of the plants listed in Table 6.2 are salt tolerant.

Information on promising trials in *Journal of Arid Environments*, 11(1986), pp. 37–59.

Sources: Sharkhov 1956; Bernstein 1962; Chapman 1960; Israelsen & Hansen 1962; Boyko 1968; National Academy of Sciences 1975; Adams *et al.* 1978; Hollaender 1979; Finkel 1982.

reduce the evaporation losses from the crop so that moisture derived from the seawater is sufficient.

Some plants like mangroves, Nipa palms or certain algae might be used to provide crops (charcoal, timber, tan-bark, alginates, fodder and so on) using seawater or brackish estuarine water.

Using waste water

Considerable amounts of sewage or industrial effluent and *irrigation return flows* (excess water from irrigation schemes, often contaminated with salts and agrochemicals) are presently unused. The treatment of effluent and the risks associated with using it for irrigating farmland have been discussed (see pp. 96–97). Certain mangrove species and certain algae can grow in heavily contaminated water; if effluent were led through shallow algae-filled channels or artificial 'mangrove swamps', biological processes might render it harmless, at the same time yielding some of the useful crops mentioned in the last section (fodder, and alginates for human consumption excluded; San Pietro 1982: 195; Hoddy 1983).

Rehabilitating and preventing the recurrence of salinized or alkalinized soil conditions

Faced with salinized or alkalinized soil, irrigation management must first assess the cause(s) and attempt rehabilitation if it is feasible. If the main cause of salinization is a high soluble salt concentration in the soil, rehabilitation is comparatively easy provided there is adequate natural or artificial drainage, and sufficient water of good enough quality to apply to the soil to leach away the salts. It should also be noted that leaching also removes soluble nutrients as well as salts. Therefore, careful application of fertilizer may be required after rehabilitation by leaching.

To rehabilitate saline–alkali soils the problem is to replace the exchangeable sodium (that attached to clay particles) with calcium or other suitable ions, to reduce the sodium to calcium ratio and then to leach out the replaced sodium and other soluble salts which are no longer bound to the clay. If the mistake is made of leaching away soluble salts leaving the exchangeable salts bonded to clay particles behind, the soil structure can deteriorate making it very difficult further to improve the soil. Exchangeable sodium can be replaced by calcium under field conditions by adding a soluble calcium salt such as gypsum (Seneviratne 1979), or a soluble acid such as sulphuric. Acids are not usually applied to the soil, but powdered sulphur, which is rapidly oxidized in most soils to sulphuric acid, has been used. Aluminium and iron sulphates have also been used as both of these hydrolyse, leaving the hydroxide behind as a precipitate and liberating sulphuric acid. Routine methods of controlling alkalinity in dryland soils may involve the regular addition of acidifying agents.

The rehabilitation of alkali soils is often difficult and expensive, largely because they are so impermeable it is difficult to leach away salts. If care is not taken, attempts to rehabilitate saline or saline-sodic soils can convert them to alkali soils.

A (slower, less-effective) alternative to *chemical amendment* (or amelioration) is to grow, harvest and remove plants which are able to draw salts from the soil. Such plants are more useful as part of a salinity control strategy and could be rotated with crops to help maintain a salt balance. This has been tried on parts of the Gezira Scheme where a fodder crop – saltbush (*Atriplex* spp.) – was grown in rotation with cotton one year in twelve with promising results; if land were badly salinized such crops might be grown more often (Withers & Vipond 1980: 118). In addition to saltbush, other promising salt-control plants include: *Tamarix* spp., *Suaeda fructicosa*, *Prosopis spicifera*, *Leptochloea fusca* and some of the rush family (*Juncus* spp.). The latter could be used to remove salt and then be harvested for thatch, strawboard or paper production or for fuel (Heathcote 1983: 89).

It may sometimes be possible to control salinization by altering vegetation cover to lower the water-table below the critical depth, so preventing capillary rise and deposition of salts. For example, in India and Pakistan deep-rooted crops – pulses, oilseed rape or wheat – are often grown after rice so that their roots can pursue the falling capillary fringe, their transpiration then lowers the water-table, reducing the risk of waterlogging and salinization after the next rainy season (Kovda *et al.* 1973: 255 lists deep-rooted plants suitable for this use). An alternative is to plant a belt of deep-rooted plants which, hopefully, do not compete with crops for soil moisture or in any other way reduce yields, alongside cultivated plots to depress the water-table enough to avoid salinization (Fig. 7.3) (Hay 1972; Downing & Gibson 1974; Holmes & Talsma 1981: 129).

Salt-tolerant crops and appropriate salinization-avoidance cultivation strategies, if they can be perfected and disseminated, will open up considerable areas of land to farming and could benefit small-scale cultivators in remote areas.

Water supply: scheduling and allocation

When sufficient supplies of water can be delivered to exceed peak demand, management merely has to maintain a constant flow to allocate supplies. Ideally *constant-flow delivery* (continuous-flow delivery) should involve some regulation to ensure users do not over-water so as to drastically exceed crop needs and overwhelm natural or artificial drainage. By monitoring factors like the hectarage of particular crops under cultivation and the rate of evapotranspiration, it should be possible to regulate constant-flow delivery to prevent gross over-irrigation. In practice, finding the skilled staff and the funds to do such monitoring may not be possible. Constant-flow delivery operates best if the rates of flow

Fig. 7.3 Use of deep-rooting, high-transpiration-rate plants (phreatophytes) to lower groundwater sufficiently for annual crops to be grown with reduced risk of salinization. The phreatophytes also help lower crop evapotranspiration by acting as wind-breaks.

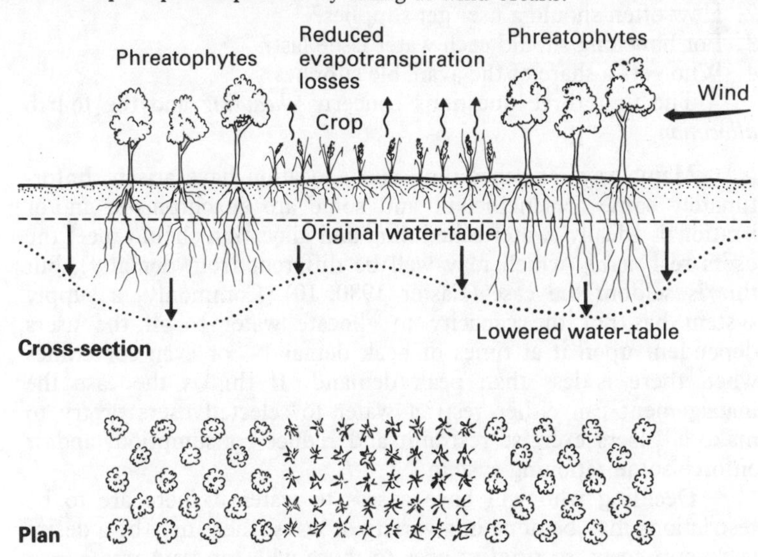

Source: based on Oliver 1972: 118.

are relatively fast; if water flows slowly the conveyance efficiency of such supply systems tends to be poor due to leakage and evapotranspiration (Kovda *et al.* 1973: 302; Easter 1980: 11; Wade 1981: 30). As demand for limited water supplies becomes greater constant-flow delivery systems are likely to become less common, although electronic devices may make monitoring water demand easier and more accurate.

Wherever water is scarce in relation to demand, individuals, communities or groups compete for it. Where water is supplied to users by an authority, one of that authority's most important functions is to oversee and manage the competition to ensure that there is efficient, and where possible fair, distribution (Jensen 1980).

User demand for water supplies is not just a function of crop needs and environmental factors; demand for water depends on the attitude of users, and greed, fear of water shortage and misinformation can cause water users to demand and apply too much. An 'adequate' water supply may cease to be, merely if users' attitudes change (Hazlewood & Livingstone 1982: 35).

Those planning and managing a water supply system face four basic questions:

1. How much water should they provide per user?
2. How often should a user get supplies?
3. For how long should each water issue last?
4. Who gets a share of the available supplies?

The first three questions concern *scheduling* and the fourth *allocation*.

Many ways of scheduling and allocation have arisen, unfortunately some are inefficient and some are anachronistic and/or irrational. Ideally, water scheduling and allocation should meet the users real *needs* (which may well be different from demand), but this is seldom the case (Easter 1980: 10). Commonly, a supply system has not the capacity to allocate water to all the users dependent upon it at times of peak demand – or even sometimes when there is less than peak demand. If this is the case the management can either restrict water to selected users or try to make all users exercise restraint and reduce consumption, and/or enforce some rationing system.

Deciding who will have access to water if there are to be restrictions may be done in a variety of ways: there may be a desire to recoup costs, so priority goes to those who can pay; water may be regarded as 'aid' and be allocated to a particular underprivileged group or to those growing a particular crop favoured by the authorities. Being selected as a user is no guarantee of being allocated water if supplies are short or interrupted.

The character of the conveyance/supply system determines the degree of control an authority may have over allocating water. Some systems can be controlled only at source and at the users' turnouts, others are far more versatile (Fig. 7.4 is an example of the latter, a typical modern canal supply system on which flows can be controlled at a number of points).

The various ways in which rate and volume of flow, duration of delivery and frequency of deliveries can be manipulated to allocate water supplies is shown in Table 7.5.

Most modern water supply/distribution systems can be controlled at the source, at the laterals, distributaries and the farmer's turnout (farmgate). A *rotation* can be organized if there are adequate controls over flow, so that one lateral gets supplies for a given time and then the water is diverted to another lateral. Along each lateral, users' turnouts can usually be locked or unlocked to operate a rotation of supplies (see Fig. 7.4).

Broadly, there are two types of rotational allocation: *rigid scheduling* and *flexible scheduling*.

Fig. 7.4 A typical canal (gravity) irrigation supply/distribution system. Topography determines the network layout: in general the main canals follow ridges, laterals are routed along lesser ridges, and drains (if there are any) follow depressions. *Command area* is the total area irrigated; this depends on the elevation of the supply canal above the lands to be irrigated. Sometimes command is not positive but negative and farmers then have to lift water to their crops.

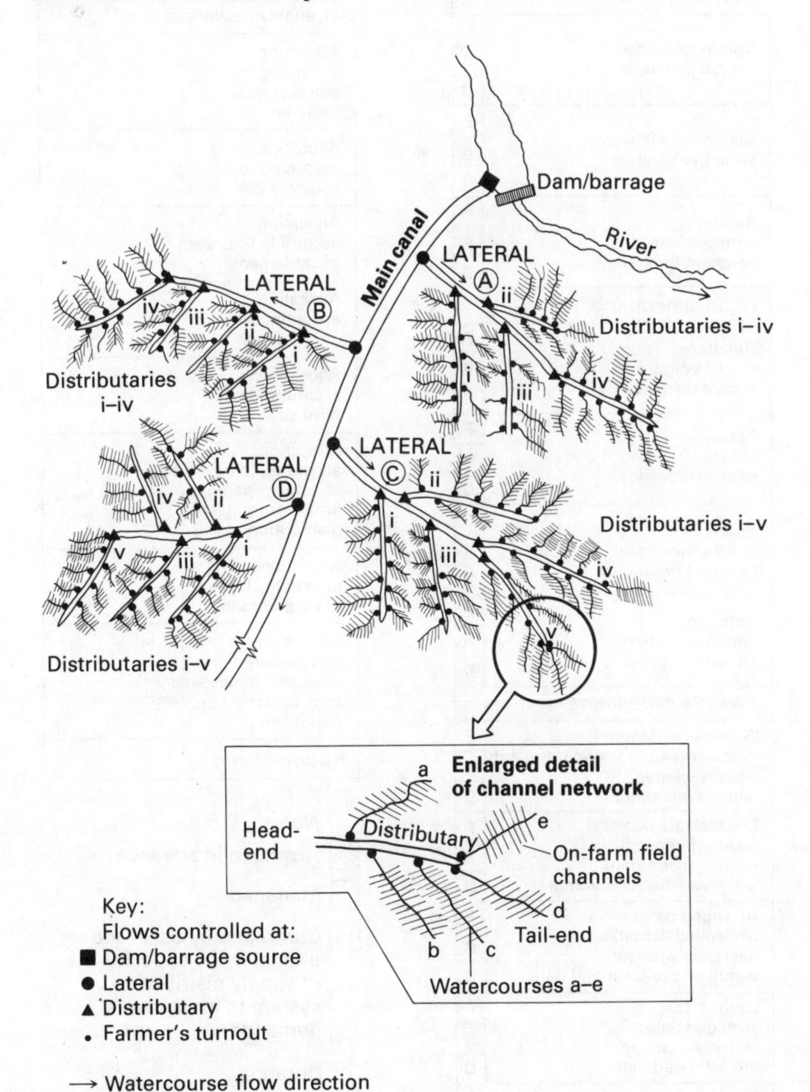

Table 7.5 Scheduling possibilities and allocation possibilities

Scheduling possibilities			
How much flow, how often, for how long			
Constant flow			
Continuous flow: constant volume	F	R	D
Continuous flow: varied volume	F	R [managed]	D
Rotation: constant rate/ constant frequency	F [managed]	R [managed]	D
Rigid scheduling			
Rotation: varied volume/ constant frequency	F [managed]	R [managed]	D [managed]
Rotation: constant volume/ varied frequency	F [managed]	R [managed]	D [managed]
Rotation: constant volume/ constant frequency	F [managed]	R [managed]	D [managed]
Rotation: varied volume/ varied frequency	F [managed]	R [managed]	D [managed]
Flexible scheduling			
Delivery on demand: (free demand or user-controlled): what is wanted, when it is wanted	F	R	D
Limited-rate demand (user-influenced): what management will give whenever wanted	F	R [managed]	D
Arranged date controlled demand: user gets what he wants on pre-arranged date	F [arranged in advance]	R	D
Limited-rate arranged date: restricted rate on pre-arranged date	F [arranged in advance]	R [managed]	D
Restricted organized: rate and duration of flow restricted on pre-arranged date	F [arranged in advance]	R [managed]	D [managed]

Allocation possibilities
Who gets the water
Allocation: according to user's location on supply network
Allocation: according to social status/ tradition
Allocation: according to ability to pay
Allocation: according to order of settlement
Allocation: according to crop priority
Allocation: according to farm size
Allocation: according to user's area cf. total irrigated area, i.e. equitable distribution
Allocation: to certain stages of crop growth
Allocation: to a favoured group– refugees, relocatees, selected target group, nepotism
No restriction

Notes

[hatched box] Arranged in advance

[shaded box] Managed

[blank box] Controlled by user (and ultimately by capacity of supply distribution system to meet user's demands)

D Duration

R Rate

F Frequency

Rigid scheduling

There are a number of ways in which water deliveries may be scheduled in a rigid manner (see Table 7.5). The simplest and most widespread form is for successive users along a lateral to take turns, the allocation to each lasting for a certain time, until a given flow has been delivered or sometimes until the user has flooded his land. If for some reason supplies to the lateral cease before a user has had his turn he should be first to get supplies when flows are restored. In practice things are seldom so fair, those nearest the main canal (*head-enders*) tend to deprive those furthest down the lateral (*tail-enders*). To try to supply the latter first is not really a satisfactory remedy, because it wastes water – conveyance losses are greater and any excess water tends to flow on to already watered land.

The locational inequality between head-enders and tail-enders is real enough when all farmers are of similar economic, social and political standing (even if all were settled on a new scheme with identical-sized plots of land this is unlikely to be the case), but in practice there is usually much more than locational inequality – the tail-enders are often weaker for a variety of reasons. It is not only water supplies which tail-enders lose out on, roads and services tend to be more accessible to those at the head-end (roads usually flank the main canals). On a big irrigation scheme tail-enders may have to wait six weeks or more for water to reach them from the main canal. A missed turn (typically because users further up the lateral have obtained more than their fair share of water by theft, bribery or threats to officials) may mean twelve weeks without irrigation in extreme cases. This delay between a user's demand and the actual arrival of water is known as the *response time* of a system. Response times can increase if maintenance of the supply/distribution system is poor, and as any silt in a system tends to get flushed down the network of channels to clog the tail-end, the tail-ender usually gets less water, longer delays and has to clear channels more than head-enders (Coward 1980: 41). If drainage provisions are poor it is commonly the tail-enders who suffer the worst waterlogging and salinization because they farm the lowest-lying land (Chambers 1981: 7).

When water is supplied to areas that are already settled, the farmers with power and money are likely to become the head-enders, and the gap between them and the poorer farmers may well widen. Some authorities have tried to compensate tail-enders by allocating them more land, better farm extension services or subsidized fertilizer, but this has had mixed results (Levine *et al.* 1976; Bromley *et al.* 1980; Wade 1982a: 46). In some cases where inequality between head- and tail-enders has been redressed there have been marked improvements in yields for both, Bottrall

(1981a: 14) cites a 5,700 ha irrigated rice scheme in the Philippines where 'modest' changes in water distribution which were more just, combined with 'minor' technical improvements, led to a 97 per cent increase in the overall production of the scheme, and a 149 per cent increase in production by tail-enders (all in a mere two years).

On another Filipino irrigation scheme head-enders were, with some difficulty, forced to reduce their water use to give the tail-enders a fairer share of available supplies, and it was found that the head-enders' crops actually increased by 23 per cent over two years, probably because regular adequate supplies had replaced excessive erratic supplies (Bottrall 1981a).

Ideally, rotation of supplies should provide farmers with water at intervals which conform as closely as possible with the irrigation requirements of the crops being grown. Most traditional rigid rotation systems try to ensure that water allocation is not pure anarchy, but seldom do they match supply to crop needs. Even with all the modern skills of the agronomist, water supply engineer and so on, working out schedules matching supply to crop need is difficult to achieve and in practice rare (Bottrall 1981b).

Until better rotation methods can be devised, traditional rotation methods are still useful. One that is widely used in North-east India and Pakistan is the *warabandi* system. Basically this is a continuous-flow, seven-day rotation rigidly controlled by a roster of turns enforced by local officers (Coward 1980: 257; Merrey 1982). The system has weaknesses, the main one being that the allocation to users is fixed so that if there is rain too much water gets applied and if there is drought too little. Another fault with *warabandi* is that if the rotation gets 'out of step', i.e. someone misses a turn because the lateral is empty, it can be very difficult to correct without individuals suffering.

In some parts of India the *karai* rotation system is practised which determines the sequence of turns according to family group or caste status and the duration of allocation by tradition. This may allow individuals to take excessive water; disruption is also caused if there are changes in size of landholdings or if farmers purchase pumps.

Most farmers want to irrigate their crops in the early morning or late afternoon, the result is that systems often have daytime water shortage. With a little organization and the issue of torches some of the farmers can be required to irrigate at night and solve the problem. This has traditionally been done in Iraq with the *waqt* system which organizes water users to take their supplies sunrise to sunset on one turn, and sunset to sunrise on the next.

Wherever rigid scheduling is practised, it is vital that corrupt practices are not allowed. If they do gain a hold, allocation becomes very inefficient and unfair. Typical forms of cheating include:

secret extraction from supply channels by hidden pipes, siphons or pumps; unlocking of turnouts; the building-up of channel beds to increase water flows over regulating wires upstream; the alteration of notched weirs; bribing or threatening of water supply officials. Sometimes water supplies are allocated according to *crop priority scheduling*, when water supplies are scarce crops deemed by the authorities to be of strategic or economic importance get priority (Easter 1980: 13).

Flexible scheduling

Flexible scheduling allows the user greater control over water supply which in turn allows him a greater choice of crops. There are some problems associated with such scheduling, however. Installation costs tend to be higher than with rigid scheduling (although operating and maintenance costs are often lower). The main weakness of flexible scheduling is that farmers tend to over-water their crops unless the supply authority takes measures to prevent this. To function efficiently, flexible scheduling really demands that the user (and the authority) has some reliable means of monitoring water need and amounts actually used. One possibility is to use *indicator plants* sown among the crops – these wilt at known levels of soil moisture and so indicate when irrigation is needed (Kovda *et al*. 1973: 245). Hopefully in the near future cheaper microelectronics will provide even better monitoring that can be used by unskilled personnel.

There are several ways in which flexible scheduling can be managed:

1. *Delivery on demand* (free demand or user-controlled);
2. *Limited rate demand* (user-influenced);
3. *Arranged date* (controlled demand);
4. *Limited rate arranged date* (controlled demand);
5. *Restricted arranged* (controlled demand).

Under delivery on demand scheduling the user gets the water he wants whenever he wants it, subject to the supplies being available. This system can suffer from users over-irrigating and there is a tendency for water supplies to be under-used at night. Delivery on demand does work quite well with sprinkler irrigation, and where mixed rather than uniform crops are grown.

Under limited rate demand scheduling the user gets what the supply authority wishes to provide whenever he wants. If arranged date scheduling is operated the user gets what he wants but only on prearranged dates. Limited rate arranged date scheduling gives the user limited flow only on prearranged dates, and restricted arranged scheduling gives the user supplies restricted both in terms of rate of flow and duration of flow only on prearranged dates. There is a sixth means of scheduling – *equitable distribution* – whereby the user gets a share of the available water based on the area he irrigates, but has no control over the quantity he gets if supplies are low nor over the timing of flows. This system is agriculturally inefficient because the user cannot match supply to crop needs (Replogle & Merriam 1981: 116).

Reducing the wastage of water during conveyance

Although most wastage of water and interruptions of supply occur in the distribution network, wastage does occur in the conveyance system as can breakdowns in supply. The most obvious way to reducing conveyance losses from canal supply systems is to route canals to take advantage of the topography and to line any which leak at unacceptable levels. Lining is only worth while when the water source is limited and the water losses exceed the cost of installing and maintaining canal linings, or if leakage causes environmental problems.

Maintenance of supply systems may be carried out on a regular basis to prevent, or at least reduce, wastage and cuts in supply and to ensure that the flows are maintained at efficient rates. Often, however, maintenance is done only when breakdowns occur or conveyance efficiency falls to a marked degree. An important point about regular maintenance is that it reduces the risks of water supply failure at critical times when crops are vulnerable.

Water application (irrigation) methods

Supplies of water, whether they arrive via a complex canal network, simply flow from springs or are pumped from streams or tubewells, have to be satisfactorily applied to crops. Water application may be subdivided into three main categories: *surface*, *subsurface* and *overhead* (Table 7.6).

Common irrigation methods and the crops which 'fit' them are listed in Table 7.7. If more than one method proves suitable for the terrain, soil and crop chosen, the final selection is made with reference to the relative merits of yield, water use efficiency, cost, ease of adoption and maintenance and risks of environmental and socio-economic impacts. A summary of the major methods of irrigation is given by Table 7.8.

Table 7.6 Water application (irrigation) methods

Categories	Methods
SURFACE	
Flood	
Wide coverage some control	Basin: check basin, padi-field, contour basin, field basin
Wide coverage poor control	Flowing water: wild flooding, spate, border check irrigation
Furrow	Furrow irrigation
Linear source releases more controlled	Modified furrow
	Corrugations Zigzig
SUBSURFACE (SUB-IRRIGATION)	
Wide coverage (Field)	Water-table manipulation
Linear source	Buried perforated tube, ditch or mole drain
Point source (local wetting)	Pitcher irrigation, buried emitter
OVERHEAD	
Wide coverage	Sprinkler: fixed, semi-portable, moving (centre-pivot, side-roll, etc.)
Limited coverage	Trickle: trickler, bubbler, dripper emitters; watering-can, hose

Sources: Kovda *et al.* 1973; Stern 1979; and others.

Some of the methods described in the following section are presently too complex and costly for general developing country use. However, there is steady improvement in reliability, in reduction of costs and the need for expensive maintenance. At present, surface irrigation methods dominate although they have quite high channel construction, field levelling and maintenance costs. Overhead irrigation (sprinkler and trickle irrigation) have no such costs but require expenditure on pipework, water emitters and pumps. Overhead irrigation methods are comparatively recent innovations and at present many developing countries can find cheap unskilled labour to construct surface irrigation systems more readily than the foreign exchange and skilled staff needed to install and run overhead irrigation. In time, however, the costs and disadvantages of overhead irrigation are very likely to decline and

Table 7.7 Irrigation methods suited to various crops

Method of irrigation	Crops
Sub-irrigation	All crops except rice
Sprinkler irrigation	All crops except rice
Furrows	Row crops Bush crops Orchards
Corrugations	Cereals Pastures Green fodder crops
Basins	Cereals Orchards Rice
Contour basins	Pastures Cereals Green fodder crops Orchards Rice
Border strips	Fodder crops Pastures Cereals

Source: Kovda *et al.* 1973: 312.

these methods can offer much greater water use efficiency and control over water application. Trickle irrigation may also offer greater advantages where irrigation has to make do with poor-quality (saline) water because salts tend to concentrate away from the emitters and crops, and because little or no water gets on to leaves.

Surface irrigation methods

Surface irrigation systems supply water to crops at ground surface level. There are six principal methods: *wild flooding, basin, border, furrow, corrugation,* and *trickle*. *Spate irrigation* might be added, but is really a form of water spreading, rather than surface irrigation (see Chapter 6). At present, surface irrigation methods are the most widespread means of applying water to the land in developing countries. Wherever there is a tradition of irrigation it is almost certainly some form of surface irrigation, commonly basin or furrow. A tradition of irrigation may not always aid agricultural improvement – where there is a strong tradition of basin or furrow irrigation farmers can sometimes be so set in their ways that they resist the introduction of new cultivation techniques or innovations in water application.

Uniform water distribution is important if uneven crop yields and/or localized waterlogging/salt accumulation are to be avoided. To obtain uniform application with surface irrigation methods it is important that the basin, field or planting plot be carefully levelled. New techniques, for example laser-levelling, have already begun to reduce costs and make this easier, even using unskilled manpower. A problem with some surface irrigation methods is that channels and banks hinder mechanization and the layout suits row crops but not field crops (i.e. those broadcast, not planted in rows) the range of crops that can be grown is restricted and rotation of crops may be difficult.

The rate at which the water distribution system can provide water is important. For all surface irrigation methods there is what is known as a *minimum unit stream* (the ratio of size of irrigated plot to rate of water delivery) necessary to achieve uniform water distribution for a given depth of irrigation (Kovda *et al*. 1973: 310). Terrain is important in determining the surface irrigation method used: if the ground slopes at less than 0.1 per cent an artificially graded slope will be needed for almost all surface irrigation methods. Furrow irrigation requires slopes of 0.5 to 1.5 per cent if it is to function well; if slopes exceed 2 per cent erosion becomes a problem.

Soils are important. Where the topography is level and basin

Table 7.8 Summary of the major methods of irrigation

Bases of comparison	Controlled flood irrigation	Furrow irrigation
1 *Total capital costs*, supply and distribution works	Low, $140–173/per ha	Low, $140–173/per ha
2 *Total annual costs*, includes interest on capital, operational costs and depreciation	Low, 0.0056–0.0064/ per m^3	Medium, 0.0064–0.0081/ per m^3
3 *Crops* to which the method is particularly suited	Pastures, hay, grain	Row crops, e.g. corn, cotton, potatoes, vegetables, orchards
4 *Soil type* most suitable	Adaptable to most soils	Adaptable to most soils having good lateral moisture movement characteristics
5 *Soil profile* most suitable	Adequate depth for necessary grading	Adequate depth of uniform soil
6 *Topography* most suitable	Slopes capable of grading to 1% max	Slopes varying from 0.5% to 12.0%
7 *Irrigation efficiency*	Good, 50–90%	Medium, 50–70%
8 *Drainage required*	Usually incorporated in the layout	Some type of waste water spillway drain
9 *Water required*	Large streams, 28.3–226.4 litres/s	Fairly large streams are sometimes necessary depending on the number and size of furrows
10 *Soil or crop damage*	Over-watering may cause salting, puddling or surface crusting	Furrow erosion is a big problem

Notes: Converted to metric by author. $US throughout.

Source: Wiesner 1970: 227 (Table 43).

Sub-irrigation	Sprinkler irrigation
High, \$741–864/per ha	Medium, \$197–222/per ha
Very low, 0.0032–0.0048/per m^3	High, 0.0097–0.0121/per m^3
Annual root crops, vegetables	Pastures, vegetables, orchards, nurseries
Must be capable of lifting moisture into the root zone. Must allow good lateral movement of moisture	Adaptable to most soils
Surface soil underlain by an impervious substrata	Adaptable to most, but useful on soils with narrow 'A' horizon
Generally adaptable to most but fairly uniform slopes preferred	Adaptable to most slopes which can be farmed
Theoretically excellent, up to 90–95% but 70–80% in practice	Very good, 65–85%
Nil, tilelines may act as drains after a storm	Very little, due to excellent control
Very little, flows as low as 75.7 litres/min have been utilized	Small flows may be utilized
Extremely unlikely	Both are possible if water drops are large

irrigation (especially padi rice production) is contemplated, slow infiltration can be an advantage, but where the ground slopes, even quite gently, applying water to such soils by almost any surface irrigation method is likely to result in erosion. When soils permit rapid infiltration, surface application of water can be ineffective, because the water tends to sink in before it can flow far from the point of release. To counter this, the release must either be of high volume or the release points must be closely spaced. If infiltration is rapid, irrigation applications are likely to require closer supervision. Sprinkler, drip or pitcher irrigation methods are better than surface irrigation when infiltration is rapid. Some soils, particularly clay soils which crack when dry, initially allow rapid infiltration but this soon decreases to stabilize at quite a slow rate as the fissures seal because of the swelling of soil particles when they become wet.

Current surface irrigation practices could be made much more water-efficient if, for example, excess water flowing off the irrigated plot (return flow) were pumped back to the water distribution system for reuse.

Wild flooding

Wild flooding consists of the delivery of water from supply channels running either along, or perpendicular to, the contours of a sloping plot. Water is released or caused to overflow from these channels to cause an advancing front of water to cover the plot. Once water is released or overflows a channel, it is uncontrolled except by the topography of the plot. This method is best suited to growing perennial, low economic return crops (especially forage or pasture) on steep slopes. Some skill is required, mainly in the selection of the water release points, but minimal land preparation is required, and labour needs are low. Water application is relatively inefficient and uneven and, particularly on light (sandy) soils, erosion may be a risk. The method is used in parts of Europe and the USA, Tibet and Nepal (Downing & Gibson 1974; Stern 1979: 51).

Basin irrigation

Basin irrigation methods are widely practised and in their more simple forms are easy to operate. Most of the world's rice is grown by basin (padi) irrigation, much of it by smallholders. To form a basin (padi) the land is carefully levelled and is surrounded with earthen bunds. The rate at which water can be supplied compared with the speed at which it evaporates/infiltrates the soil determines the size of the basin (on level ground with fairly impermeable soil they may reach several hectares), topography and land-ownership (boundaries and pathways) determine the shape of the basin. Given time, padi-fields on permeable soils tend to seal themselves and lose less water. Padi-fields are kept flooded to a depth of between 20 and 40 cm while the rice is growing, with a gentle throughflow of water.

A very different form of basin irrigation is *check flooding*. This method aims to trap ephemeral flows behind earthen bunds long enough to soak the soil thoroughly, sometimes also to accumulate damp sediments. Really it is a kind of spate or floodwater irrigation. Check basins may be useful where soils are poorly permeable, or if the soil is so permeable that other methods of surface irrigation are difficult. In the former case water is delayed so that it infiltrates, in the latter case the aim is to flood the basin fast and then allow the water to sink in reasonably uniformly all over.

Border irrigation

Border irrigation may take two basic forms: *border strip* (border check) in which a strip of land is carefully graded to an even slope of up to (2 degrees), 100 to 800 m long and up to several hectares in size bordered by low earthen bunds which check the flow laterally; and *border ditch* in which water is allowed to flow down a graded strip in just the same way as the border strip method but ditches not bunds control lateral flows. The border ditch method allows better drainage of surplus water on gently sloping land and where soil is poorly permeable. It is sometimes possible to construct much wider cultivation plots with border ditch irrigation than with border strip irrigation, the former being cheaper to construct and maintain.

The main costs of border irrigation are the land grading, banking or ditch digging and these costs are recurrent. However, weeding and harvesting are easy and as the strips drain well, moisture-related crop diseases and waterlogging are reduced. In general, border irrigation needs a water supply with high-volume flows, and is best suited to deep, medium-textured soils, deeper-rooting crops and large-scale production of grain, pasture or orchard crops (Wiesner 1970: 28; Finkel 1982: 349).

Furrow irrigation

Furrow irrigation involves the release of water along furrows, typically 25 to 30 cm wide and 15 to 20 cm deep, to wet ridges separating them. The method is best suited to deep, moderately permeable, fine-textured soils and uniform, gentle slopes. If the slope is more than 2 per cent, instead of arranging the furrows up- and downslope they must be orientated almost along the contour with just sufficient fall to maintain controllable flow. This means it is possible to use the method on slopes as steep as 14 per cent although this is rare (Johl 1980: 139).

The spacing of furrows, their cross-sectional shape and that of the ridges between are varied according to the crop, the soil character, quality of the water supply and the farming practices used. Some crops may be planted in multiple rows and broad ridges, more often crops are grown in single rows, each separated by a furrow. On permeable soils V-shaped furrows are preferable, on less permeable soils a U-shaped furrow is better, the former reduces infiltration, the latter improves infiltration. On permeable soils the furrows have to be more closely spaced than on less permeable soils.

Furrows may be supplied with water from a channel at right angles to their direction at their upslope end. Increasingly, *gated pipes* are being used to reduce wastage and they can convey water over uneven ground, and in some forms can be moved about to improve water supply. Gated pipes are plastic or metal pipes which are usually portable and have controllable outlets at intervals along their length. Lay-flat plastic piping, with perforations at intervals along its length may offer the small farmer a cheaper alternative to commercially produced gated pipes.

Furrow irrigation probably has the lowest installation costs of any water application method, but labour requirements can be high and water application efficiencies are not particularly good as the land must periodically be regraded and the furrows have to be regularly renewed. Provided the soil is not easily eroded, the skills of furrow irrigation can usually be mastered rapidly by cultivators who are unfamiliar with irrigation.

A number of crops are intolerant of standing water, or require 'earthing-up' (like potatoes). They cannot, therefore, be raised with basin irrigation but can be with furrow. On the other hand, field crops cannot be grown with furrow irrigation, only row crops. Another advantage of furrow irrigation is that it can be used where there are very strong winds. Where irrigation water contains salts, particular care has to be taken with the design of ridge and furrow and the mode of planting (see Fig. 7.2).

Corrugation irrigation

This is similar to furrow irrigation except that smaller cross-section, closer-spaced channels are used (roughly 10 cm deep and 40 to 70 cm apart), and there are no raised ridges or beds between the channels. Water fed into the corrugations wets the whole plot. The method is suitable for field crops as well as row crops and can be used on medium-textured soils and where it may be too steep to practise furrow irrigation – on slopes of up to (10 degrees). A disadvantage is that at present the method may be too costly for smallholders, and professional help with grading is usually required (Stern 1979: 48).

Trickle irrigation

Trickle irrigation (drip or dribble irrigation) applies small quantities of water at frequent intervals directly to the soil surface through various types of outlet (emitters or dribblers) lying just above, on or below the soil surface. Typically, about 2 to 10 litres/h are emitted from these outlets, the water being distributed to them through small-bore pipes (12 to 16 mm diameter) laid along the crop rows or buried just below the surface. To maintain an adequate flow in such small-bore pipes, pressures of 1 to 3 atmospheres are needed and this necessitates quite powerful pumps.

Only the immediate vicinity of the emitter is wetted, the moisture advancing through the soil until the rate of infiltration/evaporation matches the rate of emission. The result is a moist patch of ground up to 1 m in diameter (Fig. 7.5). This can have the following advantages:

1. The method is very water-efficient – typically saving 30 to 50 per cent of what other methods would use (Wolff 1977: 117; Finkel 1982: 247). Some experiments in Israel suggest it can be much more efficient than sprinkler irrigation (Postel 1985c).
2. There is scope for automation, which would closely match water supply to plant needs.
3. The method can be used with supplies containing salts, provided care is taken to ensure that if a zone of salts builds up at the limit of the wetting patterns that rainstorms do not suddenly wash this close to crops.
4. Weed control is relatively easy because weeds grow only within the wetting pattern.
5. No water is splashed on the crop leaves – this is important for some crops, particularly when the water supply is salty.

Fig. 7.5 Soil wetting patterns and salt accumulation under trickle (drip) irrigation. Salts (from the irrigation water and/or soil move outward from the emitters to the fringes of the wetting pattern.

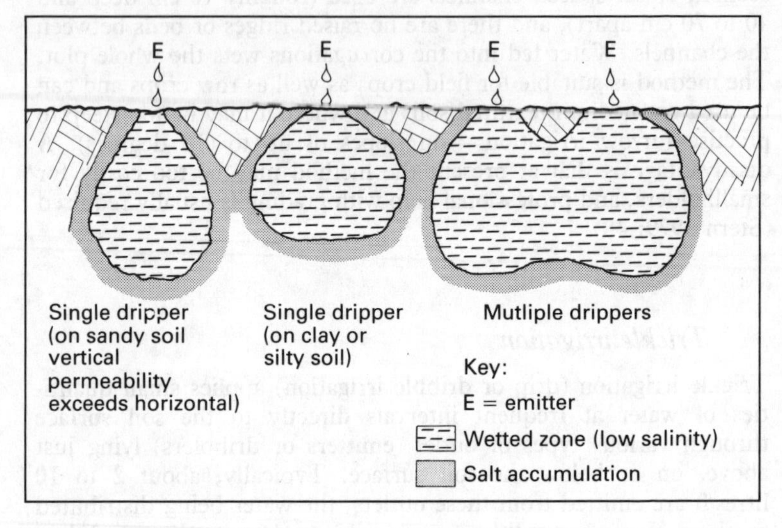

Source: Adams, *et al*. 1978: 44 (Fig. 30).

6. Fertilizers or pesticides can be efficiently and easily applied by adding them to the water supply.
7. There is little opportunity for water-related diseases to be transmitted with this method of irrigation.
8. Automation is possible (Postel 1985c: 22).

The disadvantages of trickle irrigation are:

1. The costs, according to some authorities are five to ten times the level of some of the methods so far discussed (White 1978: 17).
2. The tendency for pipes and emitters to become choked with sediment, sand or algal growths (FAO 1973; Howell *et al*. 1981).

Already some of the previously mentioned disadvantages are being countered, for example bubbler-type emitters are now available which do not require the high water pressures demanded by earlier types. Non-clog or self-cleaning emitters and better sediment filters are getting round the problem of blocked equipment, and above all the costs of installation are falling.

The technique has value where water supplies are saline, Bernstein & François (1973: 73) reported that yields from plots watered by trickle, sprinkler and furrow irrigation using slightly saline supplies showed trickle irrigation gave by far the best yields. Sceptics do question the value of trickle irrigation. Jurriens & Bos

(1980: 101) argued that it had attracted too much attention and was too expensive for developing countries: 'We estimate that for every 35 ha under drip irrigation, one article has been written. (This ratio applied to surface irrigation would yield about 5 million articles.)' Despite such scepticism there are already large areas of trickle irrigation in Mexico, Israel (10 per cent of the total irrigated land), the Middle East, Australia and, China (*ca*. 14,000 ha in 1985, some installations costing only $US200 per cropped hectare).

Subsurface irrigation methods

In some situations it is possible to irrigate crops by applying water below the ground surface (subsurface or sub-irrigation). This may be achieved by using natural features or by artificial means (Fig. 7.6).

In some localities it may be possible to manipulate the water-table, for example by constructing a barrage or barrier to cause moisture to back up and rise close enough to the surface to water crops. Another possibility where there is seasonal rainfall is to inject water into drainage pipes during the dry season to effect subsurface irrigation. Mole ploughs (where soils are prone to collapse into the ploughed channel a 'rope' of peat or other fibrous material can be laid) and buried, perforated small-bore pipes can also be used for subsurface irrigation. Care is needed with all methods of subsurface irrigation to ensure that the water-table does not rise so high as to cause salinization or kill the soil micro-

Fig. 7.6 Sub-irrigation. Natural impervious layer, plastic sheet or bitumen used to control water and cause lateral movement.

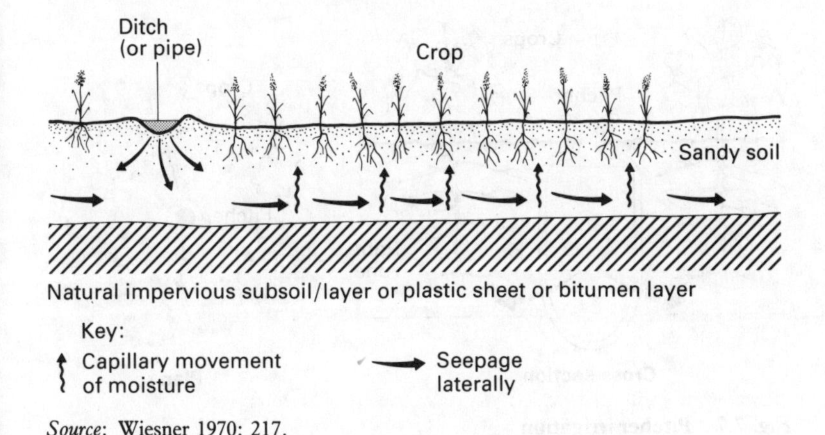

Source: Wiesner 1970: 217.

organisms which fix nitrogen. The techniques seem to perform best when the soil is fine-textured and the terrain is level.

There have been recent experiments with shallow-buried cathodes (on a net of glass fibres just below the crop roots) and an anode deeper down near the water-table. Application of an electric current to the anode 'pushes' groundwater up to the cathode where the crop roots can reach it. Although there are considerable problems with the technique at present, there may be potential for developing it as a water-efficient irrigation method where there is groundwater a little beyond the reach of crop roots (Anon. 1985b).

Pitcher irrigation

Pitcher irrigation is simple, effective and cheap, reducing water waste, suitable for free-draining soils, unlikely to promote water-related human diseases (malaria and schistosomiasis), and using materials found in most rural areas by employing easy-to-make, unglazed earthenware pitchers (gourds or bamboo with small holes pierced in them might be substituted) which are sunk into the soil (Fig. 7.7).

The pitchers leak water slowly to crops planted around them and evaporation losses are very low (Mondal 1974). The method shares many of the advantages of more costly trickle irrigation, and has another advantage that the pitchers can be filled by hand from springs with a low rate of flow, wells, or streams – sources which might not otherwise support irrigation. Production of the sun-dried or lightly fired pitchers might provide rural employment. Fertilizer can be added to the pitcher resulting in very economical applica-

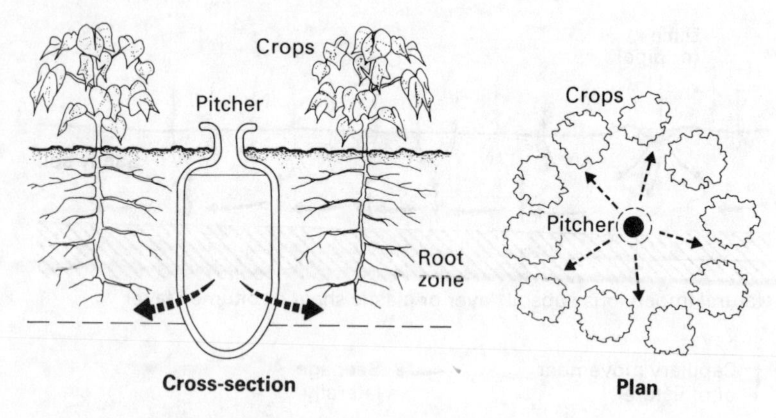

Fig. 7.7 Pitcher irrigation

tion, and the method is suitable for use with zero-tillage cultivation. The method, like trickle irrigation, is likely to be effective where the irrigation water contains salts.

Pitcher irrigation is traditional in the Punjab (India), Iran (where it is called kuzeh pot irrigation), and recently has been tried in Sahelian Nigeria, Argentina, Brazil, Bolivia and Chile. (Stern 1979: 54; Gischler & Jauregui 1984). In the Punjab the amount of water used to irrigate crops is surprisingly low – less than 2 cm/ha over an eighty-eight day growing season (National Academy of Sciences 1974: 110).

Pitcher irrigation and use of the simple watering-can have been surprisingly slow to spread, yet their potential for smallholders is considerable. Tillman (1982) estimated that one person of 'average' strength using a watering-can might irrigate around 500 m^2 effectively, provided the water source did not delay filling the can; by installing pitchers filled from the can, he/she could irrigate an equivalent of up to 3,500 m^2.

Overhead irrigation methods

Watering-can

The watering-can is the simplest form of overhead irrigation, much used for small-scale market gardening in temperate regions and a little in the Far East, but surprisingly little in tropical and subtropical countries. The area that can be watered depends on how distant the water source is from the cropped plot, a walk of 100 m might cut 500 m^2 down to 250 m^2.

Hosepipe

The coverage with a hosepipe depends upon the water pressure, and upon the diameter and length, and to a lesser degree the material of construction. One person with a hose of 12.5 mm diameter and a water supply with a 20 m head of pressure can water a plot of about 800 m^2 in nine or ten hours (Stern 1979: 56). Few rural areas have suitably high-pressure water supplies for this method to be widely used.

Sprinkler irrigation

Of the various irrigation methods, sprinkler irrigation is probably that which most closely mimics natural rainfall. Water is projected,

Fig. 7.8 A simple sprinkler irrigation system. Sprinklers mounted on laterals which can be moved down a main line when a portion of field has been watered. S = sprinklers; J = junction points where laterals can be installed.

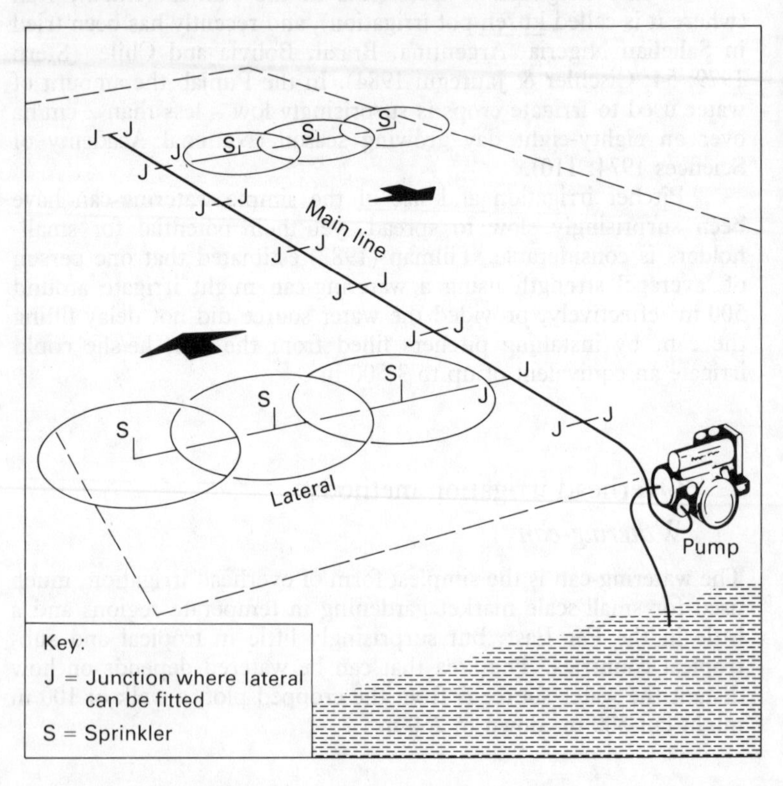

Source: Stern 1979: 59 (Fig. 21).

usually under considerable pressure (typically between 2 and 5, sometimes as high as 10, occasionally as low as 1 atmosphere) through sprinkler heads supplied by stationary, semi-portable, portable or moving pipes. The discharge of water may be over the crops or, less commonly, beneath the crop canopy which reduces evaporation loss and the damage done by water droplets lying on leaves. The simplest form of sprinkler, with fixed mains and laterals is illustrated in Fig. 7.8.

Some systems use high-pressure water 'guns' which project a spray of water over considerable distances from the gun. Some sprinkler systems have laterals (lateral pipes) which are either pivoted (*centre-pivot systems*) at the centre and sweep in a large circle to water a plot of crops. Some systems have laterals mounted on wheels and are propelled across a field watering a large rectangle of crops. Centre-pivot systems are becoming increasingly common in developing countries for large-scale commercial crop production; Libya and Egypt in particular have installed considerable areas of centre-pivot irrigation. In Egypt the Salhia Agricultural Project has centre-pivots which each water 60 ha allowing 22,260 ha of the arid Eastern Desert between Cairo and Ismailia to be cropped. Onions, potatoes, cabbages, tomatoes and fodder are produced – the latter support beef and dairy cattle and thriving chocolate and tomato purée industries, and in all employs about 1,500 people (directly involved in irrigated agriculture). Egypt has plans to expand the area of centre-pivot irrigation (largely relying on private companies) to over 2,428,20 ha by the end of 1987, financed largely from Suez Canal and oil revenues and taxation of remittances from Egyptians working in the Gulf (Hildrew 1984). In Libya the New Kufra Oasis (the Kufra Scheme) also uses large (100 ha plot) sprinklers to water barley or alfalfa (ASAE 1981: 52; Johl 1980: 143; Finkel 1982: 193).

Sprinkler irrigation, if designed and operated properly, is a very effective water application method, one which can be used over a wide range of soil conditions (the method is especially useful where the soil is coarse textured, sandy and/or has low moisture-retention capacity) and on uneven terrain where surface irrigation would be impossible. Little grading is needed, the equipment is easy to manage and there is little on the ground to hinder cultivation. Row and field crops can be grown and there is little standing water (unless application is badly managed). Pesticides and fertilizer can be applied with the water and there is little risk of increase of malaria or schistosomiasis. The control of water application is easily automated to allow a close match of crop water needs to application rates.

There are disadvantages with sprinkler application. Moisture losses can be quite high as droplets of water vaporize in flight and much is intercepted on leaves and evaporated before reaching the ground (for these reasons water application is best done in still weather conditions, at night or on a dull day). Even so, water losses are still probably less than would occur using surface irrigation (Stern 1979: 17). Some plant pathogens, especially fungal diseases, may be dispersed and helped to establish themselves by sprinkler irrigation and there is a risk to humans and livestock if sewage or other contaminated water is applied with sprinklers (aerosols formed by sprinklers may drift a long way on the wind to infect crops, livestock or people). Where conditions are windy, sprinkler irrigation may be seriously hindered – attempts to apply water when the air is not still enough leads to uneven application.

The major disadvantages of sprinkler irrigation at present are its tendency to suffer from blockage or abrasion of sprinkler heads due to sediment or algae in the water supply, the relatively high costs of equipment, the need for quite powerful pumps (which means relatively high energy costs) and the need for skilled staff for design, installation and management. It is likely that the costs of sprinkler irrigation hardware will fall, but even if they do not, today's typical installation costs of around $US2,000 per hectare (Stern 1979) is not unfavourable when compared with the costs of establishing large-scale surface irrigation schemes in sub-Saharan Africa which commonly exceed ($US8,000 per hectare (Mather 1984: 182).

Where water supplies contain salts, sprinkler irrigation may remain unsuitable because the droplets deposited on at least some crops causes burning of the leaves and moisture stress (Johl 1980: 172). Low- and medium-pressure sprinkler systems probably offer most potential for developing countries, but there is the problem that with low-pressure systems water droplets can be large and cause crop and soil damage (ASAE 1981: 53).

Chapter 8

Irrigation water supply: groundwater

Introduction

Groundwater, especially shallow groundwater (i.e. that no deeper than 200 m), is often the most practical, most rapidly developed and economical supply for irrigation. Storage and conveyance costs and water losses are negligible unless artificial recharge of aquifers is practised. Groundwater is usually available during rainless periods and sometimes in more or less rainless drylands. Other sources of water – catchments for collecting rainfall, reservoirs and canals – deny land to agriculture; groundwater exploitation needs little space.

Groundwater resources are presently relatively underexploited on the world-scale, though there may be local or even regional overexploitation in some countries. Groundwater is commonly underexploited because its presence is unknown or it is too deep for locally available technology to recover. The main problems associated with groundwater use for irrigation are:

1. *High salt content* – groundwaters tend to contain more salts than other water sources, though this is by no means always the case. As salt-tolerant crops, salt-adaptive cultivation and salt-reduction strategies become available poorer-quality groundwater will be used.
2. *Speed of exploitation plus its common resource character* – shallow groundwaters can be especially difficult to manage because many individuals are able to draw water and the authorities may encounter problems in monitoring rates of use and/or in enforcing rational management strategies.

Ideally, groundwater development should be *conjunctive*, that is, in step with surface water exploitation. The Cauvery Irrigation System in southern India practises such management; canals feed water for recharging groundwater in the wet season, and in the dry season tubewells tap the groundwater when the canals cannot meet demands. Probably preoccupation with large-scale reservoir and canal projects has been a major reason for the relative neglect of groundwater development, but also groundwater surveys in many developing countries have been inadequate. India and Pakistan are countries which have made considerable advances in groundwater exploitation, the former exploited roughly half of its proven reserves by the mid 1970s (Vohra 1975), and about one-third of the latter's irrigated land depended on groundwater supplies (Grigg 1970: 275). Despite the lower priority given to groundwater supplies, those which have been developed have had considerable effect. Pearse (1980: 225) felt that probably: '. . . the most important development in irrigation during the last decade and that which has made greatest contribution to the new technology [i.e. the spread of green revolution agriculture] is the use of groundwater.'

Exploiting groundwater resources

Groundwater often makes up a considerable part of streamflow, so that irrigators, even those using river or stream water, at least seasonally, depend indirectly upon it. Exploitation of groundwater is easiest where there are springs or shallow groundwater beneath soft rocks. Where rocks are harder and groundwater lies deeper, exploitation may require considerable skill and effort to overcome the problems of access and/or lifting.

Access: wells

Hand-dug wells

Hand-dug wells have been used for millennia for domestic, livestock and irrigation supply and are still probably the most widespread means of exploiting groundwater. Their potential is by no means fully realized, especially in regions with shallow aquifers with a low rate of yield and where geological conditions favour excavation.

Favourable conditions for excavation means rocks that are stable and not too hard, where more or less unskilled labourers can easily sink a 1.5 to 2 m diameter shaft to 20 or 30 m. If groundwater lies close to the surface but the seepage into a shaft is slow, a large-diameter shallow well (*seepage reservoir*) can be dug. Where ground conditions are favourable, unskilled labour, for example the farmers themselves during 'slack' times of the year, may construct wells (Wagner & Lanoix 1959: 71). Where conditions are less favourable specialists may be needed, although there are techniques now becoming available for shielding less skilled diggers. Well-lining not only makes digging safer and allows shafts to be driven deeper in less favourable conditions, it can also greatly extend the life of a well.

Many of the wells dug in developing countries have been unlined and typically last for about twenty years rather than the centuries which might be the case if they were properly lined. The tendency has been for people to dig new wells when the need arises rather than lining and maintaining those in use. Increasingly, the costs of new well construction far exceed the costs of lining and maintenance. In the past, brick or masonry were often used for lining and required skilled workmen; a promising replacement is locally cast concrete rings which these sink like caissons protecting the diggers and form an excellent cheap liner – above all the digging and lining does not require much in the way of skill.

Modern technology in the form of explosives and portable pneumatic drills have greatly increased the spread of well construction, specially in regions where hard rocks have hindered or prevented manual labour from reaching water. For example, in the Deccan Plateau (Central India) very hard lava flows prevented diggers from reaching groundwater: farmers or teams of labourers can now hire drill sets and reach water where twenty years ago it was virtually unknown. Powered rotary drills can also be used to drive a pattern of collector boreholes out from a hand-dug well into water-bearing layers – this can greatly increase the yield of existing wells.

Hand-dug wells have three advantages over smaller-diameter boreholes:

1. The latter can choke or run dry with little warning, hand-dug wells usually give advance warning and fail gradually.
2. It is comparatively easy for workmen to go down a hand-dug well and clear blockages.
3. Water-raising devices for boreholes are more or less restricted to modern pumps with all the attendant expense and spare-parts/maintenance problems; hand-dug wells by virtue of their greater diameter can use traditional water-lifting devices which are cheaper to run and more likely to be locally repairable.

'Horizontal' wells

There are situations where groundwater can be tapped by 'horizontal' (in practice gently sloping) shafts; the most widespread form is the quanat (karez – Pakistan, foggara – North Africa, mina or laoumi – Spain, karezes – Mexico/Chile) (Welchert and Freeman 1973). Quanats are underground passages dug to intercept one or

Fig. 8.1 'Horizontal' wells (**a**) *Cross section of a quanat.* (**b**) *Cross section of a well bored into a suitable rock formation.* Seepage or vegetation often indicates the presence of an aquifer.

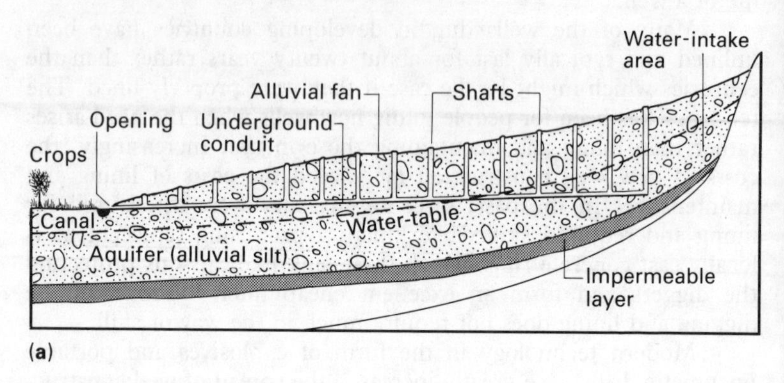

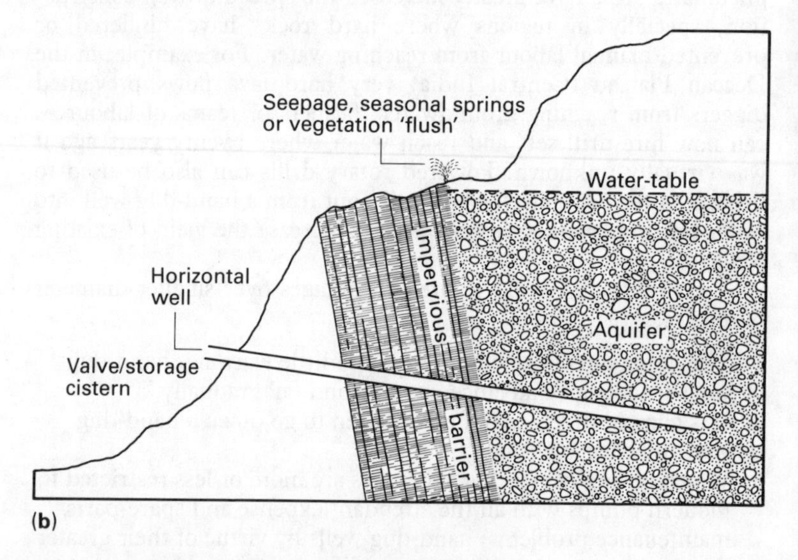

Sources: (**a**) author; (**b**) Welchert & Freeman 1973: 255.

more vertical 'mother wells' usually sunk into water-bearing alluvial fans. The passage is sloped at a gradient less than that of the ground surface and delivers water to an outlet 10 km or more distant (there are records of quanats over 100 km in length) (Fig. 8.1a).

The origin of the quanat has been attributed to the Persians around 3,000 years ago. Today the technique is used world-wide but is especially common in southern and south-eastern Afghanistan, Iran, Pakistan (especially the province of Baluchistan), the Arabian Peninsula, Palestine, Turkey, Soviet Central Asia, Spain, Sicily, China (Sinkiang Province), and in Central and South America (Cressy 1958; English 1958; Wulff 1968; Evenari *et al.* 1971: 173; Hall *et al.* 1979; Rahman 1981).

Quanats usually flow without needing pumps, they draw water from quite high up in the catchment where it is likely to be relatively salt-free. Conventional wells sunk near the quanat outlet often give saline water because by that point groundwater has leached a lot of minerals from the aquifer.

Today, maintaining existing and constructing new quanats is becoming difficult and expensive because workers can often find less dangerous and more rewarding employment. A typical 15 km long Iranian quanat cost between $US13,500 and $US34,000 in the late 1960s, the lower figure reflecting favourable geology and terrain. Such a quanat would give a 10 to 25 per cent return on investment (Cressy 1958: 36; Wulff 1968; Kamiar 1983).

Quanats are vulnerable to earthquake damage, tend to supply water to areas with poor soil (at the foot of alluvial fans) and give best flows, which also vary from year to year, during the season when it is least needed. There are innovations which could improve quanats: concrete or metal liners, pumps to boost the flow from the 'mother wells', pneumatic drills to improve excavation, valves to reduce flows and conserve supplies of water for the season of peak demand, but so far innovation has tended to cause the decline of once effective quanat systems, as the following example shows.

Case study: the breakdown of quanat irrigation

In North and North-west Baluchistan (Pakistan), quanat or karez systems, as they are known in the region, are common. They have traditionally operated in harmony with local political, social and economic conditions. Typically, a system supports one village and is either owned by one or a group of wealthy landowners or is built with money provided by 'shareholders'. While the distribution of water and village life is essentially feudal under both styles of organization, the communities do function well. The introduction of modern technology tends to help individuals not communities.

Rahman (1981) studied a typical Baluchistan shareholder-managed karez village – Achakzai. The water supply supported about 260 ha of irrigated land and there were around 130 ha of rainfed land, supporting fifty shareholder families. Each family received an agreed share of water once a week, the amount of water allocated reflecting the contribution made for building the karez.

The system worked well, binding the village together with a strong community spirit, and provided maintenance was kept up the system should have been indefinitely sustainable.

From the 1960s changes began to affect karez villages throughout Baluchistan (and it is likely in other countries). The Pakistan government encouraged individual landowners and village organizations to sink tubewells, making available loans and subsidies to help them do so. By 1981 Achakzai had seven tubewells which irrigated most of the once-rainfed land – a considerable and quite sudden change in the village economy. A few individuals with tubewells greatly increased their incomes, and owed none of this to the karez. The whole traditional karez-based cohesion of the community began to break down, there were squabbles and maintenance of the karez was neglected. Worse the tubewells around the village began to depress the water-table. Tubewell withdrawal is not easy to control and thus it is likely that the end result will be a failed karez, dried-up tubewells and a ruined village. This scenario is by no means restricted to the village of Achakzai, and is but one example of centralized decision-making which is apparently blind to the advantages of traditional water supply systems and the dangers of incautious innovation (Nir 1974: 117; Spooner 1974; Wilkinson 1977).

In Hawaii, tunnels (maui) used to be driven to intercept water-bearing fissures in the lava rock of the islands; nowadays boreholes like that illustrated in Fig. 8.1b are more usual. Where the geological structure is suitable horizontal tunnels (*infiltration galleries*) or trenches filled with coarse rubble can be placed where aquifers are too unconsolidated to allow more usual wells or boreholes (for a description of infiltration galleries see Wagner & Lenoix 1959: 103).

Utilizing subterranean flows

Low-lying alluvial land adjacent to rivers, streams or lakes, the beds of seasonal creeks or ancient buried stream channels may retain groundwater or even have water flowing through them close enough to the surface to support crops when there is little rain. Where such water is beyond the reach of crop roots, farmers may have to exploit it by digging seepage reservoirs or by lowering the fields. These *sunken fields* have been excavated in many parts of the world, and in the past were especially well developed in parts of lowland Peru (Downing & Gibson 1974: 83). Where lowering of an entire field is impractical, pits may be dug for individual trees. Date-palms are planted in such pits in some parts of North Africa so that their roots can reach groundwater. The disadvantage with

Fig. 8.2 Use of an 'underground' dam to raise flowing groundwater to within reach of the roots of crops

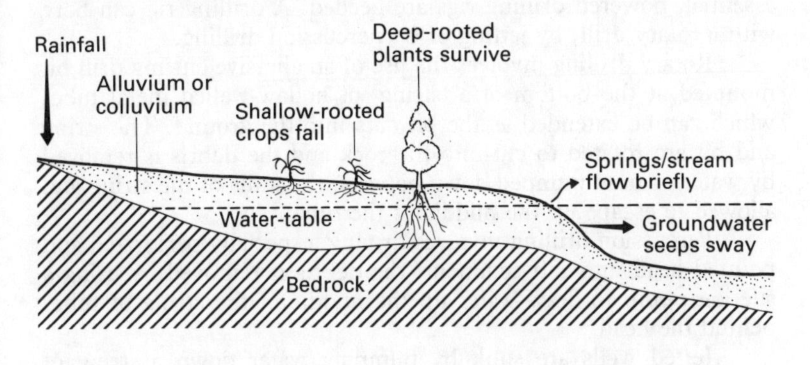

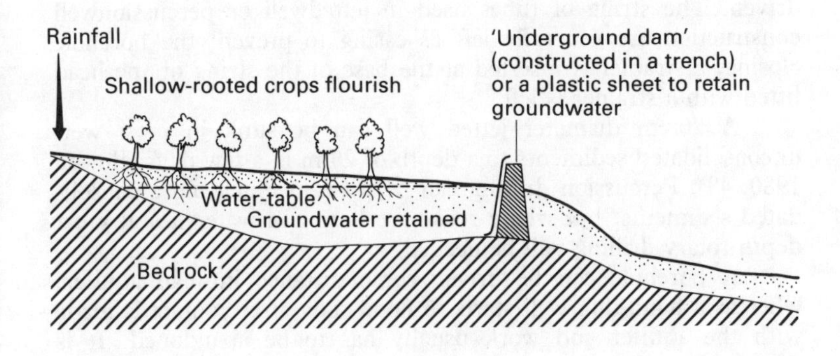

this method of groundwater exploitation is that unless great care is taken the field or pit may reach down too near the water-table and salinization may result.

In some regions valleys have been eroded and then infilled with alluvial deposits. Groundwater may flow in these buried valleys and may be conserved/exploited by constructing *underground barrages*, waterproof barriers of concrete or plastic sheet installed in a trench. This strategy is quite widely used in Morocco and has been tried in North-East. Brazil (Gischler & Jauregui 1984: 15–16) (Fig. 8.2).

Bored, driven or jetted wells

Hand-dug wells deeper than 20 m, become increasingly difficult and expensive to construct. In soft but firm sediment hand-augering can cut a 10 to 20 cm diameter bore to 60 m, depth quite fast and cheaply. In hard rock, if groundwater lies deep or if speed is essential, powered drilling rigs are needed. A drilling rig can bore with a rotary drill, by jetting or by percussion drilling.

Rotary drilling involves the use of an abrasive/cutting drill bit mounted at the bottom of a 'string' of hollow-walled metal tubes which can be extended as the bit cuts into the ground. The string and bit are rotated to cut into the rock and the debris is removed by water or mud pumped down between the walls of the string and allowed to escape up the middle of the tube.

Percussion drilling involves driving a specially perforated and pointed head into soft sediment; the debris is removed by using a suction pump to draw water up the middle of the string of tubes behind the head.

Jetted wells are sunk by pumping water down a string of tubes to wash away the sediment so that the string can sink or be driven. The string of tubes used in jettedwell or percussionwell construction are generally left as casing to prevent the borehole closing, a strainer is inserted at the base of the string or the head fitted with a strainer is left.

A 20 cm diameter jetted well can be sunk in most wet, unconsolidated sediments to a depth of 20 m in a few days (Bowen 1980: 48). Percussion drilling can be done in much more consolidated sediments, but where rocks are hard and water lies at great depth rotary drilling will be needed.

A borehole can be driven some distance into an aquifer, whereas a hand-dug well starts to flood soon after contact is made with the aquifer and work usually has to be abandoned. It is therefore possible for an 8 cm borehole extending several metres into an aquifer to collect and yield more water than a 2 m diameter hand-dug well which only just touches the aquifer. A problem with small-diameter boreholes (and tubewells) is that water can only be withdrawn by pump, and at present these require a power source which can be costly and/or prone to interruption, the pumps often suffering mechanical failure, especially if they are not regularly maintained. Boreholes and tubewells are also notoriously prone to sudden failure – usually caused by the strainer becoming blocked with fine sediment. In one study of north-east Kenya and Central Botswana, countries with higher-than-average standards of government efficiency, if one considers sub-Saharan Africa as a whole, 75 per cent of boreholes in the former and 85 per cent in the latter were broken down (Sandford 1983: 63). Simply sending out teams to drill wells makes little sense. Money would probably be better spent ensuring that those wells that are sunk continue to function.

Tubewells

Tubewells basically have two parts: a 'string' of pipes jetted or driven into unconsolidated sediments and a perforated strainer through which the water (and hopefully not sediment) enters the bore. Tubewells can be sunk fast and cheaply in response to demand for water supplies. Between 1958 and 1968 about 50,000 tubewells were installed in West Pakistan alone (Carruthers 1968: 67). In India between 1961 and 1971 the number of tubewells increased from around 2,000 to roughly 500,000 (figures vary considerably either side of the latter figure; Arnon 1981: 33).

A number of factors have played a part in the expansion of tubewells, not least in India the widespread desire to reduce dependence on seasonal rainfall after the poor monsoons of 1965–66. Also important have been the granting of loans and subsidies for tubewell development, increased rural electrification, improved opportunities for purchasing or hiring pumps, better pumps and pump maintenance and an awareness among farmers of the benefits that tubewells can bring.

Generally tubewells are sited on the land that is to be irrigated so there is little need for expensive distribution systems and less water wasted than with canal supplies. The transition from rainfed cultivation to irrigated farming may be a little easier with tubewell supplies than with surface water supplies because the farmer does not have to worry too much about water delivery schedules or maintenance of a distribution system. In areas with seasonal rainfall tubewells not only improve security of harvest they may also make a dry season crop possible.

Tubewells have a relatively high initial cost plus recurrent costs of pumping and maintenance. Nevertheless, in India they typically pay for themselves in two to four years and may last for as long as twenty years. For maximum return on investment, a tubewell ideally needs to irrigate quite a large area (by developing country standards) – say 20 ha. In India, Pakistan and Mexico where tubewells have been widely adopted larger landowners have profited more than small.

In India, diesel-pumped tubewells seem to be cheaper to install than electric but cost more to run (Mishra 1981: 345). Few rural area of developing countries have electricity; cooperatives or large-scale farming schemes may use diesel-electric generators and electric pumps at each tubewell, but for most smaller operators diesel or petrol pumps are used to lift irrigation water.

Two problems tend to affect farmers using tubewells. There is a risk that groundwater will be overexploited (this is discussed later in this chapter), and there is a need to maintain and repair pumps. As pumps become more reliable and as more parts or whole pumps are made in developing countries, this latter difficulty should decrease. Corrosion has been a real problem affecting pumps, strainers and tubewell casings. This is being countered by the use of modern plastics which are also much cheaper than, say, mild steel, are easier to machine into parts and more easily transportable.

Governments can promote tubewell installation by either investing in public tubewells or by encouraging individuals to install private tubewells. There has been a lot of debate over the relative merits of these two approaches. Where groundwater is deep, where there is a need carefully to regulate the depth of the water-table or if there is a danger of over-exploitation, government-funded public tubewells are probably most appropriate because they allow for more coordinated control of supplies (Bokhari 1976; Sigurdson 1977).

The bamboo tubewell

Tubewells were introduced into Bihar State (India) in the late 1950s and early 1960s. Farmers who wanted to install them became frustrated by the delays and high costs involved in using government drilling teams to sink tubewells and with the pace of rural electrification (Moyes 1979: 10). High-yielding wheat varieties and tubewell irrigation was a combination which offered considerable potential for farmers to make profits, and not surprisingly by the early 1960s some farmers and tubewell contractors were looking for faster and cheaper ways of sinking tubewells and for pumps which did not depend on state electricity (Clay 1980).

By the mid 1960s the bamboo tubewell had emerged initially in the Kosi area of Bihar where wet, sandy alluvium was widespread and easy to penetrate. The *bamboo tubewell* as it became known spread fast; in 1965 there were 300, by 1972 this had increased to 19,496 (Carruthers & Clark 1981: 104). The technology was relatively cheap, typically somewhere between $US8 and $US10 compared with the $US300 charged by a government team to sink a traditional tubewell (Moyes 1979). After 1972 the spread of bamboo tubewells was further accelerated when government subsidies were made available for their installation.

The bamboo tubewell consists of hollowed bamboo jointed to fit together and sunk using a simple hand-powered drilling rig. Bamboo tubes are not only at least three times cheaper than conventional steel casings, they are also not subject to the delay in supply. In place of expensive brass, steel or plastic strainers bamboo and coir rope ones are used, and although not quite so durable they do function longer than the time needed for the tubewell to pay for itself.

Under favourable circumstances smallholders with well under 0.5 ha could afford to install a bamboo tubewell (Arnon 1981: 68; Bottrall 1981a: 230). More often farmers with about 4 ha are near the lower limit of those able to afford bamboo tubewells and access to a pump. The relationship between landholding size/ability to afford to adopt bamboo tubewells is not clear cut because a number of smallholders may cooperate to sink a tubewell where their lands adjoin and hire or buy a pump set. Diesel or petrol pump sets mounted on a bullock cart and available for hire have helped to make groundwater available to smallholders and they have also been used by larger farmers who own more than one tubewell.

There has been much interest in bamboo tubewell development because the approach seems to have considerable potential for adoption elsewhere in the tropics as well as on a larger-scale in the Indian subcontinent. One attractive aspect of the technology is that it appears to have generated quite a lot of off-farm employment. In the Kosi area of Bihar alone, between 1972 and 1973 the fabrication, sinking and maintenance of bamboo tubewell systems is believed to have generated about 600,000 man-days of employment and, in addition, irrigation which resulted probably created another 4 or 5 million man-days of employment (Clay 1980). The unemployed and the rich seem to have had the greatest benefit from the bamboo tubewell. In the Kosi area larger farmers seem to have benefitted more from the innovation than small farmers, it was the larger farmers who were mainly behind the development initially. Cheap tubewells enable a larger farmer to water his land without expensive levelling or water distribution systems. Even if they obtain water, smallholders still have to compete with larger landowners for access to fertilizers, seeds, pesticides and the market.

Water lifting

Water-lifting devices, especially pumps can be costly to install and run and are prone to failure. If groundwater is quite close to the

---

Fig. 8.3 Traditional water-lifting devices (**a**) *the bahari (Middle East/Africa)*; (**b**) *the Archimedean screw (Asia/Middle East)*; (**c**) *the shadouf (Egypt/Middle East/Africa/South Asia)*; (**d**) *the dall (India)*; (**e**) *the Persian wheel – sakiya or saguya (Middle East/West and South Asia)* – inset shows details of bucket-chain pump. (**f**) *the mot (India)*; (**g**) *the noria or norya (Middle East/South Asia/Far East)* (See pages 256, 257 and 258).

	Lift (m)	Power	Area that can be irrigated (ha/day)
Goatskin bucket; Man; Crops; Man; Ropes; Ropes; Shallow well or canal **(a)**	1.0–3.0	2 men	—
Water source **(b)**	1.0–1.5	1 man	0.3
Counter weight **(c)**	2.0–5.0	1 man	0.1–0.3

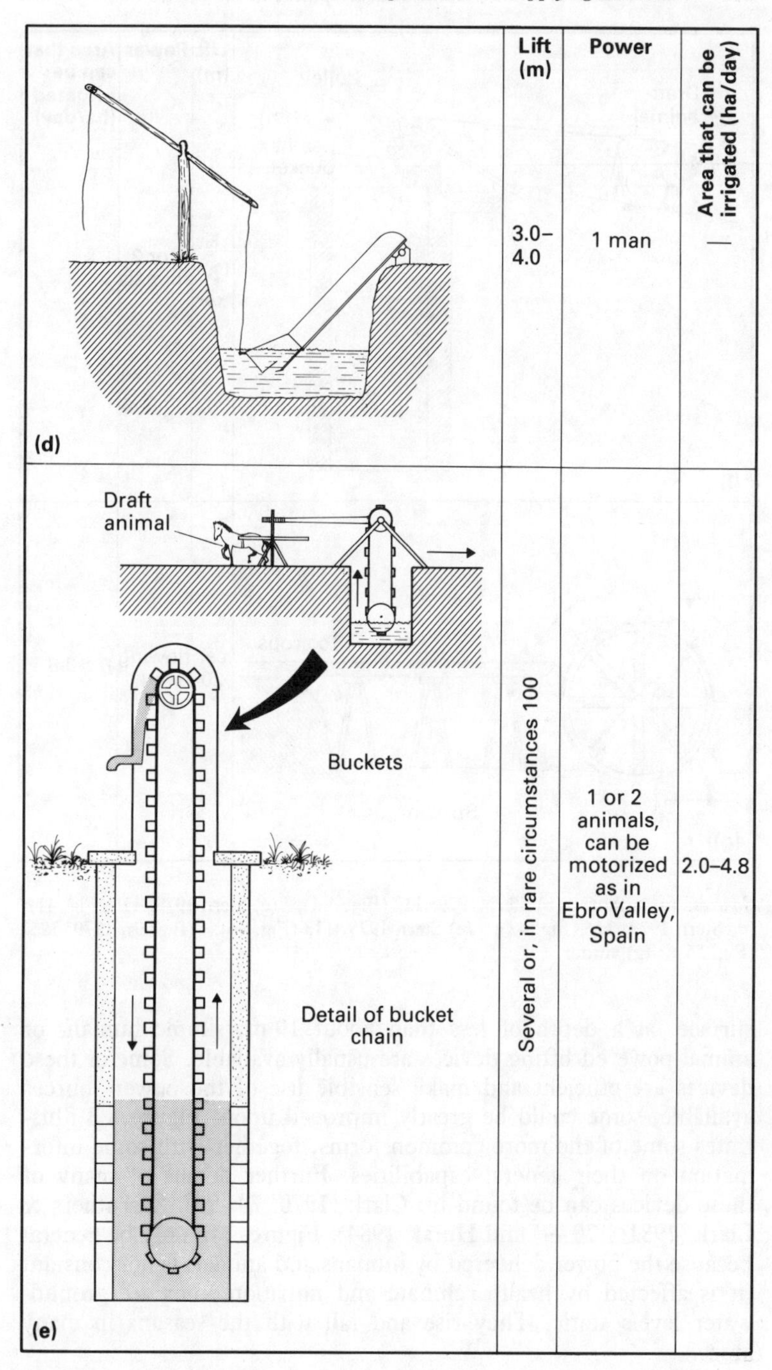

	Lift (m)	Power	Area that can be irrigated (ha/day)
(d)	3.0–4.0	1 man	—
(e)	Several or, in rare circumstances 100	1 or 2 animals, can be motorized as in Ebro Valley, Spain	2.0–4.8

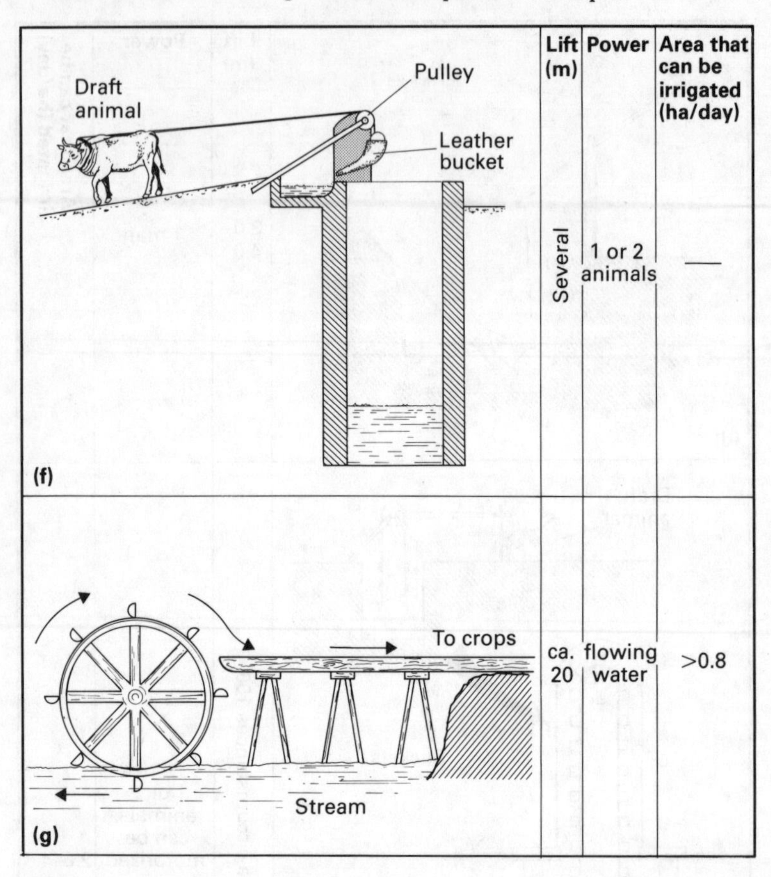

Sources: **(a)** author; **(b)** Stern 1979: 117 (Fig. 4.4); **(c)** Stern 1979: 116 (Fig. 41); **(d)** Stern 1979: 116 (Fig. 42); **(e)** Stern 1979: 118 (Fig. 46); **(f)** Stern 1979: 125 (Fig. 55); **(g)** author.

surface, at a depth of less than about 10 m, simple human- or animal-powered lifting devices are usually available. Some of these devices are efficient and make sensible use of the power sources available, some could be greatly improved upon. Figure 8.3 illustrates some of the more common forms, together with some information on their general capabilities. Further details of many of these devices can be found in: Clark (1970: 74–93), Carruthers & Clark (1981: 120–4) and Hurst (1984). Figures can only be general because the power delivered by humans and animals is not constant (it is affected by health, climate and nutrition); nor are groundwater levels static. They rise and fall with the seasons in many areas.

Pumps

Correct choice of a pump which is appropriate for the situation, is not likely to break down and for which spares are available, is essential or irrigation projects may lie idle and crops be lost (Bateman 1974; Logenbough 1980). Recently, considerable attention has been directed towards perfecting village-level operated and maintained hand-pumps for domestic water supply and small-scale irrigation. Cast iron or mild steel has traditionally been used for pump construction and has made local repair or manufacture difficult. In Malawi, Sri Lanka, Thailand, Ethiopia and rural Malaysia there have been encouraging experiments with pumps made all or in part of polyvinylchloride (PVC) or polyethylene plastics (although in Malawi hyenas were found to chew some of the PVC installations).

Developing countries can adopt one of the following strategies:

1. Total self-reliance – the village craftsman or the farmers make, service and mend pumps.
2. Partial self-reliance – specialized firms make the pumps in the developing country and village craftsmen or farmers maintain and repair them.
3. Purchase pumps from a developed country and hope that they will need little or no maintenance or can be serviced with parts and skills locally available.

New materials and new pump designs are making maintenance and repair less of a problem than it used to be and may even reduce the installation costs. Some of the more common types of pump are listed in Table 8.1. There are two main types of pump: those which use a rotating impeller to move the water (*turbine pumps*), and *reciprocating pumps* which have a piston with valves that pumps water on the up-stroke and a check or foot-valve assembly to stop water escaping on down-strokes. If water lies within about 8 m of the surface the impeller or piston/foot valve can be housed in a pump body cylinder above the ground. Such pumps draw water up by suction and are therefore often called *suction pumps*. If the groundwater lies deeper the impeller or piston assembly can be installed down the well – this type of pump forces the water up and is therefore often called a *lift pump*. On the whole, electrically-powered turbine pumps are most suitable for installation down deep wells.

Pumping costs may be divided into *overheads* (the initial investment, depreciation and possibly interest charges on loans) and *running costs* (fuel or electricity charges and maintenance). Low-performance pumps generally have lower overheads but higher running costs. The cost of high-performance pumps is

Table 8.1 Types of pump

Mechanical principle	Pump type	Power
Displacement	Reciprocating (piston)	H, W, E, M, A, S
	Rotary	H, E, M, A, S
	Chain	H, E, M, A, S
	Turbine	E, M[a]
Velocity	Centrifugal	E, M[a]
	Jet	E, M
Buoyancy	Airlift	E, M
Impulse	Ram	Falling water

Key: H = hand, W = wind, E = electricity, M = diesel, petrol or gas, A = animal, S = solar.

Note: About 8 m is the limit of operation for surface cylinder pumps, if water is deeper then the pump cylinder or the whole pump must be underground within 10 m of the water.

[a] Relatively costly to run.

usually more than low-performance pumps, but they are likely to be cheaper to run and can exploit deeper groundwater (German Foundation for International Development 1977: 28). Smaller farmers are most likely to use low-performance pumps.

Manual pumps

For many small farmers the only power available is their own or their families' labour. Even if they cannot support full irrigation, they can often raise enough water to ensure the survival of their rainfed crops during a period of poor rainfall or boost yields when conditions are less pressing. The potential of hand-pumps or foot-pumps should not be underestimated, for they offer the small farmer a cheap, robust and nowadays usually easily repairable means of exploiting shallow groundwater (Lal 1972; Karala 1982).

A number of improved manual pumps have been developed, for example there is the *India Mark II* – a reliable, moderately priced (about $US200 per installation in 1981), 'low-maintenance' hand-pump, first developed in India about twenty years ago and now widely adopted in Africa and Latin America as well as southern Asia. Unfortunately, low-maintenance does not mean zero-maintenance, and there is also a need to see that pumps are carefully installed. In Maharashtra State (India), where an India Mark II hand-pump programme began in the 1960s, a recent United Nations Children's Emergency Fund (UNICEF) survey found 36 per cent of pumps had broken down and 70 per cent had inadequate foundations (Anon. 1985c). Hopefully, low-maintenance pumps now being manufactured are an improvement on those of the 1960s.

The *Rower pump* a simple, very cheap (about $US14 in 1981) hand-pump, made largely out of PVC plastic is being produced in Bangladesh. A PVC hand-pump which appears to need virtually no maintenance (provided the PVC is dark-coloured to deter hyenas!) is the *Malawi shallow well PVC handpump* (in 1981 these cost roughly $US28). An improved version of a shallow well reciprocating-type hand-pump produced by the Battelle Columbus Laboratories, USA – the *No. 6 Pump* – has been produced in large numbers in Bangladesh and is promoted elsewhere by USAID (Anon 1981).

Since the late 1950s UNICEF have supported a project for installing shallow tubewells and No. 6 Pumps in Bangladesh – the Manually Operated Shallow Tubewell for Irrigation Projects. Over 60,000 had been installed by 1970, each watering an average of 0.2 ha of rice or 0.25 ha of other crops. On average, each installation cost (including the tubewell and the pump) $US70 (Commonwealth Secretariat 1978: 17; Stern 1979: 121).

Few hand-pumps can at present reach water below 30 m depth, but there have been reports of the India Mark II hand-pump extracting water from as deep as 80 m and where the water-table lies at less depth delivering as much as 750 litres/h. Foot-operated pumps have received less attention than hand-operated, which is surprising because foot-operated pumps are potentially more efficient. Even an only moderately fit individual should be able to deliver about 0.1 horsepower for quite long periods using his/her feet, something unlikely with a hand-pump.

Hand- or foot-operated pumps have considerable potential where the groundwater is at shallow depth. But, because they deliver relatively small quantities of water, in drier regions at least, they would be best used in conjunction with irrigation and moisture-conservation methods which make optimum use of water – for example pitcher irrigation.

Motor-driven pumps

Two types of motor are most commonly used to power pumps: *internal combustion* (diesel, petrol, gas or kerosene), and *electric*. In the future there may be others, for example closed-cycle engines (like the Stirling engine) which might function cheaply and reliably on local fuels like vegetable oil or biogas.

Pumps may be either fast- or slow-running; if water is at considerable depth, high-speed (high-performance) pumps are virtually obligatory. These are usually electrically-powered, and because of their speed of running tend to be expensive to buy and maintain. Slow-running pumps are in general more robust. Diesel motors generally require less maintenance than petrol motors and have no electrical systems which could be upset by humid tropical conditions or storms, however, their spares have been costly and more difficult to fit than those of petrol motors. Diesel motors can be converted to run on gas produced by the gasification of wood, coal or on biogas.

With both petrol and diesel motors, a farmer (provided he can afford to) can buy reserves of fuel in advance of need. Electric pumps leave a farmer much more vulnerable, he can be affected by sudden, sometimes protracted power cuts due to storm damage, breakdowns or industrial strikes. Electricity supplies are also prone to voltage fluctuation which can damage a farmer's pump motor(s).

Access to electricity supplies is a problem for most of the South's farmers, for example, in 1978 only 38 per cent of India's villages had electricity despite considerable progress; settlements away from roads, larger towns and hydroelectric schemes, even if not particularly remote, are likely to be without supplies for some time to come. In Africa and Latin America the electricity supply picture is generally worse. Also, diesel and petrol pump sets are more readily moved than electric ones and are therefore more flexible. Hurst (1984) weighed up the costs and benefits of various pumping methods in North-east India and concluded that, at least in that part of the world, diesel pump sets presently have the edge over electric-, petrol- or animal-powered pumps for medium and larger farmers.

A major advantage of the electric pump is that it can be designed to be installed down a tubewell, linked to the surface by power cables. This configuration can raise water from depths practically inaccessible to pumps powered by internal combustion engines or animals. Where groundwater lies deep, or where there are large, well-organized irrigation projects or in villages which have access to electricity supplies electric pumps may be the best choice. Large projects or cooperatives may be able to run quite large numbers of electric pump sets from a diesel-electric generation plant.

It should be noted that, compared with petrol, diesel and electric pumps, biogas and animal-powered pumps have been relatively little developed technologically. In the future these power sources may become more competitive.

Wind-, solar- and water-powered pumps

Small, metal-vaned wind-pumps have been produced in the UK, USA, South Africa and Australia since the early part of this century. These robust, simple pumps can raise water from great depths – usually to serve livestock or domestic water supply needs. The Dutch, Cretans and Chinese developed fabric-sailed windmills and pumps for irrigation and drainage purposes, indeed the Netherlands from the thirteenth century has in large part been shaped by such devices which played a vital role in land reclamation. It is surprising in view of the early development of wind-powered pumps and the success had with them that only recently has any real effort been made to improve designs and adapt them for use in developing countries. A review of low-cost, practical designs developed in India, Peru, Thailand, Ethiopia, Kenya, Sri Lanka and the Philippines was provided by Van Veldhuizen (1982).

The disadvantages of wind power are that initial costs are high and that there is always some risk that a lull in the wind will deprive crops of water at some critical moment in their growth. Despite such problems there does seem to be considerable potential for wind-powered pumps (Tenari 1978; Anon. 1983d).

At low latitudes the hourly solar energy receipt of horizontal surfaces is 6 to 8 kWH/m^2, certainly a power source worth considering. Recently the Intermediate Technology Group (London) reported developing a solar-pump capable of raising water 20 m for very low running costs (Anon. 1983c).

Animal-powered pumps

Draught animals are important in many regions for raising water. In parts of the Punjab (India) the average size of landholding reflects the amount of water a pair of bullocks can raise from a well (Clark 1970: 74). The beast used for water lifting can also be used for other farm tasks or for transport of materials or people – a distinct advantage over other means of pumping. The relative power outputs of various animals is listed in Table 8.2. In most cases harness design could be improved so as greatly to increase efficiency. A real breakthrough would be the perfection of an animal-powered high-speed turbine pump, because at present animal power is applied almost entirely to reciprocating or bucket-chain-type pumps or simple lifting devices, all of which are able to reach only quite shallow groundwater and need relatively large-diameter wells rather than cheaper boreholes (Hurst 1984: 147).

Table 8.2 The power output of various draught animals

Animal	Weight (kg)	Draft force (kg)	Average speed (m/s^{-1})	Power	
				kg/s^{-1}	hp
Horse	400–700	60–80	1.0	75	1.00
Bullock	500–900	60–80	0.6–0.8	56	0.75
Cow	400–600	50–60	0.7	35	0.45
Mule	350–500	50–60	0.9–1.0	52	0.70
Donkey	200–300	30–40	0.7	25	0.35

Source: German Foundation for International Development 1977: 31.

The impacts of groundwater exploitation

If groundwater reserves are depleted too much, the aquifer may be damaged or the quality of the groundwater may suffer. Hydrostatic pressure within an aquifer carries much of the weight of overlying rocks; if the water is pumped out the aquifer may be collapsed. Sometimes it can recover, but often it is permanently damaged and can never again store water. Collapse of a dewatered aquifer may cause the land above it to subside as much as 10 m; considerable areas have been affected in this manner, for example in Argentina (Frederick 1975; El Hinnawi & Hashmi 1982: 184; Cooke & Doornkamp 1974: 165–71).

In coastal regions where seawater can enter an aquifer, fresh water accumulates as a floating lens-shaped body. It is generally assumed that little mixing takes place, and that if a well is sunk salt water will be encountered, not at sea-level but at a depth below sea-level equivalent to about forty times the height of the (fresh) water-table above sea-level. The relationship between the fresh and saline water is known as the *Ghyben–Herzberg relationship*. This represents the condition of approximate hydrostatic equilibrium between the lighter, fresh groundwater and the denser, saline groundwater. If a well or borehole is pumped incautiously this relationship can be upset. For every unit depth the water-table is lowered around the well/borehole the fresh/saline groundwater interface is raised towards the surface by approximately forty times more. A well may thus suddenly yield salt water or cause salt-water intrusion which leads to other wells becoming saline with little warning or even salinization of the land (Ward 1967: 286–8; Goudie 1981: 110). These problems have been apparent in Hawaii, near Dakar

(Senegal), along the Caspian Sea shores of Iran, the United Arab Emirates and Bahrain.

The use of groundwater for livestock in drylands requires particular caution. In general, such developments have brought more problems than benefits, yet aid agencies and governments are still promoting them. Cattle can range up to about 19 km from water, goats about 20 km and sheep 5 to 8 km. Even a low-yield borehole, spring or water-hole can support quite large numbers of livestock, but the surrounding vegetation often cannot stand up to the grazing. Seldom are livestock numbers adequately controlled and so, as cattle, sheep and goats graze into the wind, a tear-shaped pattern of overgrazing forms with the watering-point at the 'point' of the 'tear' (UNESCO 1962; Dasmann *et al.* 1973: 90; Walker 1979: 130). During a recent drought in Niger, Bororo herdsmen were so afraid that wells being sunk by the government and aid agencies to help them would attract outsiders and lead to grazing conflicts (something which had happened in other parts of Niger) that they actually asked for an end to well-sinking (Swift 1977b; Sandford 1983). Grainger (1982: 69) commented that he knew of no instance in which sinking wells or boreholes had any positive effect on semi-arid or arid land pastoralists. Yet in spite of the numerous problems $US7.5 million was spent between 1975 and 1982 on sinking wells for pastoralists in The Gambia, Mauritania and Burkina Faso by aid agencies and governments (Grainger 1982: 70).

Overgrazing might be reduced by sinking a specific pattern of wells, and then locking wellheads or moving the pump set(s) between wellheads so that stock are kept off some of the grazing long enough for it to recover. In practice, however, the herders may not be very cooperative.

Groundwater management

Equitable distribution of groundwater

There is a tendency for groundwater to be either monopolized by a few rich landowners, or to be over-exploited by many, depending on the abundance of the groundwater, ease of access and the effectiveness of government controls. If supplies cannot meet the demands made upon them water-tables fall. In addition to the risks of collapsed aquifers or surface subsidence already described, falling water-tables make pumping more expensive which favours richer landowners.

Where there is a mix of larger and smaller landowners using

the same groundwater (particularly if water-tables are falling) and a new, improved, but somewhat expensive, means of extraction becomes available, there is a risk that the larger farmers will be able to adopt it and will be able to buy out the smaller. Rich farmers who obtained new, more powerful pumps depressed the water-tables of one Saharan oasis by 16 m, consequently the date-palms of smaller farmers could not reach the groundwater and died. The rich farmers were able to take their share of water and, because there were fewer growers, were able to push up the price of dates to make greater profits (Fauchon 1980). Rural electrification may cause a 'patchy' pattern of access to groundwater – one village may get power and improve its irrigation, another may not. The adoption of tubewells and electric-pumps in North Arcot District (southern India) was studied by Farmer (1977: 201), who suggested that access to water was made easier for optimum-size holdings: 2 to 3 ha for one well/pump, and 4 to 6 ha for two wells/pumps. Farms larger and smaller than these two size ranges were slower to adopt the technology. The optimum landholding size according to Farmer will vary according to depth of water-table, the crop water requirement and the power of the pumps that are available.

In recent years the Agricultural Development Bank of Pakistan has lent money to farmers with landholdings between roughly 5 and 10 ha, a lending criterion which excluded at least 80 per cent of the nation's farmers (Griffin 1979: 27). In India, Bottrall (1981a: 229) observed that too much stress had been placed on 'high-technology' tubewells which only the rich could afford, and that comparatively little attention had been given to the shallow tubewell (including the bamboo tubewell) which could be more easily adopted by smaller farmers.

Controlling groundwater exploitation

In most, but not all, developing countries groundwater exploitation is still very much a *laissez-faire* process in need of much tighter control (Bottrall 1981a: 232). The legal framework is often inadequate. In India, for example, a country which is rapidly developing its groundwater, there was no special groundwater law in 1982 (Carruthers & Stoner 1982: 16). Attempts at public control often infringe traditional private rights, and 'equitable distribution' is not easy to achieve. Even when legislation has been passed its enforcement is difficult because farmers can withdraw groundwater from wells on their own lands at night, can falsify estimates of water taken or practice a plethora of other ploys to avoid close regulation. More often than not those who first get control of groundwater in a locality retain that control (Frederick 1975).

There is a need for better data collection and assessment. Without information on groundwater reserves, rates of renewal, rates of withdrawal and so on, little can be done to rationalize exploitation. Modern aerial surveys, satellite imagery and geophysical techniques are improving the data base necessary for rational groundwater exploitation, but many mistakes have been made because developers have failed to seek basic information. The Kufra Irrigation Scheme (southern Libya) is a good example of this. According to Walker (1979: 220) when this project was implemented it was expected that the water-table would fall only about 35 m in forty years of exploitation, yet in the first year of extraction it fell 15 m in parts of the well-field. Debate over the sustainability of the Kufra Basin's water supplies continues, but planners seem to pay little attention to it (Pearce, 1984b).

There continues to be considerable disagreement over whether public or private sector control of groundwater development is preferable. The public sector will only improve equity if the needs of all users and not just those of special-interest groups are needed; often it is the special-interest and pressure groups which have their way. The public sector may be better able to pioneer groundwater exploitation than the private (Carruthers & Clark 1981: 106), but once aware of the potential, private developers can exploit resources very swiftly. Overall control by private developers is generally poorer than that of public authorities, so a degree of public control is desirable if groundwater reserves are limited or if environmental or other difficulties are likely to accompany unwise groundwater use (Carruthers & Stoner 1982: 15).

Chapter 9

Irrigation water supply: large impoundments and diversion of streamflow

If conditions preclude safe, productive, rainfed cultivation, and if there is no suitable groundwater and rainfall collection and concentration is impractical, it may still be possible to develop streamflow to support irrigation. Streams seldom flow constantly year-round, and in some cases they may fluctuate from year to year; also they may not pass close to land which is suitable for irrigated cultivation. The problem of inconstant flows may be solved by engineering (the construction of dams or barrages) or by watershed management or by both. To date, engineering approaches have usually been adopted and this seems likely to remain so for the immediate future. The solution when rivers do not run near the land it is desired to irrigate is to build a canal, pipeline or aqueduct and divert the water to where it is needed; this is increasingly being done and on an ever-growing scale and over longer distances.

Provided flows are not low for extended periods, a *barrage* with controllable floodgates can often maintain adequate supplies for irrigation, but a dam impounding a large storage reservoir is necessary where poor flows are prolonged. Large dams hinder streamflow much more than large barrages and so tend to generate more environmental, social and economic problems. Storage reservoirs may be divided into *intra-year storage* reservoirs, designed to save water from wet months for use in dry months of the same year, and *inter-year storage* reservoirs, designed to reduce the risks of crop failure due to years of low rainfall that cause streamflows to cease to be adequate for long periods (Hazlewood & Livingstone 1982: 60).

Since the Second World War many large reservoirs have been created in developing countries; over 40 exceed 1,000 km^2 and at least 320 are of more than 100 km^2 surface area (Table 9.1). Sometimes streams or rivers are impounded solely for irrigation supply, but commonly there are other primary or secondary goals, for example the generation of hydroelectricity or flood control. There have been impoundments created for hydroelectricity generation or flood-control purposes which offer irrigation opportunities, but these are not realised or are under-exploited, for example the drawdown areas of many tropical reservoirs could be farmed, but seldom are.

A number of developing countries are heavily dependent on large dams or barrages for irrigation supply, notably Egypt, India, Pakistan, Malaysia and the People's Republic of China; large and medium-sized dams increased India's irrigation potential from 9.7 million ha in 1950 to 26.6 million ha in 1979 (Gribbin 1982: 439). To put an actual figure on the number of people in the tropics who are dependent on large dams or barrages for irrigation water is not easy, it is considerable, although not anywhere near as great as the number who practice smallholder irrigation and who obtain water from small streams, flood water or groundwater. A high proportion of irrigation development expenditure in the last twenty to thirty years has been spent on large dam-supplied schemes though.

Vast sums of money have been spent, and will continue to be spent on building dams to provide water for irrigation and to generate electricity. In Latin America and the Sudan, about 50 per cent of irrigation development expenditure in the early 1980s went on schemes supplied with water by large dams or barrages, in the Near East and Africa (excluding the Sudan) about 70 per cent, in South Asia (Bangladesh, India and Pakistan) about 40 per cent and in Asia and the Far East (excluding South Asia) about 60 per cent (Mather 1984: 182).

Some have argued that improved crop storage or increased application of artificial fertilizers to rainfed farmland would be a better investment than building dams and large irrigation schemes, but this is unlikely to halt dam building. To ensure effective application of fertilizer or better crop storage by numerous smallholders would almost certainly require road building, the establishment of extension services and other facilities; more important, it may be difficult to overcome institutional difficulties. Engineering routes to improved crop production can be more practical and quicker, especially where it is difficult to reach and motivate people to alter their practices.

There is a considerable literature on the benefits and costs of large dams/reservoirs, but only a fraction provides careful assessment, a considerable portion portrays large dams as environmental, economic and social 'disasters', and some portray them as wholly beneficial. There is still debate as to whether one of the first, the Aswan High Dam, is on balance a success or failure, in spite of about twenty years of study (Fahim 1981). Large dams/reservoirs are costly and they have frequently generated severe environmental and social problems; sometimes these problems have eclipsed their benefits, but it seems unlikely that this has been the case with more than a portion of those built.

Table 9.1 Some large tropical impoundments

Scheme	Country	Approximate date full	Approximate surface area (km^2)
Volta	Ghana	1969	8,482–9,000
Kariba	Zambia (S. Rhodesia)	1963	4,300–5,250
Kainji	Nigeria	1969	1,500
Kossou	Ivory Coast	1978	1,600
Aswan High	Egypt/Sudan	1974	4,000–6,216
Cabora Bassa	Mozambique	1975	2,700
Kafue	Zambia	1972	800–3,100
Ubolratana	Thailand	1966	410
Lam Pao	Thailand	1971	230
Pa Mong	Thailand	Projected	4,000–6,216
Sirithorn	Thailand	—	292
Upper Pampanga	Philippines	1973	—
Ord River	Australia	1972	3,089
Brokopondo	Surinam	1967	1,560
Sobradinho	Brazil (North-East)	—	4,500
Tucurui	Brazil (Amazonia)	1982	2,400
Maduru Oya	Sri Lanka (Mahaweli)	1983	—
Victoria	Sri Lanka (Mahaweli)	1985	—
Kotmale	Sri Lanka (Mahaweli)	1986?	—
Randenigla	Sri Lanka (Mahaweli)	1987?	—

Notes: H = hydroelectricity, C = flood control, F = fisheries development, I = irrigation, Scheme = the most frequently used title which may refer to dam, reservoir or river.

Sources: Lowe-McConnel 1966; SCOPE 1972; Freeman 1974; Petr 1978; Welcomme 1979; Madeley 1983.

Environmental impacts associated with large dams and reservoirs

Judged merely on technical and economic grounds, not by the overall costs borne by the environment or people, most large dams/reservoirs have attained their planner's goals. Yet environmental and socio-economic impacts have frequently overshadowed benefits (for bibliographies see: Lagler 1969; Freeman 1974; Hunter & Munroe 1978; Barrow 1981; Herren *et al*. 1982; Interim Mekong Committee 1982; Madeley 1983; Petts 1984; Goldsmith & Hildyard 1984 and 1986; Canter, 1986: 77–110).

The impacts of the Aswan High Dam and its reservoir – Lake

Regional vegetation	Drawdown (approximate maximum, m)	Ratio of inflow to outflow	Original planned purpose (if known)
Savanna/woodland	3–4	1 : 4	H
Savanna/woodland	9–14	1 : 9	H
Savanna/woodland	10–11	4 : 1	H (C, F)
Savanna	3	—	H
Desert	10–20	1 : 2	I, F, H
Savanna/forest	31	—	H
Savanna	—	—	H
Monsoon forest	—	—	H
Monsoon forest	—	—	H
Monsoon forest	—	—	I
Monsoon forest	—	—	H
Tropical forest	—	—	I, H
Forest/savanna	Considerable	—	I
Tropical forest	3	4 : 3	H
Tropical forest	36	—	—
Tropical rainforest	12	—	H
Tropical forest	—	—	H, I, C
Tropical forest	—	—	H, I, C
Tropical forest	—	—	H, I, C
Tropical forest	—	—	H, I, C

Nasser have probably been the most discussed. Table 9.2 sets out most of the documented costs and benefits together with an indication of their significance. Overall the contribution made by the Dam to Egyptian agriculture is very considerable and outweighs the disadvantages. The Aswan High Dam was one of the first large dams to be constructed (1968); mistakes are therefore hardly surprising. What is unfortunate is that similar ones are still being made now, despite over two decades of hindsight experience which should enable planners to avoid them. If the planning and implementation of large dams could be improved there would be fewer problems and more benefits.

Figures 9.1 and 9.2 indicate the impacts (beneficial and

Table 9.2 The impacts of the Aswan High Dam

Impact	Status[a]	Assessment of significance
Control of potentially devastating floods	+	Important. Saved lives in 1964, 1967 and 1975
Power generation	+	Important. Supplies *ca.* 58% of Egypts' energy needs. Energy generated by the Dam has enabled rural electrification
Irrigation supply	+	Important. Two or even three crops a year now possible instead of only one with flood agriculture. Between 1968 and 1979 roughly 450,000 ha of new land has been brought into production with water supplies from L. Nasser
L. Nasser fisheries	+	Significant but enough to compensate for downstream catch reductions?
L. Nasser drawdown cultivation	+	Considerable potential but not exploited
Waterlogging and salinization	−	Significant
Loss of silt deposited during floods	−	Minor. In view of increased output of perennial irrigation, also fertility of silt probably over-stated. Manufacture of 13,000 tonnes of calcium nitrate would compensate and could be done using Aswan electricity. Traditional brick-making affected
Increased incidence of water-related diseases	−	Significant. Schistosomiasis has increased since 1968
Possible route for spread of new diseases	−	Possibly significant. Mosquitoes and snails may spread
Reduction of marine fish catches	−	Role of pollution in reduction of catches?
Reduction of estuarine fisheries	−	Significant
Increased erosion of Nile Delta	−	Significant. Could be controlled

[a] + = Beneficial impact; − = Harmful impact.

Sources: Abu-Zeid 1979; Zonn 1979; Fahim 1981; Biswas *et al*. 1983.

Fig. 9.1 Potential environmental impacts of a large tropical dam due to obstruction of streamflow

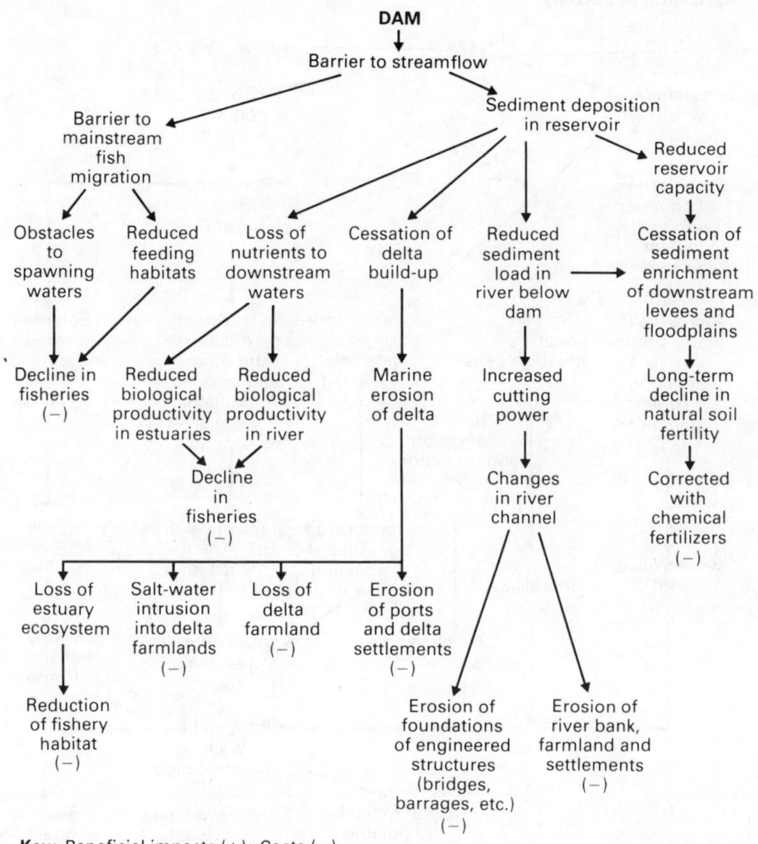

Source: Petr 1978: 372.

harmful) of dam construction, Fig. 9.3 indicates the impacts which result when a reservoir is formed. In practice, avoiding unwanted impacts has been difficult for a number of reasons:

1. The tropics are relatively poor in natural lakes so scientists and planners have had largely to learn by trial and error (Fernando 1980b).
2. Tropical biogeochemical processes have repeatedly proved more complex and swift-operating than expected.
3. Each stream that is impounded differs in water quality, flow

Fig. 9.2 Potential environmental impacts of a large tropical dam due to reduction of flooding, regulation of downstream flow and increased agricultural activity

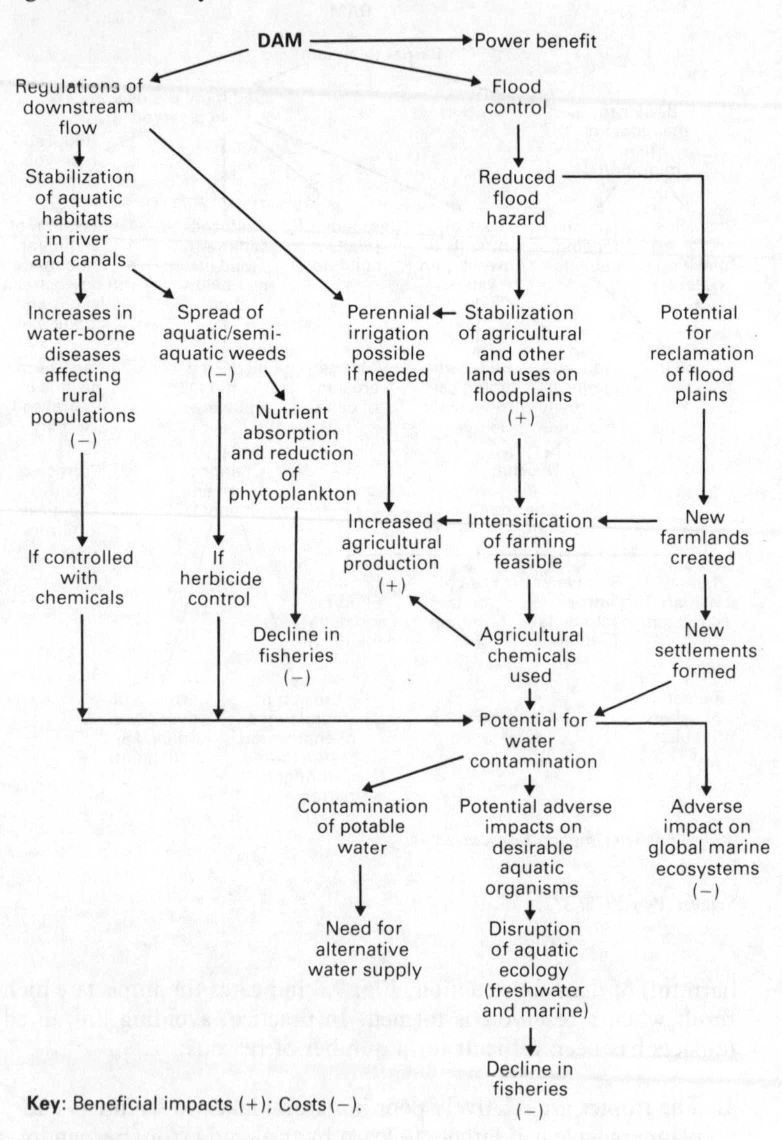

Key: Beneficial impacts (+); Costs (−).

Source: Petr 1978: 373.

Fig. 9.3 Potential environmental impacts resulting from the creation of a large tropical reservoir

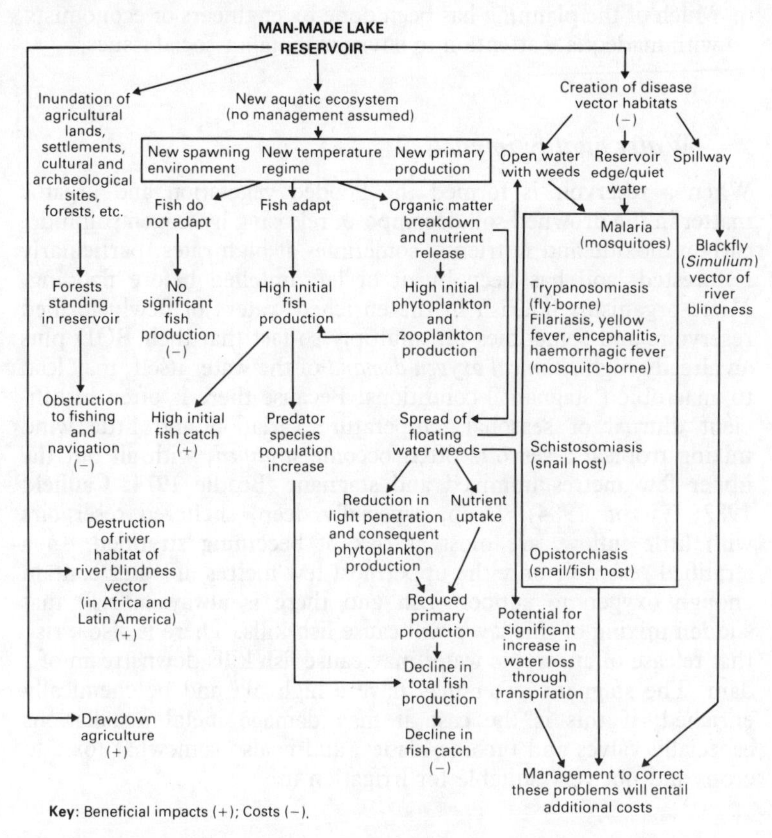

Source: Petr 1978: 373.

regime, vegetation and fauna and so on; to interpolate from previous experience is therefore difficult.

4. Reservoirs are especially prone to problems for roughly ten years after the dam is closed. Real 'stability' may not be attained for twenty-five years, and even after that changes in catchment management or the arrival of a new plant or animal may upset the homoeostasis. Twenty-five years is a long time for a country to have to run expensive monitoring programmes, and so data collection has not often been complete or comprehensive enough for the length of time necessary to help planners draw up guidelines for reservoir planning.

5. Many of the impacts of reservoir development are indirect and therefore difficult to forecast.
6. Much of the planning has been done by engineers or economists with inadequate attention to environmental or social issues.

Water quality impacts

When a reservoir is formed the flooded vegetation and organic matter in the drowned soil decompose, releasing hydrogen sulphide, carbon dioxide and nutrients, sometimes at high rates, particularly if forested land has been burnt or left unfelled before flooding, Micro-organisms flourish in the enriched waters of newly formed reservoirs and sometimes can multiply so fast that their BOD plus an already high *chemical oxygen demand* of the water itself, may lead to anaerobic ('stagnant') conditions. Because there is often insufficient diurnal or seasonal temperature variation and little wind mixing tropical reservoirs often become *stratified*, with all but the upper few metres unmixed and stagnant (Beadle 1974; Caufield 1982; Garzon 1984). Deep, especially deep, sheltered reservoirs with little inflow, are most at risk of becoming stratified. In a stratified reservoir only the uppermost few metres of water contain enough oxygen to support fish and there is always a risk that sudden mixing of this layer will cause fish kills. There is also a risk that release of anaerobic water may cause fish kills downstream of a dam. The stagnant water may have a high pH and be chemically enriched; if this is the case it may damage metal installations especially valves and turbine blades, and is also somewhat toxic to crops and not very suitable for irrigation use.

Weed growth impacts

Aquatic plants have caused problems in many tropical reservoirs and the canals and irrigation schemes they supply. If the water becomes nutrient-rich, algal 'blooms' may occur which can poison fish, livestock and people or may choke filtration equipment used for trickle irrigation or purifying domestic supplies.

On many reservoirs problems have been caused by higher plants (macrophytes). These aquatic 'weeds' are either spread by birds, floodwaters, arrive on fishing equipment or are already in the stream system before impoundment. The most common nuisance plant is the water hyacinth (*Eichhornia crassipes*); other frequent pests are *Salvinia auriculata*, *Pistia stratiotes* and *Scirpus cubensis*. These weeds and others slow the flow of water through spillways or channels; if flows are slowed enough, water deliveries are reduced and siltation of the channel may occur. They can hinder navigation, fishing and power generation; they harbour insects or snails which carry human diseases, and in smaller reservoirs can seriously reduce the storage capacity and increase water losses (this is because aquatic weeds can transpire more water than would evaporate from the equivalent area of unvegetated water; Hammerton 1972; Thomas 1979).

A broad pattern of succession of weeds can be recognized in man-made lakes, although in detail it differs from reservoir to reservoir because local environmental conditions are not the same. In general, however, floating weeds develop first, these may form a floating mat (sudd) colonized in time by semi-aquatic weeds like sedges, often the floating sudd vegetation decreases in extent as nutrients in the lake decline. A little while after the reservoir is established, submerged weeds increase in the shallows and then an emergent flora ('drawdown flora') becomes established on land periodically exposed by fluctuating water levels.

In general weed problems are more pronounced in lakes with nutrient-rich water; those with acidic, nutrient-poor waters suffer least. Wave action and great depth also tend to discourage weed growth (Gangstad 1978; Gaudet 1979).

Altered sedimentation rates

Too often dams have been built only to find that the rate at which the reservoirs they impound silt up has been grossly underestimated. For example, the Ksob Dam (Algeria) silted up ten years before the irrigation project it supplied could have paid for its construction (Arnon 1981: 120). In Pakistan, the Tarbela Reservoir will probably silt up within fifty years and the Mangla Dam (Pakistan) will lose roughly 30 per cent of its reservoir storage capacity in the same period (Carruthers & Clark 1981: 139). Circumstances change during and after construction of a dam – a catchment may be deforested by shifting cultivators which leads to increased siltation of a reservoir – settlement or agricultural development may also alter rates of siltation. For the former and a range of other reasons, predicted siltation rates often prove to be far too optimistic.

Impacts on wildlife

When a reservoir is formed it provides organisms with a new, different, unoccupied ecological niche. There are also considerable movements of people, livestock, materials, the construction of roads, transmission lines or canals which increase the chances of dispersal of plants and animals.

Not all introductions into reservoirs and the irrigation systems they supply are by chance, a number of fish species have been introduced to establish new or improve existing fisheries, or to control mosquitoes, weeds or snails. Diamant (1980) listed over 265 species of freshwater fish introduced into tropical reservoirs or irrigation systems for mosquito-control purposes.

Newly created reservoirs often have sparse fish stocks away from the shore because river fish are seldom adapted to life in open waters. In some cases suitable species have been introduced to fill such niches (Balon 1978). Great care is needed to ensure that any introduction will not later be a nuisance in the way the predatory Nile perch (*Herotis niloticus*) has proved to be in some African reservoirs. As agricultural development becomes more intense every opportunity should be taken to ensure there are some habitats for wildlife which may often be useful predators of pest insects in the farmers' fields. The margins of reservoirs and the banks of large irrigation canals can offer such habitats if managed in the right way. Reservoir development also has the potential to destroy wildlife; the richest wildlife, many reserves and national parks are to be found in river valleys, precisely where dams flood the land. In a number of countries dams have damaged nature reserves, for example, in India, the Silent Valley and Vishnuprayag hydroelectric projects have stirred up considerable protest because of their destruction of wildlife, there have been similar problems in Malaysia, Brazil and Indonesia (Rubeli 1976; UNAPDI 1979; Gribbin 1982). Reservoirs act as silt-traps and reduce the amount of silt passing downstream. This, together with the regulation of flooding (which affects floodland ecosystems) can have tremendous impacts on wildlife, even beyond the point where the river flows into the sea (Barrow 1983: 448).

Groundwater: geophysical and climatic impacts

Downstream of a dam water-tables may well fall; upstream of a dam they are more likely to rise fed by leakage from the reservoir, irrigation channels and irrigated land and because the river is 'backed-up'. In the case of the Aswan High Dam rise of water-tables has been as much as 12 m in some localities and this has damaged soil organisms, archaeological remains and the foundations of buildings (Fahim 1981: 31).

Large reservoirs can cause earthquakes (Rothe 1968; Baxter 1977; Balon 1978) – possibly because the weight of impounded water stresses rocks, or water penetrates faults or permeable rock layers and lubricates them. Whatever the process involved, the shocks which result can be severe; in the case of the Kariba (Zambia and Zimbabwe), Cabora Bassa (Mozambique) and Koyna (India) dams they measured 6 or higher on the Richter scale.

It is likely, but unproven, that large reservoirs moderate the diurnal range of air temperature in their vicinity by supporting onshore and offshore breezes, and increase fog, mist and cloud cover downwind (Ackermann *et al.* 1973: 398; Baxter 1977: 275).

Socio-economic impacts associated with large dams and reservoirs

Large dams tie up considerable amounts of investment capital and often there are few local beneficiaries. Hydroelectricity tends to be transmitted to cities or foreign aluminium processing companies. Irrigation is likely to be for large-scale commercial crop production with the profits channelled to the cities or multinationals. Dam and reservoir management is often orientated towards meeting the demands of electricity generation and it may be difficult to agree trade-offs which can help local farmers or fishermen.

Well before a dam is finished the arrival of construction workers, changes in local communications and employment opportunities can have profound socio-economic impacts on the local 'host' population, they are beset by numerous new challenges to their established ways.

Resettlement

Even in regions where population is sparse, a reservoir which floods thousands of square kilometres displaces many thousands of people (Table 9.3). Those forced to move, even if compensated, suffer considerable trauma and frequently have great difficulty re-establishing themselves. Resettlement of forced relocatees has become one of the least satisfactory and costly aspects of large reservoirs (for example, see: Lawson 1968; Butcher 1971; Chambers & Moris 1973: 522–4; Hart 1980; Barrow 1981).

Reservoirs drown much good land in rich alluvial valleys. Alternative good lands for resettlement are difficult to find. Although cash compensation is an alternative, it is difficult to distribute fairly and there is the risk that relocatees will misspend or be relieved of the money by the unscrupulous before they can use it to establish a new livelihood and home – but if land is unavailable there may be little alternative to cash compensation.

Table 9.3 Some large impoundments which have dislocated people

Impoundment[a]	Approximate surface area (km^2)	Cost of resettlement ($US per capita)[b]
Volta 1965	78,000–84,000	650
Kariba 1963	50,000–57,000	220 Zambia 160 (S Rhodesia)
Kainji 1969	42,000–50,000	500
Kossou 1971	75,000–100,000	
Aswan (L. Nasser) 1968	120,000	1,900 Sudan 2,000 UAR
Nam Pong 1965	25,000–30,000	—
Lam Pao	30,000	—
Pa Mong	200,000–480,000	—
Nam Ngum 1971	3,000	—
Upper Pampanga 1973	14,000	—
Brokopondo 1971	5,000	—
Bhakra 1963	36,000	—
Damodar 1953	93,000	—
Gandhi Sagar	52,000	—
Tarbela 1974	86,000	—
Nanela 1967	90,000	—
La Angostura	20,000	—
Presidente Miguel		—
Aleman	22,000	—
Sobradinho 1983	70,000	—
Itaipu	20,000	—
I and II	20,000	—
Yacireta	12,000	—
Maduru Oya 1982	10,000	—
Victoria 1984	45,000	8,400

[a] Date of flooding where known.
[b] Exchange rates at time flooding began.

Sources: Lowe-McConnel 1966; Jennes 1969; Ackermann *et al.* 1973; Lightfoot 1978; Madeley 1983.

The Volta Project necessitated the relocation of nearly 1 per cent of Ghana's 1968 population – about 67,500 were resettled and roughly 9,000 were cash compensated (HMSO 1956; Chambers 1970: 48; Lumsden 1973). The resettlement plans were ambitious, involving attempts to improve traditional agriculture and the

Total cost resettlement ($US)[b]	Nations involved
54,400,000	Ghana
7,600,000	Zambia
3,600,000	S. Rhodesia
25,200,000	Nigeria
	Ivory Coast
95,100,000	Sudan
100,200,000	UAR (Egypt)
—	Thailand
—	Thailand
—	Thailand/Laos
—	Laos
—	Philippines
—	Surinam
—	India
—	India
—	India
—	Pakistan
—	Pakistan
—	Mexico
—	Mexico
—	Mexico
—	Brazil
—	Brazil/Paraguay
—	Argentina
—	Argentina/Brazil
—	Sri Lanka
—	Sri Lanka

creation of fifty-two new villages where the once-scattered people were to enjoy the advantages of improved housing, health care and other public services. Although much effort and money went into house construction and establishing villages (probably too much, compared with that directed towards re-establishing agriculture) it can hardly be said to have had satisfactory results for by 1968 of the 65,500 relocatees transferred to the resettlement villages only 25,900 remained (Hart 1980: 80).

The Volta Resettlement Scheme experience and subsequent relocation attempts in many developing countries indicates that relocatees require several years to overcome the trauma of their move and re-establish their previous levels of productivity (Colson 1971; Montgomery *et al.* 1977). Planners have tended to overlook this resettlement trauma, and have overestimated the innovation capacity of relocatees (Afriye 1973). They have also tended to underestimate the risks of conflict between 'host' populations and the relocatees and the relocatee community disillusionment which tends to follow relocation (Scudder 1973).

Health impacts

The damming of a river and the formation of a reservoir initiates two broad categories of impact on the health of people in the vicinity. Firstly there is a short-duration 'stress period' which may last several years, during which health problems are generated by the relatively sudden disruption of biogeophysical conditions or from the movement of people (resettlement and inmigration). Secondly, longer-term health problems may result from the presence of a larger body of water and people's increased contact with it because they are fishing, irrigating or watering livestock. Raised water-tables, altered diets (people may be newly able to fish after impoundment or they may get poorer catches compared with river fishing), better access to medical care and improved communications all have long-term health impacts.

The short-term health impacts tend to be relatively unpredictable, longer-term health impacts are in part determined by project design and management, so are often more easily foreseen.

The range of human diseases associated with impoundment is rather similar to that associated with irrigation development, as is to be expected. Schistosomiasis, a debilitating urinary or intestinal disease affects at least 250 million people in developing countries and is commonly associated with impoundment and irrigation development. (Stanley & Alpers 1975; Obeng 1978; Harriss 1980; Deacon 1983). One of the best ways to prevent or control the disease is to design reservoir and irrigation systems to ensure that the breeding and spread of the host snails is prevented or discouraged. Provided flows are maintained above 30 cm^3/sec and channels are kept mud and weed free, snails can be discouraged. There may also be prospects for growing plants which could be steeped in reservoir shallows, canals or ditches so as to release snail-killing compounds, for example: *Ambrosia maritima*, *Phytolacca dodecandra* or *Euphorbia ingens* (Walker 1979: 226; Joblin 1978; Ruddle & Manshard 1981: 166; Deacon 1983: 17).

The shallows of reservoirs, canals and water bodies in irrigated fields are potential breeding sites for mosquitoes which can transmit malaria, filarial infections and arboviral diseases like dengue haemorrhagic fever, yellow fever and viral encephalitis (Donaldson 1978). Over 100 species of mosquito can carry malaria; all vary in habitat preference and behaviour so there can be no simple, effective, generally applicable control. Instead malaria-control authorities must acquire local knowledge and remain vigilant to apply appropriate measures when they are required.

Blackfly-transmitted river blindness (onchocerciasis) can be a problem in West and Central Africa, Mexico, Colombia, Venezuela and Brazil (Asibey 1977). The blackfly vector may breed where weirs, spillways or fast-flowing water in channels oxygenates the water; careful design of dam spillways and irrigation supply canals can do much to reduce such opportunities. Sleeping sickness, caused in Africa by tsetse-fly-borne *Trypanosoma brucei* spp. *gambiense* and in Latin America Chagas's disease caused by *Trypanosoma cruzi* (which is spread by bugs that live in substandard housing) may be a problem when reservoirs are built and movements of people increase. In Africa the tsetse-fly is attracted to vegetation fringing reservoirs or irrigation canals where it comes into contact with humans and livestock. In Latin America migrant workers employed on dam construction or land clearance have helped spread the bugs which transmit Chagas's disease (Bucher & Schofield, 1981).

Disruption of fisheries and farming downstream of dams and barrages

In addition to upsetting fisheries by reducing the silt/nutrient load of streams which damages fishing by reducing food supplies of the fish stocks, and by regulating flooding, upsetting the livelihoods of fishermen and farmers in floodlands, impoundment may also lead to increased erosion of deltas and salt-water intrusion up the lower reaches of rivers (Beadle 1974; Baxter 1977). Prawn and other shellfish fisheries are quite important in the lower reaches of many Southeast Asian, African and Latin American rivers and these are very vulnerable to disruption by upstream impoundments (Petr 1980).

Development opportunities associated with large dams and reservoirs

The opportunities offered by large impoundments are rarely fully exploited; either people do not realize the possibilities or they are somehow prevented from doing so. Sometimes those who do grasp the opportunities are not the local people but opportunistic immigrants.

Reservoir fisheries

There is considerable under-exploitation and mismanagement of reservoir fisheries (Fernando 1980b), and where there are apparently valuable catches these should be weighed against the losses which have probably been suffered downstream (Zonn 1979; Abu-Zeid 1979). Developing reservoir fisheries is not easy, due partly to the difficulty in forecasting the maximum sustainable yield. In part this is because catches are largely made by scattered fishermen who seldom give accurate catch returns. Another problem is that reservoir catches decline after initially being high. The final yield depends on what species are introduced, whether the reservoir area was properly cleared of trees which might otherwise foul nets and many other factors difficult to predict. Even if there are fish to be caught and the local people can catch them, the problem of transportation and marketing those perishable goods remains.

It is likely that tropical reservoir fish yields can be greatly increased if suitable strategies and suitable varieties of fish can be found for reservoir pisciculture and if local people can successfully adopt them. There has already been considerable research on tropical freshwater pisciculture, for example, the *Tilapia* spp. have attracted much attention.

Drawdown and reservoir margin agriculture

Those areas periodically exposed as a reservoir fills and empties – the *drawdown areas*, offer considerable potential for agriculture, especially smallholder rice production (Scudder 1980). For example, it is estimated that Lake Volta has between 84,000 and 90,000 ha of such land (Ackermann *et al.* 1973) and Lake Nasser has roughly 84,000 ha (Fahim 1981), and many other reservoirs have such areas.

To realize fully the potential of these areas will require the cooperation of reservoir managers, who may wish to keep water levels high to ensure that electricity can be generated in the dry season, and those who wish to farm drawdown lands and who want the water levels as low as possible for as long as it takes their crops to mature. Plant breeders might help if they could develop with a crops short growing season or flood-resistant crops. There may also be conflicts of interest between those wanting to grow rice and those wanting to fish the shallows (Thomi 1984: 125).

Large reservoirs may control floods and lower river levels sufficiently to allow areas of swampland to be drained and farmed, for example, considerable areas downstream of the Volta Dam have been reclaimed for arable cultivation.

Usually irrigation development associated with reservoir storage consists of large-scale gravity-fed schemes below the dam; many reservoirs could support pump irrigation of lands around their margin, but it has seldom been encouraged probably because managers are afraid of silt from the farmland entering the lake, and because engineers are familiar with the configuration of gravity-fed irrigation down-valley from dams and are slow to change. Ironically, a belt of agriculture around the margins of reservoirs and in the drawdown areas, **if it were carefully managed**, could act as a silt-trap to prevent some of the detritus from the catchment washing in. It would also help reduce the risk of eutrophication because drawdown area crops and weeds if harvested and removed would extract nutrients from the reservoir ecosystem. Aquatic weeds like *Eichhornia* are used in China and South East Asia for livestock feed, and may be used to make charcoal, biogas, alcohol or paper (Land 1980; Wolverton & McDonald 1979). Nuisance weeds might also be used as compost for farmland around a reservoir or alongside larger irrigation canals (Junk 1980). The problem is harvesting and processing the weeds, but it should be possible to organize this and the time may come when troublesome weeds play a part in establishing rural industries (Gangstad 1978).

Tourism

Reservoir-based tourism is quite important in Europe, USA, Australia and New Zealand. Sport fishing, sailing and other aquatic activities are likely to be growth areas, and already there has been some development at Kainji and Lake Nasser (Barrow 1981: 146). There is a need for caution, however, in attempting to establish sport fisheries. The stocking of some African reservoirs and streams with fish like the Nile perch has resulted in damaging predation of the local fish, in some cases markedly reducing commercial fish catches and endangering wildlife.

The promotion of regional economic growth

Regional economic growth may be stimulated by the creation of a dam/reservoir. There were hopes that Lake Volta (Ghana) would improve the transport of goods, especially foodstuffs from the north for sale in southern cities, hopes which have largely failed to materialize (Lumsden 1973; Thomi 1984). In Brazil, if planned dams on the Tocantins River create a navigation route along which produce could be shipped to and from the relatively remote lands south of the Amazon forests to the Atlantic, considerable economic development of the cerrados and parts of the Matto Grosso might result.

That many large tropical dams/reservoirs cause problems and fail to promote regional economic development does not mean all impoundment projects fail; some like the Muda Scheme (Malaysia) have given excellent returns. With hindsight, fewer problems should arise with future dams/reservoirs and with the development of pisciculture, drawdown farming, reservoir-margin irrigation, reservoir-based industries and tourism projects benefits should increase. If planners and administrators learn from past mistakes the beneficiaries of future impoundments should outnumber the 'losers'.

Large-scale diversion and transfer of water supplies

Technology is now available for diverting large volumes of water to almost anywhere it is wanted in a catchment, or even to convey it out of a catchment considerable distances from water-surplus areas to those with a deficit. Advances in earth-moving, channel, pipeline, tunnel and pump engineering have greatly reduced the obstacles of rugged terrain, hard rock and porous soils, making long-range *inter-basin transfer* increasingly practical (Howe 1982). Channel construction techniques range from highly mechanized (e.g. the Jonglei Canal, Sudan) to largely hand-dug (as is often the case in India and China).

In the past the diversion of water from turbid rivers was a problem, intakes and channels became silted up and soft banks gave way, constantly damaging installations. Modern intakes (the *headworks* of a canal or pipeline) are generally sited on the convex side of a river curve with baffles that deflect high-velocity flow. Using modern headworks the intake of silt is drastically reduced and the likelihood of flood damage is decreased (Askochensky 1962).

Fig. 9.4 Canals and tunnels planned as part of India's National Water Grid which will transfer supplies from the Ganges and Brahmaputra to areas of moisture deficit

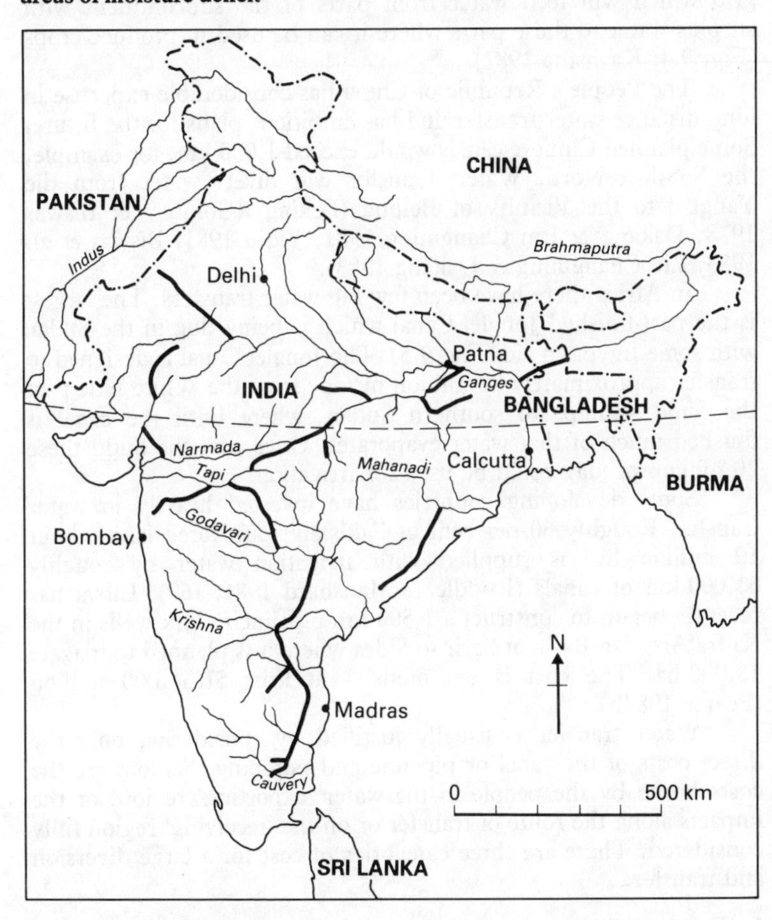

Source: Falkenmark & Lindh 1976: 101.

India, China, Pakistan, the Sudan, Israel, Mexico, Peru, Chile and Libya have built or are constructing impressive water conveyance schemes (Cummings 1974; Stone 1984). Already millions of Indian, Chinese and Pakistani farmers depend for irrigation on large-scale diversion and transfer by canal. By 1979 India had invested Rs 105,000 million (roughly $US900 million), nearly 14 per cent of the country's total planned expenditure, on dams, barrages and major supply canals (Gribbin 1982). In the long term, India plans to progress from the development of selected river basins to establish regional and ultimately a national water grid which will feed water from parts of the subcontinent with surplus water to those parts where it can be used to produce crops (Fig. 9.4; Kayastha 1981).

The People's Republic of China has considerable expertise in long-distance water transfer and has ambitious plans for the future. Some planned Chinese canals would exceed 1,000 km, for example, the South-to-North Water Transfer will divert water from the Yangtze to the Vicinity of Beijing (Peking) (Golubev & Biswas 1978; Dakang & Liu Changming 1981; Yiqiu 1981; Biswas *et al.* 1983; Liu Changming & Dakang 1983).

In Africa there have been few big water transfers. The largest is the part-finished Jonglei Canal which is being dug in the Sudan with some Egyptian aid (Fig. 9.5). The Jonglei Canal is designed to transfer approximately 20 million m^3/day from the White Nile past the Sudd swamps of southern Sudan, where until the canal is finished much of that water evaporates. Once past the Sudd these 20 million m^3/day could be used for irrigation.

Some developing countries have invested heavily in water transfer. Roughly 60 per cent of Pakistan's cultivated land (about 10 million ha) is supplied with irrigation water by roughly 61,000 km of canals (Ruddle & Manshard 1981: 169). Libya has recently begun to construct a 1,500 km pipeline to link wells in the Kufra Artesian Basin at Sarir to Sidra where it is planned to irrigate 18,000 ha. The cost is enormous – at least $US4,000 million (Pearce 1984b).

Water transfer is usually justified by considering only the direct costs of the canal or pipeline and pumping. Seldom are the costs borne by the people in the water 'exporting' region, or the impacts along the route of transfer or on the 'receiving' region fully considered. There are three categories of cost for a large diversion and transfer:

1. The construction, operation and maintenance costs. Weed clearance alone can cost a lot, for example in India in 1970 it was $US600 to $US1,200 per km.
2. The opportunity costs of the land occupied by the canal or pipeline (were it not used in this way it might have yielded crops).
3. External costs borne by parties sometimes far removed from the actual water transfer project.

In view of the vast sums of money that are tied up in large water transfer projects, better pre-project CBAs and environmental impact assessments are needed (Howe & Easter 1971; Herren *et al.* 1982).

Fig. 9.5 The Jonglei Canal

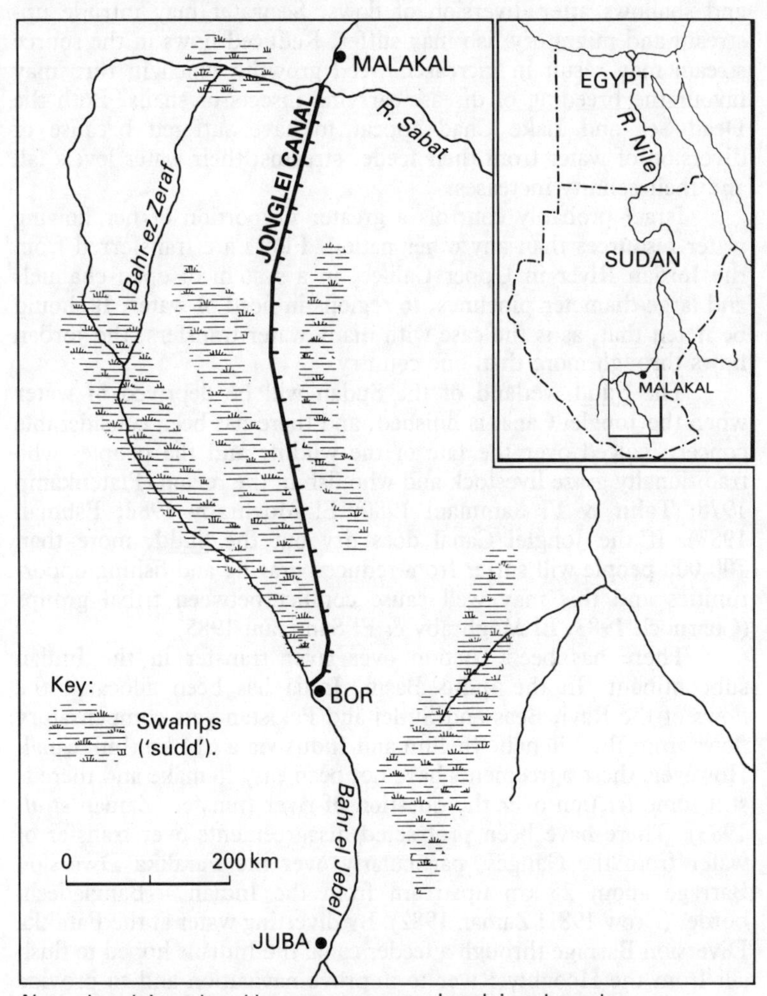

Note: Jonglei rendered by some maps as Junglei or Junqoley

Source: Charnock 1983: 286.

Environmental impacts associated with large-scale water diversion and transfer

Impacts below the point of diversion

Diversion reduces the flow of the source stream below the point of extraction. This may reduce the ability of the source stream to cope with pollution. Shipping may have difficulties negotiating rapids and shallows after diversion of flows. Seawater may intrude upstream and migratory fish may suffer. Reduced flows in the source stream may result in increased weed growth, which in turn may favour the breeding of disease-carrying insects or snails. Both the Dead Sea and Lake Chad appear to have suffered because of diversion of water from their feeder streams; their water levels fall and their salinity increases.

Israel probably controls a greater proportion of her flowing water resources than any other nation. Flows are transferred from the Jordan River in Upper Galilee, in a system of open channels and large-diameter pipelines, to regions in need of water. It should be noted that, as is the case with many water transfers, the Jordan flows through more than one country.

The Sudd wetland of the Sudan will be deprived of water when the Jonglei Canal is finished, and there has been considerable concern voiced over the fate of the wildlife and the peoples who traditionally graze livestock and who fish in the region (Platenkamp 1978; Tahir & El Sammani 1980; El Moghraby 1982; Eshman 1983). If the Jonglei Canal does dry out the Sudd, more than 700,000 people will suffer from reduced grazing and fishing opportunities and this may well cause conflict between tribal groups (Charnock 1983; El Moghraby & El Sammani 1985).

There has been friction over river transfer in the Indian subcontinent. In the Indus Basin, India has been allocated the flows of the Ravi, Beas and Sutlej and Pakistan gets compensatory flows from the Chenab, Jhelum and Indus via a 614 km link canal. However, these agreements have not been easy to make and there is still some friction over the question of river transfer (Zaman *et al.* 1983). There have been protracted disagreements over transfer of water from the Ganges, particularly over the Farakka Diversion Barrage about 25 km upstream from the Indian – Bangladeshi border (Crow 1981; Zaman 1982). By diverting water at the Farakka Diversion Barrage through a feeder canal the Indians hoped to flush silt from the Hooghly River to improve navigation and to provide irrigation supplies (Fig. 9.6). Bangladesh objected that between 1975 and 1978 her farmers had suffered because the diversion reduced flows allowing seawater to penetrate into the Ganges Delta ricelands and had reduced irrigation supplies in the dry season. In due course an agreement over the diversion was reached and functions today (Zaman 1982).

Impacts along the course of transfer

Even major supply canals are often unlined and can leak vast

Fig. 9.6 The Farakka Diversion Barrage and Feeder Canal

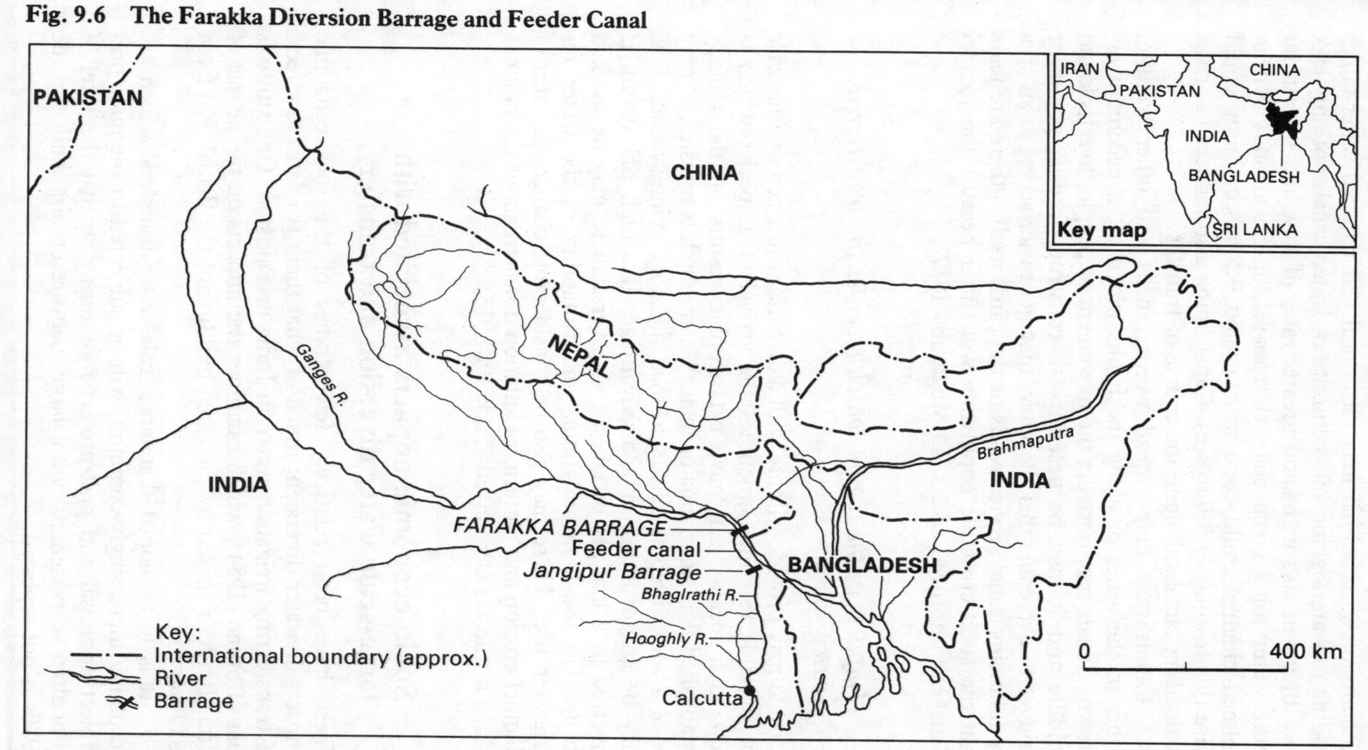

Source: Zaman 1983: 30 (Fig. 1).

amounts of water which may raise local and regional water-tables leading to waterlogging and salinization. Lined canals not only leak less, they can also withstand greater rates of flow and so transfer more water for a given channel cross-section (in a sandy soil an unlined channel could cope with around 0.5 m/sec, lining could raise this to roughly 3.0 m/sec). Canal lining also reduces the risks of flooding farmland along the course of transfer.

Canals may help spread weeds, snails and other wildlife, either in the water or along the banks which often become overgrown. Canals may obstruct the movement of people, livestock and wildlife and it may be necessary to construct fording points or bridges to overcome this. Poachers may get easier access to an area by following canal service roads or the canal itself – there are fears that the sudd crocodile population will suffer because the Jonglei Canal has improved access (El Moghraby 1982).

Impacts in the region which receives the transferred water

A canal may provide wildlife with a dispersal route across otherwise inhospitable terrain. Sometimes the arrival of a species which is new to a locality can have catastrophic results on the already established flora and fauna or man. Water which is transferred may be of a very different quality to that in the receiving system. This may be an advantage or a disadvantage. If a high pH stream is diverted to a low pH or lake or vice versa there may be marked changes in the flora and fauna and suitability of the water for irrigation use. There may also be considerable changes in rates of channel erosion and patterns of siltation downstream of the point at which a transfer is received (Park 1981: 13).

Socio-economic impacts associated with large-scale water diversion and transfer

There have been relatively few studies of the socio-economic impacts of water diversion; much of what there is relates to effects of transferring irrigation water in India or Pakistan. One study is that of Stone (1984), which examines the impacts on the people of Uttar Pradesh caused by the canal building of the British Raj from 1817 onward.

It is to be hoped that large-scale water transfers, which are becoming increasingly common, will be subjected to adequate pre-project appraisal and post-project monitoring so that the level of difficulties experienced with large dams/reservoirs built in the 1970s is not repeated.

Chapter 10

The biogeophysical and human consequences of irrigation development

Even with the best planning, factors will be overlooked and unexpected environmental and socio-economic impacts will sometimes be generated Canter (1985: 135–136). Careful review of past experiences should reduce the risk of this happening and should give planners a chance to prepare contingency plans for problems which might arise.

Impacts on the physical environment

The more common physical impacts of irrigation development are listed in Table 10.1. Of all the problems, however, waterlogging and salinization are the most important.

Soil erosion, soil structure changes and siltation problems

Not all irrigation impacts are harmful. The preparation of land for irrigation may sometimes have a beneficial effect in that it reduces erosion; gently graded or level plots of irrigated land with carefully constructed storm drains, or flooded padi-fields, are likely to suffer less erosion than would be the case if the same land were used for pasture or rainfed cultivation.

Irrigated soils may suffer *suffosion* – the washing out of soluble salts which results in subsidence; loess soils and gypsum- or salt-rich soils are especially prone. The humus content of some soils may decrease after a period of irrigation or the installation of drainage and this *dehumification* may alter the soil nutrient status and structure. The clay content of clay/loam soils can be washed down the soil profile by irrigation causing the soil texture to alter,

Table 10.1 Physical impacts of irrigation development

	Causes/effects	Remedy
Soil		
Salinization		
Secondary salinization (man-induced)	High water-table often due to poor drainage. Use of poor-quality water. Insufficient leaching	Install adequate drainage/try to meet leaching requirement. Rotate crops with salt-reduction crops/ leave land fallow for rains to wash out salts
Erosion		
Sheet	Poor levelling of plots/ poor water application	Level land/better water application. Use cover crops/ mulch
Gully		Level land/better water application. Use cover crops/ mulch
Structural changes		
Removal of soluble salts (especially loess soils and gypsum or salt-rich soils)	Leaching-out of soluble compounds or fine particles. Salts may contaminate groundwater/ streamflow	
Piping or tunelling	Concentration of subterranean streamflow	Improve drainage and water application
Texture changes	Use of saline water	Better water/chemical amelioration
Dehumification	Intermittent-wetting/ drying and entry of oxygen. Sometimes follows installation of drainage (peaty soils esp. vulnerable) – can shrink 2 m in few years	Add humus (green manure, etc.)/raise soil moisture
Decrease in clay content of clay–loam soils	Clay particles get carried down the soil profile. Texture alters	

Table 10.1 (continued)

	Causes/effects	Remedy
Structural changes (continued)		
	– crumb structure alters, more easily compacted	
Decrease in proportion of humic to fulvic acids	Leaching by irrigation water, and addition of fertilizers. Changes in pH and soil nutrient status and structure deteriorate	
Crust formation	Especially in humid and subhumid tropics. Impermeable crusts on surface and/or impermeable pans at varying depth, formed by gypsum, carbonates or silicates. Problem greatest when soils are rich in iron or aluminium compounds. Ferricrete formation (most common crust form)	
Groundwater		
Aquifer collapse	Removal of water from voids in aquifer which then collapse under pressure of overlying rocks	Usually irreversible, high-pressure injection of water may help
Seawater intrusion	Excessive extraction	May be difficult to reverse, pressure injection of water may help
Raised water-tables (local and regional)	30 m in 10 years not uncommon. Seepage from reservoirs, channels irrigated land raises groundwater	Reduce seepage by lining canals, cutting over application/ install deep drains. Grow plants that have high evapo-transpiration rates and deep roots

Table 10.1 (continued)

	Causes effects	Remedy
Groundwater (continued)		
Contamination	Salts or agro-chemicals leached from irrigated land. Nitrates from fertilizer a problem (levels above 45 mg/litre can harm children who drink water). Pesticide pollution of groundwater increasing. Pathogenic organisms pollute groundwater less often than agro-chemicals. Some agro-chemicals persist for a long time	Removal of contaminants from groundwater is very difficult, must be left to nature and that can be very slow Control use of agro-chemicals
Surface streamflow and lakes		
Return flows contaminate streams and/or water bodies	Direct runoff/return flows and/or contaminated groundwater for safe disposal	
Siltation of streams	Silt eroded from irrigated land or in the supply system providing water	Silt traps, better design of channels, leveled plots and better application methods
Eutrophication	Excess nutrients lead to micro-organisms and sometimes aquatic macrophytes, making excessive growth which deoxygenates water	Reduce pollution
Climatic		
Altered albedo (reflectivity)	Change in plant cover or planting pattern	
Increased atmospheric moisture downwind	Increased evapo-transpiration from irrigated crops	

Sources: White 1978: 33; Kovda 1980: 172; Bull 1982.

reducing its resistance to compaction, and causing a deterioration of crumb structure and fertility.

Particularly in the humid and subhumid tropics, impermeable crusts may form on the surface of some soils; hard, impermeable layers (*pans*) may also form within the soil. The majority are caused by the concentration, consolidation and hardening of silicates, but carbonates or gypsum may also be responsible. If pans form just below cultivation depth they are called *plough soles* (Webster & Wilson 1966: 32).

Impacts on groundwater

The problems of aquifer collapse and seawater intrusion as a consequence of groundwater extraction have been discussed earlier. Another major problem associated with irrigation is seepage from reservoirs, canals, distribution channels and the irrigated land itself. The result can be marked rises in local or even regional water-tables (White 1978: 33). Sometimes groundwaters are contaminated by agro-chemicals, salts or disease organisms from sewage effluent applied to irrigated fields (Bull 1982). Once an aquifer is contaminated it may take a very long time for it to run clean again.

Salinization

Irrigation-caused salinization is widespread, very important and by no means restricted to developing countries or the tropics. The USA, the USSR, Syria and Australia for example, have vast areas of land that was once irrigated and is now salinized and unproductive (Askochensky 1962; Holmes & Talsma 1981: 335; Yaron 1981: 1; Khan 1982). The extent of salinization in ten badly affected countries is shown in Table 10.2.

At no time can irrigation management be allowed to falter; the ancient civilizations of Mesopotamia successfully irrigated with the waters of the Tigris and Euphrates for centuries (roughly between 2500 and 1700 BC); then management seems to have degenerated and salinization ruined the land. How long modern irrigation systems can function under good management without salts building up, is in practice uncertain; few projects are more than thirty years old, most are much younger, what **has** been established is that poorly managed projects can become salinized very fast.

Some of the salinized land listed in Table 10.2 has been coursed by mismanagement of the land, some is natural. Estimates of the areas and rates of increase in salinized land vary and should

Table 10.2 Distribution of salt-affected soils (areas in thousand hectares)

Country	
Australia	357,240
Argentina	85,612
China	36,658
Iran	27,085
India	23,796
Paraguay	21,902
Ethiopia	11,033
Chile	8,624
Chad	8,276
Iraq	6,726

Sources: White 1978: 35; Kovda 1980: 251–3; Carpenter 1983: 163.

be treated with caution. World-wide there are probably about 91 million out of a total of roughly 756 million ha of cultivated land which are irrigated, at least a third (roughly 30 million ha), possibly even half the irrigated land (roughly 46 million ha) is in a poor state due to salinization (Goldsmith & Hildyard. 1984: Appendix 2; Hollaender 1979: 47; Hoffman *et al.* 1980: 148; Kovda 1983: 92).

A conservative estimate is that at present land is lost to agriculture through salinization at a rate of 160,000 ha/y world-wide (Buringh 1978); Kovda (1980: 124) put the figure higher at 200,000 to 300,000 ha/y, and more recently at 1 to 1.5 million ha/y (Kovda 1983: 92).

Irrigation-related waterlogging and salinization/alkalinization has been particularly common in the Indus Basin, where the World Bank (1982: 62) estimated there were over 8 million ha of water-logged soils and 40 per cent of these were severely salinized. Pakistan, according to one recent report, will spend about $US317 million in the next few years to combat salinization and improve irrigation (Anon. 1985d).

Salinization can be a problem not only for an irrigator but also for his neighbours, as salts may leach from one plot of irrigated land to another or to unirrigated land or may get into streams and affect people even further removed.

Climatic impacts

It seems likely that large irrigation projects, particularly in arid regions, can affect local, and possibly regional, climates by altering the albedo of the area cropped. This might change local air movements and cause convection which leads to rain clouds. There may also be increased atmospheric humidity downwind from an irrigated area due to the increased evapotranspiration.

Impacts on biota

The impacts which tropical water storage and distribution systems have upon biota have been discussed in Chapter 9 – the focus here is upon the impacts of water application to the land and the effects of water draining from that land. The more common impacts of irrigation upon organisms are listed in Table 10.3.

Irrigation development causes relatively sudden, often quite drastic, environmental changes; inevitably these affect organisms. Sometimes natural wetland ecosystems are already present close to the irrigation development. If this is the case it is easier to forecast which organisms might colonize the irrigated land and water supply channels. If, however, there are no natural wetlands, or none with moisture regimes similar to the irrigated land, biological impacts may be more difficult to predict. The forecasting of biological impacts is more difficult than the prediction of physical impacts because of the sheer diversity of organisms which might colonize, because organisms can change their behaviour and/or tolerances (sometimes in ways never before recorded), and because the biological impacts are often indirect.

In general it is the more highly specialized organisms which suffer most from irrigation development. They may be sensitive to altered moisture availability, increased human disturbance, use of agro-chemicals, the replacement of natural vegetation with crops and many other changes.

Impacts of agro-chemicals associated with irrigation development

Most of the pesticide, herbicide and artificial fertilizer used in developing countries is used on irrigated land. There have already been serious problems in many parts of the world and it seems likely that such problems will increase. It should not be forgotten that man is part of the biota, i.e. humans as well as wildlife are affected by agro-chemicals (Bull 1982). Great care is needed in storing, handling and applying agro-chemicals, especially pesticides and herbicides. High tropical temperatures may make agro-

Table 10.3 Biological impacts of irrigation

Impact	Problem	Remedy
Agro-chemical pollution	Pesticides/herbicides and chemical fertilizers. Affects soil flora/fauna, stream, lake and even marine plants and animals. Pesticide may be used on the crop or may be used on disease-carrying organisms which become established on irrigation schemes	Control use of agro-chemicals, where possible use 'safe' compounds or alternative pest control methods (integrated pest management)
Dispersal of organisms	Across reservoirs or along their margins, in channels or along banks of channels. Can drastically affect native species. May spread crop/livestock/human diseases. Can aid wildlife conservation and migration of wildlife	Monitoring, chemical treatment, use of screens in channels
Pest organism breeding sites	Insect, rodent, bird pests breed on irrigated land, in canal banks or field bunds or in vegetation by canals, etc.	Pesticide, scrub clearing, encourage predatory species
Increased incidence of water-related diseases	Due to presence of water carrying pathogenic organisms, animals that carry diseases in the water or in vegetation beside canals, etc.	Chemical treatment, health education, drug treatment of affected people, livestock, engineering solutions to pest and disease problems

chemicals unstable and prone to decompose into harmful compounds. Locally produced compounds may be impure and there is a history of European and American chemical companies promoting products in developing countries which have been banned as unsafe in their own nations. Workers are often badly-equipped with safety clothing, ill-informed and poorly-trained. It is often forgotten that some organisms can concentrate poisons and if consumed expose their consumer (man or beast) to dangerous levels of contamination. For example, in 1975, health authorities in the Malnad area of Karnataka (South India) noted the increasing incidence of an apparently new disease; over forty villages were affected and the victims were all poor people. In due course it became apparent that at certain times of the year when food supplies were scarce, the poor resorted to catching and eating freshwater crabs from padi-fields. Richer landowners were treating their padi-fields with pesticides and the crabs were accumulating enough to poison those who ate them (Bull 1982: 63). Similar cases to that of Malnad can now be found throughout the developing countries and, as if that were not bad enough, many more people must face long-term exposure to low levels of contamination. The effects of such exposure are unknown but are likely to include mutagenesis and carcinogenesis.

One manifestation of careless and/or excessive use of pesticides is the increased resistance of insect pests to them. In the Cameron Highlands of Peninsular Malaysia, over-use of insecticides by smallholders has led to pesticide-resistant strains and to increased use of pesticide in response – the end result is that crops have a dangerously high content of poisons and the insects gradually get the upper hand. Much more effort needs to be devoted to farmer education and the search for safer agro-chemicals and control of those presently in use (or, better still alternatives to agro-chemicals) if the impacts are not to get out of hand.

If chemical fertilizer application is timed to coincide with maximum crop need, less should 'escape' from the irrigated farmland to contaminate streams, groundwater or lakes. Chemical fertilizers might be engineered to have a controlled rate of nutrient release so that they slowly and steadily dissolve close to plant roots, feeding the crops but tending not to get leached away to cause contamination. By rotating deeper-rooting crops with shallow-rooting crops some of the fertilizer which leaches down into the soil (when the latter are grown) can be used and removed.

Impacts on soil flora and fauna

In general, irrigation tends to reduce the number of species of soil microflora, but the populations of those which remain tend to increase (White 1978: 43). Changes in soil moisture (at its most extreme, flooding and saturation of the soil), alterations in salinity or pH, tillage and the use of agro-chemicals alone or in combination may be to blame for species reduction.

Non-symbiotic (nitrifying) bacteria capable of fixing atmospheric nitrogen tend to be sensitive to salt and saturation of the soil. Any decline in these bacteria will cause a reduction of soil fertility. Organisms active in oxidation and reduction can be affected by irrigation, for example bacteria of the genus *Desulphvibro* or *Thiobacillus thio-oxidans* may become abundant in saturated soils and may accumulate insoluble iron sulphides which reduce the soil fertility. A bacterium which plays an important role in the sulphur cycle, *Microspora desulphuricans* may also be affected by soil saturation, as may organisms which decompose cellulose, for example the Actinomycetes which are aerobic and dislike salts (Kovda *et al.* 1973: 412).

Irrigation water can disperse plant pathogens, for example, the tobacco fungus (*Phytophthora parasitica* var. *nicotianae*) and the cotton root rot fungus (*Phymatotrichum omnivorum*) can be spread by sprinkler irrigation. If sprinkler irrigation is used on vines there is an increased risk of downy mildew (*Peronoplasmopara viticola*), if used on citrus *Xanthomonas citri* (a mould) may be a problem (Finkel 1983: 43). Worthington (1977: 41) reported the beneficial impact of surface irrigation on the banana fungus *Fusarium oxysporum* in Central American plantations; Arnon (1981: 321), however, noted an increase of *Fusarium* infections of beans after the introduction of irrigation in north-west USA. Surface irrigation which saturates the soil seems to favour fungi of the genus *Phycomycetes* which can cause the 'damping-off' of seedlings, especially of legumes and solanaceous crops (which include potatoes and tomatoes) (Finkel 1983: 43).

Larger burrowing animals are quickly displaced to the margins of irrigated fields or are quickly destroyed. Of those which manage to remain, some may tunnel into banks and channel sides and cause considerable damage. Tanks in India and Sri Lanka sometimes fail because of the attentions of rodents, and losses of water from unlined channels is increased when there are rodent or other animal burrows. Rodents consume crops and insecurely stored grain, and may carry diseases (e.g. leptospirosis). In South and South East Asia it is not uncommon to lose a third of a rice crop to rodents; in Africa especially, birds like the quela also take their toll of crops.

Earthworms usually increase in moistened soil and are likely to be beneficial, improving soil aeration and fertility (White 1978: 42); termites generally seem to dislike irrigation development, but ants seem less affected (Kovda *et al.* 1973: 413). A common problem is an increase in nematode (eelworm) infestations in irrigated fields and in some cases this has been serious enough to stop the cultivation of certain crops. Regular crop rotation can often help to reduce fungal and nematode infestations and deserves more attention.

Organisms which have caused human health problems or which have damaged crops on or near irrigation schemes

Irrigation development provides new opportunities for organisms, some of these may become a nuisance or worse. Table 10.4 lists some of these 'problem organisms'. Table 10.5 lists irrigation development.

Irrigation generally involves the replacement of mixtures of hardy, traditional crop varieties with potentially higher-yielding but less hardy, more uniform varieties. The result can be the increased vulnerability of crops to pests, especially insects and plant diseases. Pests have been known to become established on irrigated land and then to spread out to surrounding rainfed farmland causing problems of a greater magnitude than were known before the introduction of irrigation (Rivnay 1972). Locusts and grasshoppers have done this in parts of Africa. A major pest of irrigated agriculture in Africa is the quela (*Quela quela*), a weaver-bird which forms vast flocks, each individual of which can consume about 100 g of grain a day. The Sudan, Ethiopia, the Sahelian states and Kenya have had great difficulties controlling this pest.

Beneficial biological impacts

The literature tends to report only the harmful impacts of irrigation; but there are also some beneficial ones, expected and unexpected. Fish and other edible organisms usually colonize irrigation channels and padi-fields and, provided these are not contaminated with agrochemicals, they can provide a valuable source of food for man and wildlife.

Increasingly, irrigated land is the only wetland left as natural swamps and marshes are drained. Migrating birds increasingly follow routes from irrigated area to irrigated area so that they can feed and rest. The farmer may benefit from such visitors if they feed upon the insects on his crop, so there is a good chance that irrigators will willingly cooperate to ensure wildlife conservation. The International Union for the Conservation of Nature and Natural Resources recognized this and has made efforts to educate and help irrigators conserve wildlife (Worthington 1977: 43).

Table 10.4 Organisms which have caused problems following irrigation development

Organism	Location	Nature of impact[a]	Source
Weeds			
Unspecified	Iraq and Egypt	Crop damage L	Johl 1980: 89
Nutgrass (*Cyperus* spp.)	Kenya	20% crop reduction L	Arnon 1981: 22
Invertebrates (not insects)			
Nematodes	Widespread, especially lowland tropics	Crop damage L	Author; Kovda *et al.* 1973: 220
Aquatic snails		Human health L, R	
Freshwater crabs/ prawns	Widespread in lowland tropics	Damage to channels and tank embankments L	
Insects			
Locusts (*Schistocera gregaria* and *Nomadacris septem-fasciato*)	Africa, Middle East, South Asia, Australia	Crop damage L, R	
Grape moth (*Polychrosis botrana*)	Iraq, Egypt	Crop damage L	Johl 1980: 89
Corn stalk borer (*Chile agamemnon*)	Iraq, Egypt	Crop damage L	Johl 1980: 89
Cotton leafworm (*Spodoptera littoralis*)	Iraq, Egypt	Crop damage L	Johl 1980: 89
Brown plant hopper *Nilaparvata ingens*)	South and South East Asia	Rice damage L	
Whitefly (*Bemisia gossypiperda*)	Sudan (Gezira)	Cotton damage L	
Mosquitoes	World-wide	Human health L, R	Bull 1982
Blackfly (*Simulium* spp.)	Africa, South and Central America	Human health L, R	

Table 10.4 (continued)

Organism	Location	Nature of impact[a]	Source
Vertebrates			
Rodents, e.g. ricefield rat (*Rattus rattus argentiventer*)	World-wide	Damage to banks, channels, crops, stored produce and human health L, R	
Quela birds (weaver or dioch *Quela quela*)	Sub-Saharan Africa	Crop damage L, R	

[a] L = local, i.e. immediate field area where crops are grown; R = affects considerable area around irrigated fields. Some organisms may respond not so much to irrigation as pesticide use destroying predators.

The socio-economic impacts of irrigation

The impacts of irrigation on human health

Most of the literature on the health impacts of irrigation dwells on the negative aspects; while these are often very serious there are sometimes beneficial effects. The provision of irrigation water often means improved, safer drinking-water supplies (especially if deeper groundwater resources are developed). On large irrigation schemes, funds and facilities may be available for health care and health education where previously there was little or none. Irrigation can (if successful) increase crop yields, and can ensure that harvests fail less often, and this in itself improves nutrition and reduces under-nutrition on malnutrition and related diseases. As previously mentioned, fish, crabs, ducks and so on may become a part of the diet after irrigation development and this improves nutrition (Deom 1976; Worthington 1977: 45).

Water can act as a medium for the direct transmission of human disease or harmful chemical pollutants. Water can allow disease-carrying organisms to breed, and water attracts man to localities where they become infected. People settled on irrigation schemes may come from a variety of environments and/or from

Table 10.5 Diseases, and disease vectors associated with irrigation development

Disease	Transmission/vector organism
Schistosomiasis	Direct in water (skin penetrated by parasite released from aquatic snails)
Typhoid	Direct in water (ingestion)
Cholera	Direct in water (ingestion)
Hepatitis (infectious)	Direct in water (ingestion)
Diarrhoea/enteritis	Direct in water (ingestion)
Poliomyelitis	Direct in water (ingestion)
Amoebic dysentery	Direct in water (ingestion)
Bacillary dysentery	Direct in water (ingestion)
Arboviruses (encephalitis, etc.)	Habitat favours mosquitoes (mosquitoes are vector)
Malaria	Habitat favours mosquitoes (mosquitoes are vector)
Filariasis	Habitat favours mosquitoes (mosquitoes are vector)
Onchocerciasis	Habitat favours blackflies (blackflies are vector)
Trypanosomiasis	Habitat favours tsetse-fly/increased human contact (tsetse-fly vector)
Chagas's disease	Increased human activity/poor housing favours reduviid bugs (reduviid bugs are vectors)
Flukes (various)	Fish in schemes become infected (ingestion of fish)
Schigella	Helminth infestation (contact with infested soil)
Dracunculosis	Contact with moist soil, or ingestion (intermediate host waterfleas)

well-dispersed, formerly isolated settlements or households; concentrated together the result can be catastrophic as people come into contact with diseases against which they have little immunity. Careful monitoring and appropriate health measures are required to overcome such problems (Worthington 1977: 47).

Irrigation and malaria

As early as 1869 markedly increased outbreaks of 'fever' (it is debatable that all cases were malaria, due to poor diagnosis) were associated with waterlogging caused by leakage of the Ganges (irrigation) Canal in India (Stone 1984: 138, 144–58). Malaria is nowadays one of the major irrigation-related diseases (Farid 1977; Finkel 1983: 47). The relationship between establishing an irrigation project, reservoir or irrigation water supply system and outbreaks of malaria (and other mosquito-transmitted diseases) is not straightforward. Different malaria-carrying mosquitoes may have very different breeding habits, so avoidance of one does not necessarily mean another will not take over as disease vector. The population levels of mosquitoes that transmit malaria depend on both the presence of suitable breeding sites and contact with warm-blooded hosts. It is possible for a population of mosquitoes which do not transmit malaria to be present and thereby to exclude malaria-carriers. It is also possible for cultivators to have daily routines which to a considerable extent keep them away from mosquitoes.

Irrigation does tend to increase the contact between people and mosquitoes, and there is also a tendency for pesticide use to lead to increased pesticide resistance of mosquitoes which in turn makes their control more difficult. Added to the latter problem the causative organism of malaria (*Plasmodium* spp.) is in many regions becoming resistant to anti-malarial drugs. If anything, malaria control is becoming more of a problem and is thus an irrigation impact growing in seriousness.

Irrigated land and the margins of the canals which supply them with water can provide dispersal routes for mosquitoes and other organisms (Diamant 1980; Jayaraman 1982). The practice of irrigation can also cause mosquitoes to alter their feeding pattern. In Kenya one irrigation scheme led to a reduction of cattle-herding and increased human settlement. Before these changes roughly two-thirds of the local malaria-carrying mosquitoes fed upon cattle; after the scheme was finished these mosquitoes fed on humans and a major malaria outbreak occurred (Carpenter 1983: 318).

There have been attempts at drug control of malaria on irrigation schemes. On the Mwea Scheme in 1976 reasonably good results were obtained in controlling the disease among the farmers and their families at the cost of $US1.2 per irrigated hectare (Carruthers & Clark 1981: 208).

Irrigation and schistosomiasis

Schistosomiasis has commonly increased when irrigation is developed (Eckholm 1977; 1982). The disease is serious and widespread, although some claims, for example that the Aswan High Dam and associated irrigation development has increased the incidence of the disease so much that one in every five deaths in Egypt are attributable to it, are possibly excessive (Carpenter 1983: 319).

Control of schistosomiasis on irrigation schemes using drugs has been tried. The Mwea Scheme found it to be cost effective in 1976 at $US1.7 per irrigated hectare (Carruthers & Clark 1981: 208). However, on a regional- or national-scale, drug treatment is too costly for most developing countries and reinfection would constantly occur. Both malaria and schistosomiasis, and many other irrigation-related diseases, must be dealt with using an integrated approach combining drugs, health education, sanitation, disease-avoidance engineering and a sound understanding of the nature of the disease (Joblin 1978; Deacon 1983: 24).

Other mosquito-transmitted diseases associated with irrigation development

Outbreaks of mosquito-borne encephalitis have been associated with irrigation schemes. The virus is transmitted from wild birds to man by mosquitoes, and changes in local bird populations often result from irrigation development. The factors leading to encephalitis outbreaks do not appear to have been very well studied and at present control of outbreaks is not easy (Carpenter 1983: 319). Filarial and arbovirus infections are also transmitted by mosquitoes which may multiply on irrigation schemes.

The economic impacts of irrigation

'The full impacts of irrigation and associated drainage development upon water, soil, aquatic and terrestrial ecosystems and human health extend through the social and economic fabric of local and national societies' (Worthington 1977: 53).

The success or failure of irrigation development must not be judged by techno-economic criteria alone, but also by less easily quantifiable impacts on people. Also, various socio-economic impacts may not affect the same segment of society at the same time (ASAE 1981: 22).

The economic costs and benefits of irrigation development can be divided into:

1. Direct benefits and costs;
2. External benefits and costs;
3. Indirect income effects;
4. Employment.

Costs and benefits may be short-term or long-term. The first category just described (direct benefits and costs) affect the farmer. Category 2 includes third-party damages (e.g. problems caused by saline return flows or waterlogging). Category 3 covers impacts stemming from irrigation, and are either forward-linked (for example, processing, marketing or transport of produce) or backward-linked, i.e. induced effects which could include irrigation inputs such as seed, fertilizer, farm machinery or pump servicing (Sudeglin 1978; ASAE, 1981: 22).

Impacts of irrigation development on the local economy

Irrigation has the potential greatly to increase agricultural output; whether or not the individual cultivator benefits from any such increase depends very much on the style of irrigation management. Cultivators may be tenants, own their own land or work on large irrigation schemes as landless labourers. They may get fixed wages or share in total profits; they may shoulder all risks or be to some degree protected from crop failure. Local peoples not actually working irrigated land may benefit from access to fallow plots for grazing livestock, may share community services (schools, roads, health centres and so on), may fish irrigation channels or reservoirs, sell and service machinery – or may get no benefit at all.

Impacts of irrigation development on the regional economy

Young (ASAE 1981: 28) was sceptical that irrigation development stimulated much regional growth, at least in the USA. However, a recent study of the Muda Irrigation Scheme (a large gravity irrigation, padi rice scheme in Kedah State, Peninsular Malaysia, opened 1973) attempted to quantify the indirect effects of investment on the surrounding region and further afield – in particular on incomes, services and demand for goods. The goals of the Muda Scheme were to make Malaysia self-sufficient in rice, and to raise the income of small farmers in what was the one of Malaysia's poorest regions. The Muda Scheme would seem to have met its goals, and the study (Bell *et al.* 1982: 9) concluded that there have been marked indirect benefits on the surrounding region: '. . . for every $M of value added, generated directly by the project, another 80 cents or so were generated downstream . . .' (in the economic sense).

Impacts of irrigation development on national economies

At the national level, clearly a number of developing countries have been markedly affected by irrigation development. Nations once importing food grains at considerable cost have, thanks to irrigation, become partly or fully self-supporting, and in a few cases now export grain. Some, however, have run up debts purchasing irrigation equipment and must now find interest payments and money for spare parts and other costly foreign – manufactured inputs.

Experience has shown that planners need to ask whether irrigation development plans are scale-neutral, i.e. will they differentially benefit larger and smaller farmers – if larger farmers benefit more than the small then there is likely to be increased disparity of incomes, rising debt among the poor, discontent and possibly calls for reform. Irrigation development can easily alter the costs of agricultural inputs, labour, transport and market prices for produce.

Commonly, irrigation projects benefit far fewer people than the planners had hoped would be the case (the Muda Scheme has it seems been more successful than most), and those who do benefit may not be the most deserving, but merchants controlling supplies of fertilizer and other inputs, rather than the poorer farmers. To combat inequity, an agency or state may act to control the size of landholdings, may fix the price of inputs, aid smaller farmers with grants, direct social services (health care, education and so on) to those who most need it, control wages and the price of products irrigated farming produces. In short, irrigation development may well require social welfare innovations and careful management of its economic effects, or the poor will suffer.

Case-study: the Kano River Project

An example of irrigation development which has absorbed a considerable proportion of a developing country's funds, but which has done relatively little to help people even in its immediate vicinity is the Kano River Project. This is a large-scale, capital-intensive irrigation project initiated in 1971 – the first of several schemes which, it was hoped, would irrigate up to 59,000 ha of northern Nigeria.

Traditionally, farmers in the region had grown a single crop of guinea-corn, millet, cowpeas and sometimes groundnuts between May and October. The Kano River Project was to improve the living standards of the peasant farmers by allowing them to grow irrigated wheat in the dry season (October to May). Wheat was not traditionally part of the diet of Kano State's rural people. It is consumed in Kano City and other Nigerian townships and when the Kano River Project was planned was an increasingly expensive import – which leads one to suspect that there were hopes that the project would grow wheat for urban Nigeria.

A massive share of the total crop subsector budget was allocated to the project under Nigeria's *Third Development Plan*. Today a mere 22,733 ha is irrigated, and whatever benefit this has generated has probably affected less than 1 or 2 per cent of the population of Kano State. To put it crudely, a single rather mediocre performance project took a major share of Kano State's agricultural development funds for some years.

The project has had a number of adverse impacts. When the Tiga Dam which supplies water to the project was built, 13,000 peasant families had to be relocated. Some received cash compensation, but in the main fared badly at resettling themselves. The rest of the relocatees obtained new landholdings and a house in one of seven new resettlement villages, but not all of the land was good enough (ODI 1980: 16; Wallace 1979, 1981).

Although production on the project was running at 1.9 tonnes/ha in 1981, which was better than traditional yields, that was still well below the planned 4 tonnes/ha (Wallace 1981: 291). Low yields would seem to have been caused by the cultivators giving too much time to subsistence crops and insufficient to the dry season wheat. Larger farmers seem to have been able to make a reasonable living growing dry season wheat, but smaller farmers, especially those with poorer water supplies, have often been forced to sublet their land to Kano businessmen. The project has created few jobs, and inmigration of skilled labour has filled many of those that were created. A technically complex, part-mechanized and very costly project seems to have done little to improve rural welfare.

Impacts of irrigation development on the global economy

Roughly 10 per cent of the world's total arable cropland is irrigated and supports about two-thirds of developing country farmers (World Bank 1982: 63). The remainder depends upon rainfall, ephemeral streams or upon periodic flooding (Oregon State University 1979). The FAO's estimate is that in AD 2000 about 84 per cent of the world's total farmland will still be rainfed and will yield roughly 51 per cent of crops (Biswas *et al.* 1983: xiii). There are already signs that the expansion of irrigation is slowing because the best, more cheaply irrigated land is already in use (World Bank 1982: 92).

It is probable that without the irrigation development that has occurred since the Second World War there would have been much greater food shortages, certainly in the Indian subcontinent, China and parts of Africa, especially in the 1960s and early 1970s. Irrigation development removed the dependence many countries would otherwise have had on American, Canadian or Russian grain. On the other hand by taking up so much of total development funds, large-scale irrigation may well have delayed improvement of rainfed farming and crops grown by the poor.

The socio-cultural impacts of irrigation

Although it is often social or cultural issues rather than problems of water supply or cultivation which jeopardize the functioning of irrigation schemes, such aspects have received only a mere fraction of the attention and resources directed to improving water supply engineering. White (1978: 58) commented: '. . . The response of human society and especially peasant farmers to irrigation has not been studied nearly as thoroughly as has the response of soils or agricultural crops. Only in recent years has the social change provoked by irrigation schemes become a vigorous subject for study.'

Sociological interest in irrigation development has largely focused on two questions: (1) Does it result in food surpluses which may lead to population increase? (2) Does organization and management of irrigation, particularly large-scale irrigation, determine the pattern of political structure? The latter interest was partly triggered by the work of Wittfogel (1957) a Marxist economist who saw evidence of a link between the stratified irrigated societies of the past and the totalitarian organization of some modern states. Wittfogel's theories have two main thrusts. One concerns the role of irrigation in the evolution of the state – does it lead to centralized and bureaucratic 'hydraulic societies'? The other concentrates on elucidating a model of how certain kinds of states, in particular what Wittfogel termed 'oriental despotisms' are structured. Where sufficient water and land is available Wittfogel suggested irrigation grew to demand large-scale, unified organization. This, he suggested, was because labour has to be organized, jobs and duties delimited and then careful central supervision develops (Chambers & Moris 1973: 513 give a bibliography on literature dealing with Wittfogel's ideas).

More recently Downing & Gibson (1974) and Hunt & Hunt

(1976) have examined the influence of irrigation on social organization. Lees (in Downing & Gibson 1974: 125) made the observation that if an agency or government installs irrigation devices which in construction and maintenance requirements are beyond the capabilities of the local community, then that community's autonomy is reduced and non-local forces affect it more.

Irrigators often adapt themselves to irrigation schemes, seldom have planners adapted their schemes to suit farmers (White 1978: 58). In the Philippines, at least, this observation seems to have been accepted and has resulted in some attempts to involve irrigators in the planning of irrigation schemes from the earliest proposals stage. In the past Filipino farmers, like their counterparts elsewhere, were usually ignored during planning and implementation, but were then often suddenly asked on project completion to become involved in user management. Obviously when this happens the farmer-managers feel little commitment to the project and there are likely to be faults in it that they could have warned the planners to avoid (Korten 1982).

Resettlement and irrigation development often go hand-in-hand. Planners, however, seldom seem to appreciate that people adapting to being moved are unlikely readily to master new farming/irrigation techniques. These problems are especially marked when attempts are made to sedentarize nomadic or semi-nomadic peoples, and are well illustrated by the experiences with the Khasam el Girba Project (Sudan) and the Helmand Project (Afghanistan), where people simply drifted away from the schemes (Worthington 1977: 53). An irrigation project near Sansanding on the Niger River failed in the 1960s, not through technical problems but because the farmers did not want to settle in that locality (Arnon 1981: 29). Similar troubles have been frequently repeated elsewhere, for example, a discussion of the social impacts of irrigation in Senegal may be found in Adams (1977). On one Nigerian irrigation scheme recruitment of irrigators was left to village chiefs who, anxious to improve their own established communities, directed the inept, the misfits and malcontents to the scheme – which consequently failed.

The role of women in agriculture has often been overlooked: in The Gambia during the 1970s, for example, rice schemes failed because the planners only consulted the menfolk, who in reality did little farming. Another problem is that when irrigation is introduced to rainfed farming communities the farmers can sometimes become so infatuated with irrigation that they neglect rainfed crops which should supply their subsistence needs. Sometimes the opposite situation arises, farmers neglect the irrigated crops to tend rainfed subsistence crops (Rao 1978). When irrigation/resettlement schemes run into difficulties, for example delays over clearing or levelling the farmland or in finishing the supply channels, there is a risk that the younger, more dynamic settlers will get fed up and move elsewhere, leaving only the inept and the old – as happened on some of the ambitious irrigation schemes along the lower Senegal River.

Irrigation development can evoke a sense of national or community achievement and pride – something which is difficult to quantify, but which may be very beneficial. Irrigation development may help to eliminate seasonal unemployment and/or underemployment. It can also provide a channel through which 'modernization' and urban influence can reach rural folk, and this may be a good or a harmful thing.

Conclusions

There is a need for careful and unbiased appraisal to identify which approaches to irrigation development and which management strategies offer the greatest benefits at the lowest cost. More care is needed to assess in advance the environmental and socio-economic impacts of irrigation development – even imperfect prediction of problems would permit better avoidance of mistakes and would be better than the usual, present situation where assessments, thorough or otherwise, are made too infrequently.

The majority of people in tropical countries derive their living directly or indirectly from agriculture. How might water resources development best help these people, and with which group or groups, if any, should priorities lie? Much money has been spent on large-scale commercial irrigation schemes which have often proved inefficient. The efficiency of such irrigation must be improved, to obtain better value for money and to ensure that the development is sustainable and not temporary. On most large irrigation projects, management, maintenance and repairs could be vastly improved. In the past, these aspects of irrigation have not had sufficient funding, and have been seen as less important and less prestigious than irrigation engineering.

The style of irrigation management and scale of project would seem to be less important than the genuine involvement of farmers in planning, implementation and management.

Communications between irrigators and those supplying inputs, especially water supplies, are important and in general could be greatly improved. Motivating the irrigators is also important. In Japan the motive force which drove the peasantry to uprate their agriculture was, according to Pearse (1980: 225), feudalism – they were forced; in the People's Republic of China sense of duty plus a revolutionary government seems to be the motivating force; South Korea and Taiwan appear to have been driven by occupying forces which coerced the peasants to produce more. Nowadays feudalism or colonial domination are unlikely to drive the development of agriculture, other incentives are required. Profit is a strong motivating force. However, ensuring that peasants can adequately profit from the introduction or improvement of existing irrigation may not be easy. Another motivation might be to offer the peasant more secure ownership of his land and the right to pass that ownership on to his children. Lack of security of tenure is certainly an important disincentive to agricultural development.

In remote and impoverished regions, more productive rainfed cultivation, possibly with supplemental irrigation, must be promoted to feed growing populations and to ensure that the land does not degenerate as traditional farming strategies come under pressure. Techniques and crops which could be used to build up low-cost, productive and sustainable agricultural systems even in harsh environments are already available or are being developed. The need now is for various components to be 'assembled' into appropriate 'packages' and then disseminated to farmers.

In view of the uncertainty over rainfall in many regions, plus the increasing deforestation of catchments and competition for available water, it is important that planners developing water resources exercise caution and establish irrigation only where there seem to be adequate long-term supplies. It should not be assumed that rainfall is a constant: precipitation might well deteriorate in parts of the tropics in the future, or it might improve. If finite groundwater resources have to be developed, then ideally there should be investment of some of the profits to ensure that the farmers have alternative livelihoods when the supplies run out.

Hopes of 'technological fixes' to the problem of water supply should not run too high. Breakthroughs leading to cheap desalination, towing icebergs to Chile, Australia or the Middle East and the development of seawater irrigation are all possible, but far from certain (Husseiny 1978; ASAE 1981). Agriculture in the tropics on the whole must make better use of what water resources it now has rather than expect new supplies.

Improving irrigation, introducing new irrigation or rainfed farming techniques is often difficult and is likely to remain so because human attitudes and institutions are a brake on change. Such attitudes and institutions can be very difficult to overcome. It is attitudes and institutional problems rather than technical and faulty cultivation or overgrazing which are at the root of much wastage of water. Considerable effort and funding has to be directed towards training those who control water supply systems, those who manage irrigated and rainfed agriculture developments and the cultivators/pastoralists themselves. Schumacher (1974: 64) stressed the importance of training in developing countries when he noted: 'All history – as well as current experience – points to the fact that it is man, not nature, who provides the primary resource: that key factor of all economic development comes out of the mind of man. . . . In a very real sense, therefore, we can say that education is the most vital of all resources.'

References

Ackermann, W. C., White, G. F. and Worthington, E. B. (eds) (1973) *Man-made Lakes: their Problems and Environmental Effects*, American Geophysical Union, Washington, DC. (Monograph 17).

Abu-Zeid, M. (1979) 'Short and long term impacts of the River Nile projects', *Water Supply and Management*, **3** (3), 275–83.

Adams, A. (1977) 'The Senegal River Valley: what kind of change', *Review of African Political Economy*, **10**, 33–59.

Adams, A. (1981) 'The Senegal River Valley', in Heyer, J., Roberts, P. and Williams, G. (eds), *Rural Development in Africa*, Macmillan Press, London, pp. 325–53.

Adams, R., Adams, M., Willens, A. and Willens, A. (1978) *Dry Lands Man and Plants*, Architectural Press, London.

Adefolalu, D. O. (1983) 'Desertification in the Sahel', in Ooi Jin Bee (ed.), *Natural Resources in Tropical Countries*, Singapore University Press, pp. 402–38.

Afriye, E. K. (1973) 'Resettlement agriculture an experiment in innovation', in Ackermann, W. C., White, G. F. and Worthington, E. B. (eds), *Man-made Lakes: their Problems and Environmental Effects*, American Geophysical Union, Washington, D.C., pp. 726–9. (Monograph 17).

Agarwal, A. (1980) 'A decade of clean water', *New Scientists*, **88** (1226), 356–9.

Agarwal, A. (1981) 'Supplying water: maintenance problems and community participation', *Development Digest*, XIX (4), 74–85.

Ahmad, Y. J. and Sammy, G. K. (1985) *Guidelines for Environmental Impact Assessment in Developing Countries*, Hodder and Stoughton, London.

Allan, P. (1965) *The African Husbandman*, Oliver and Boyd, Edinburgh.

Almeyra, G. (1979) 'Mexico's new frontier: the humid tropics', *Ceres*, **12** (5), 27–33.

Ambroggi, R. P. (1980) 'Water', in Scientific American, *Economic Development*, W. H. Freeman, San Francisco, pp. 37–49.

Anderson, J. M. (1981) *Ecology for Environmental Sciences: Biosphere, Ecosystems and Man*, Edward Arnold, London.

Anderson, T. L. (ed.) (1983) *Water Rights: Scarce Resource Allocation, Bureaucracy, and the Environment*, Ballinger Publishing, Cambridge, Mass.

Andreae, B. (1981) *Farming Development and Space: a World Agricultural Geography*, Walter de Gruyer, Berlin.

Anon. (1977a) 'The creeping deserts (a special report)', *Time*, 12 Sept., pp. 14–31.

Anon. (1977b) 'The World Water Conference: a suggested programme for irrigation management', *ODI Review*, **1**, 106–10.

Anon. (1980) *New Irrigation Schemes: Planning, Design and Social Impacts*, Overseas Development Institute (ODI), London, 24 pp., (ODI Irrigation Management Network Paper).

Anon. (1981) *Development Forum*, **9** (1), 11.

Anon. (1982a) 'New stress put on improvements for rainfed rice', *Ceres*, **15** (5), 6–8.

Anon. (1982b) 'Will polymers turn the desert green?', *New Scientist*, **95** (1316), 305.

Anon. (1983a) 'Third World review: why the desert is winning', *The Guardian*, 28 Jan., p. 10.

Anon. (1983b) 'Contour bunds save water in semi-arid Ethiopian highlands', *United Nations University Newsletter*, **7** (2), 7.

Anon. (1983c) *New Scientist*, **97** (1348), 663.

Anon. (1983d) 'The Poghil Windmill pumps water for Indian village projects', *United Nations University Newsletter*, **7** (2), 7.

Anon. (1984) 'New hope in farming the dry lands', *Mazingira*, **8** (3), 7.

Anon. (1985a) 'The green advance of millet', *International Agricultural Development*, 5 (3), 8.

Anon. (1985b) 'Crop irrigation without wasting water', *New Scientist*, **108** (1478), 38.

Anon. (1985c) 'Maharashtra's handpump programme in turmoil, *World Water*, **8** (9), 6.

Anon. (1985d) 'Pakistan acts to halt salinization', *World Water*, **8** (9), 13.

Arnon, I. (1972) *Crop Production in Dry Regions*, vol. 1: *Background and Principles*, Leonard Hill, London.

Arnon, I. (1981) *Modernization of Agriculture in Developing Countries: Resources, Potentials, Problems*, Wiley, Chichester.

ASAE (1981) *Irrigation Challenges of the 80s*, American Society of Agricultural Engineers P.O. 410, St Joseph, Michi. (Proceedings of the Second National Symposium, 20–23 Oct., 1980, Lincoln, Nebr. ASAE Publication No. 6–81).

Asibey, E. A. (1977) 'The blackfly dilemma', *Environmental Conservation*, **4** (4), 291–5.

Askochensky, A. N. (1962) 'Basic trends and methods of water control in the arid zones of the Soviet Union', in UNESCO, *Arid Zone Research, XVIII*, UNESCO, Paris, pp. 401–10 (The Problems of the Arid Zone: a Symposium).

Balek, J. (1977) *Hydrology and Water Resources in Tropical Africa*, Elsevier, Amsterdam.

Balek, J. (1983) *Hydrology and Water Resources in Tropical Regions*, Elsevier, Amsterdam and New York.

Balon, E. K. (1978) 'Kariba the dubious benefits of large dams', *Ambio*, VIII (1), 40–8.

Bangladesh Rice Research Institute (1975) *Proceedings of the International Seminar on Deep-water Rice, 21–26 Aug.*, 1975, Bangladesh Rice Research Institute, Dacca.

Barrow, C. J. (1981) 'Health and resettlement consequences and opportunities created as a result of river impoundment in developing countries', *Water Supply and Management*, 5 (2), 135–50.

Barrow, C. J. (1983) 'The environmental consequences of water resource development in the tropics', in Ooi Jin Bee (ed.), *Natural Resources in Tropical Countries*, Singapore University Press, pp. 439–76.

Barrow, C. J. (1985) 'The development of the várzeas (floodlands) of Brazilian Amazonia', in Hemming, J. (ed.), *Change in the Amazon Basin*, vol. 1: *Man's Impact on Forests and Rivers*, Manchester University Press, Manchester, UK, and Dover, NH, USA, pp. 108–28.

Bateman, G. H. (1974) *A Bibliography of Low-cost Water Technologies* (3rd edn), Intermediate Technology Publications, 9 King Street, London WC2E 8HN.

Baxter, R. M. (1977) 'Environmental effects of dams and impoundments', *Annual Review of Ecology and Systematics*, **8**, 255–83.

Bayliss-Smith, T. P. (1982) *The Ecology of Agricultural Systems*, Cambridge University Press, London and New York.

Bayliss-Smith, T. P. and Feacham, R. G. A. (eds) (1977) *Subsistence and Survival: Rural Ecology in the Pacific*, Academic Press, London.

Beadle, L. C. (1974) *The Inland Waters of Tropical Africa: an Introduction to Tropical Limnology*, Longman, London.

Beckford, G. L. (1973) 'The economics of agricultural resource use and development in plantation economies', in Bernstein, H. (ed.), *Underdevelopment and Development: the Third World Today (2nd edn)*, Penguin Books, Harmondsworth, pp. 115–51.

Beets, W. C. (1982) *Multiple Cropping and Tropical Farming Systems*, Gower Publishing, Amsterdam and New York.

Bell, C., Hazell, P. and Slade, R. (1982) *Project Evaluation in Regional Perspective; a Study of an Irrigation Project in Northwest Malaysia*, published for the World Bank by Johns Hopkins University Press, Baltimore and London.

Belmares, H., Barrera, A., Castillo, E., Monjaras, M. and Tristan, M. E. (1979) 'Industrialization of plant resources of arid and semi-arid zones of North America', *Interciencia*, **4** (6), 320–4 (in Spanish).

Bernstein, L. (1962) 'Salt-affected soils and plants', in UNESCO, Arid Zone Research, XVIII UNESCO, Paris, pp. 139–74, (The Problems of the Arid Zone: a Symposium).

Bernstein, L. and François, L. E. (1973) 'Comparisons of drip, furrow, and sprinkler irrigation', *Soil Science*, **115** (1), 73–86.

Biswas, A. K. (1979a) 'Management of traditional resource systems in marginal areas, *Environmental Conservation*, **6** (4), 257–63.

Biswas, A. K. (1979b) 'Water development in developing countries: problems and prospects, *Geojournal*, **3.5**, 445–6.

Biswas, A. K. (1980) *Labour-based Technology for Large Irrigation Works: Problems and Prospects*, International Labour Organization (ILO), Geneva, (ILO World Employment Programme Research Working Papers).

Biswas, A. K. (1981) 'Long distance transfer of water', *Water Supply and Management*, **5** (3), 245–51.

Biswas, M. and Biswas, A. K. (1982) 'The devastating cost of environmental neglect, *South*, No. 20, 57–59.

Biswas, A. K., Zuo Dakang., Nickum, J. E. and Liu Changming (1983) *Long Distance Water Transfer: a Chinese Case Study and International*

Experiences, Tycooly International Publishing, Dublin.
Blackie, M. J. (ed.) (1984) *African Regional Symposium on Small Holder Irrigation*, 5–7 Sept., University of Zimbabwe, Harare, Overseas Development Unit of Hydraulics Research, Wallingford, UK.
Bolton, P. and Pearce, G. R. (1984) *Report on African Regional Symposium on Small Holder Irrigation*, Hydraulics Research, Wallingford, Oxfordshire OX10 8BA UK, 27 pp. (Report OD/TN10).
Boserüp, E. (1965) *The Conditions of Agricultural Growth: the Economics of Agrarian Change Under Population Pressure*, George Allen and Unwin, London.
Bottrall, A. F. (1981a) *A Comparative Study of the Management and Organization of Irrigation Projects*, World Bank, Washington, DC. (World Bank Staff Working Paper No. 458).
Bottrall, A. F. (1981b) 'Improving canal management: the role of evaluation and action research', *Water Supply and Management*, 5 (1), 67–79.
Bottrall, A. F. (1981c) 'Water, land and conflict management', *ODI Review*, No. 1, 73–83.
Bottrall, A. F. (1985) *Managing Large Irrigation Schemes: a Problem of Political Economy*, Overseas Development Institute, 10–11 Percy Street, London. (ODI Agricultural Administration Unit, Occasional Paper 5).
Bouwer, H. (1978) *Groundwater Hydrology*, McGraw-Hill, New York.
Bowen, R. (1980) *Ground Water*, Applied Science Publishers, Barking, Essex, UK.
Box, T. W. (1971) 'Nomadism and landuse in Somalia', *Economic Development and Cultural Change*, **19** (2), 222–8.
Boyko, H. (ed.) (1968) *Saline Irrigation for Agriculture and Forestry*, Dr W. Junk, The Hague.
Brady, N. C. (1981) 'Food and resource needs of the world', in Manassah, J. T. and Briskey, E. J. (eds), *Advances in Food-Producing for Arid and Semi-arid Regions*, Part A, Academic Press, New York, pp. 3–31.
Brandt Commission. (1983) *Common Crisis North–South: Co-operation for World Recovery*, Pan Books, London.
Breckle, S. W. (1982) 'The significance of salinity', in Spooner, B. and Mann, H. S. (eds), *Desertification and Development: Dryland Ecology in Social Perspective*, Academic Press, London and New York, pp. 277–92.
Bromley, D. W. (1982) *Improving Irrigated Agriculture: Institutional Reform and the Small Farmer*, World Bank, Washington, DC. (World Bank Staff Working Paper No. 531).
Bromley, D. W., Taylor, D. C. and Parker, D. E. (1980) 'Water reform and economic development: institutional aspects of water management in the developing countries', *Economic Development and Cultural Change*, **28** (2), 365–87.
Brown, L. R. (1970) *Seeds of Change: the Green Revolution and Development in the 1970s*, Overseas Development Corporation, London.
Brown, L. R. (1974) *By Bread Alone*, Praeger, New York.
Bruins, H. J., Evenari, M. and Nessler, U. (1986) 'Rainwater-harvesting agriculture for food production in arid zones: the challenge of the African famine', *Applied Geography*, **6** (1), 13–32.

Bucher, E. H., and Schofield, C. J. (1981) 'Economic assault on Chagas' disease', *New Scientist*, **92** (1277), 321–4.
Bull, D. (1982) *A Growing Problem: Pesticides and the Third World Poor*, Oxfam (Public Affairs Unit), Oxford.
Bunting, H. (ed.) (1970) *Change in Agriculture*, Duckworth, London.
Bunting, H. (1979) 'Towards a better age', *New Scientist*, **81** (1148), 1043–5.
Buringh, P. (1978) 'The natural environment and food production', in McMains, H. J. and Wilcox, L. (eds), *The Alternatives for Growth: the Engineering and Economics of Natural Resources Development*, Ballinger Publishing, Cambridge, Mass., pp. 99–129.
Butcher, D. P. (1971) *An Operational Manual for Resettlement (a Systematic Approach to the Resettlement Problem Caused by Man-Made Lakes with Special Reference to West Africa)*, FAO, Rome.
Butcher, W. S. (1979) 'Salinity management and water resources developments', *Geojournal*, **3.5**, 457–60.
Byres, T. J., Crow, B. and Mae Wan Ho. (1983) *The Green Revolution in India*, Open University Press, Walton Hall, Milton Keynes. (U204 Third World Studies: Case Study 5).
Canter, L. W. (1986) *Environmental Impacts of Water Resources Projects*, Lewis Publishers Inc., Chelsea, Michigan 48118.
Caponera, D. A. (1954) *Water Laws in Moslem Countries*, FAO, Rome.
Caponera, D. A. (1983) 'International river law', in Zaman, M. *et al*. (eds), *River Basin Development*, Tycooly International Publishing, Dublin, pp. 173–84.
Carpenter, R. A. (ed.) (1983) *Natural Systems for Development: what Planners Need to Know*, Macmillan Publishing, New York, Collier-Macmillan Publishing, London.
Carruthers, I. D. (1968) *Irrigation Development Planning: Aspects of Pakistan Experience*, Economics Department, Wye College, Ashford, UK, (Agrarian Development Studies Report No. 2).
Carruthers, I. D. and Clark, C. (1981) *The Economics of Irrigation*, Liverpool University Press.
Carruthers, I. D. and Stoner, R. (1982) What water management needs: a legal framework in the public interest, *Ceres*, **15** (5), 12–20.
Caufield, C. (1982) 'Brazil, energy and the Amazon', *New Scientist*, **96** (1329), 240–3.
Chambers, R. (ed.) (1970) *The Volta Resettlement Experience*, Pall Mall Press, London.
Chambers, R. (1977) 'Men and water: the organization and operation of irrigation', in Farmer, B. H. (ed.), *Green Revolution: Technology and Change in Rice Growing Areas of Tamil Nadu and Sri Lanka*, Methuen, London, pp. 340–55.
Chambers, R. (1981) 'In search of a water revolution: questions for canal irrigation management in the 1980s', *Water Supply and Management*, 5 (1), 5–18.
Chambers, R., Longhurst, R., and Pacey, A. (eds) (1981) *Seasonal Dimensions to Rural Poverty*, Frances Pinter, London, Allenhead Osmum, Englewood Cliffs, NJ, USA.
Chambers, R. and Moris, J. (eds) (1973) *Mwea: an Irrigated Rice Settlement*

in Kenya, Weltforum-Verlag, Munich.
Chapman, V. J. (1960) *Salt Marshes and Salt Deserts of the World*, Leonard Hill, London.
Charnock, A. (1983) 'A new course for the Nile', *New Scientist*, **100** (1381), 285–8.
China Facts and Figures (1985) Foreign Language Press, Beijing.
Chow, Ven Te. (ed.) (1964) *Handbook of Applied Hydrology: a Compendium of Water-resources Technology*, McGraw-Hill, New York.
Christiansson, C. (1979) 'Imagi Dam – a study of soil erosion, reservoir sedimentation and water supply at Dodoma, Central Tanzania', *Geographiska Annaler*, **61** (ser. A), 113–45.
Clark, C. (1970) *The Economics of Irrigation* (2nd edn), Pergamon Press, Oxford.
Clarke, A. C. (1977) *The View from Serendip*, Pan Books, London.
Clay, E. J. (1978) 'Genetic evaluation and utilization: deep-water-yields of deep-water rice in Bangladesh', *International Rice Research Institute Newsletter*, **3** (5), 11–13.
Clay, E. J. (1980) 'The economics of the bamboo tubewell', *Ceres*, **13** (3), 43–7.
Clay, E. J., Catling, H. D., Hobbs, P. P., Bhuiyan, N. I. and Islam, Z. (1978) *Yield Assessment of Broadcast Aman (Deep-Water Rice) in Selected Areas of Bangladesh, 1977*, Bangladesh Rice Research Institute, published by Agricultural Development Council, USA.
Clayton, E. (1983) *Agriculture, Poverty and Freedom in Developing Countries*, Macmillan Press, London.
Colson, E. (1971) *The Social Consequences of Resettlement*, University of Manchester Press, Manchester, (Kariba Studies, vol. 4).
Commonwealth Secretariat (1978) *Proceedings of the Commonwealth Workshop on Irrigation Management, Hyderabad, India, 17–27 Oct. 1978*, Food Production and Rural Development Division, Commonwealth Secretariat, Marlborough House, Pall Mall, London SW1Y 5HX.
Conklin, H. C. (1961) 'The study of shifting cultivation', *Current Anthropology*, **2** (1), 27–30.
Cooke, R. U. and Dornkamp, J. C. (1974) *Geomorphology in Environmental Management: an Introduction*, Clarendon Press, Oxford.
Cooper, A. (1983) 'Hydroponics for the small farmer', *Shell Agriculture*, May 1983 issue, p. 3.
Council On Environmental Quality and Department Of State (1982) *The Global 2000 Report to the President: Entering the Twenty-First Century*, vol. 1, Penguin Books, Harmondsworth and New York.
Coward, E. W. Jr (1976) 'Indigenous organization, bureaucracy and development: the case of irrigation', *Journal of Development Studies*, **13** (3), 92–105.
Coward, E. W., Jr (ed.) (1980) *Irrigation and Agricultural Development in Asia: Perspectives from the Social Sciences*, Cornell University Press, Ithaca, NY and London.
Cox, G. W. and Atkins, M. D. (1979) *Agricultural Ecology: an Analysis of World Food Production*, W. H. Freeman, San Francisco.

Cressey, G. B. (1958) 'Quanats, karez, and foggaras', *Geographical Review*, **48** (1), 27–44.
Cross, M. (1983) 'Last chance to save Africa's topsoil', *New Scientist*, **99** (1368), 289–93.
Cross, M. (1985) 'Africa needs to change its choice of grain, *New Scientist*, **107** (1473), 32.
Crow, B. (1981) 'Logjam in Ganges talks', *The Guardian*, 1 April, p. 9.
Cummings, K. G. (1974) *Interbasin Water Transfers: a Case Study in Mexico*, Johns Hopkins University Press, Baltimore.
Dahlberg, K. (1979) *Beyond the Green Revolution, the Ecology and Politics of Global Agricultural Development*, Plenum Press, New York.
Dakang, Z. and Liu Changming. (1981) 'East China Water Transfer: its environmental impact', *Mazingira*, 5 (4), 48–57.
Darch, J. P. (1985) *Drained Field Agriculture in Central and South America*, (British Archaeological Reports, International Series, No. 189).
Dasmann, R. F., Milton, J. P. and Freeman, P. H. (1973) *Ecological Principles for Economic Development*, Wiley, London and New York.
Deacon, N. H. G. (1983) 'A review of the literature on schistosomiasis and its control', *Tech. Note OD/TN 3 Sept, 1983*, Hydraulics Research Institute, Wallingford, UK, 36 pp.
DeCachard, M. and Balligand, P. (1982) 'Water, heat, cooling and food in one system', *Development Forum*, **10** (6), 8–9.
De Donnea, Fr. X. (1982) 'The role and limits of normative models in water resources planning and management', in Laconte, P. and Haimes, Y. Y. (eds), *Water Resources and Land-use Planning: a Systems Approach*, Martinus Nijhoff, the Hague, pp. 7–19.
Denevan, W. M. (1970) 'Aboriginal drained field cultivation in the Americas', *Science*, **169** (39), 647–54.
Denevan, W. M. (1980) 'Review article', *Geographical Review*, **70** (1), 106–7.
Dennel, R. W. (1982a) 'Archaeology and the study of desertification', in Spooner, B. and Mann, H. S. (eds), *Desertification and Development: Dryland Ecology in Social Perspective*, Academic Press, London, pp. 43–60.
Dennel, R. W. (1982b) 'Dryland agriculture and soil conservation: an archaeological study of check-dam farming and wadi siltation', in Spooner, B. and Mann, H. S. (eds), *Desertification and Development: Dryland Ecology in Social Perspective*, Academic Press, London, pp. 171–200.
Deom, J. (1976) *Water Resources Development and Health: a Selected Bibliography*, World Health Organization, Geneva.
Diamant, B. Z. (1980) 'Environmental repercussions of irrigation development in hot climates', *Environmental Conservation*, **7** (1), 53–8.
Dickenson, J. P., Clarke, C. G., Gould, W. T. S., Prothero, R. M., Siddle, D. J., Smith, C. T., Thomas-Hope, E. M. and Hodgriss, A. G. (1983) *A Geography of the Third World*, Methuen, London and New York.
Dixey, F. (1950) *A Practical Handbook of Water Supply, (2nd edn)*, Thomas Marby, London.

Dobby, E. H. G. (1973) *Southeast Asia (11th edn)*, University of London Press.

Dobyns, H. F. (1967) 'Blunders with bolsas', in Borton, R. E. (ed.) *Getting Agriculture Moving*, Agricultural Development Council, New York, pp. 252–67 (Case Studies vol.).

Domroes, M. M. (1979) 'Monsoon and landuse in Sri Lanka', *Geojournal*, **3**,2, 179–92.

Donaldson, D. (1978) 'Health issues in developing country projects', in Gunnerson, C. G. and Kalbermatten, J. M. (eds), *Environmental Impacts of International Civil Engineering Projects and Practices*, ASCE, New York, (Proceedings of a Session sponsored by the Research Council on Environmental Impact Analysis of the American Society of Civil Engineers, Technical Council on Research, San Francisco, 17–21, Oct. 1977.) pp. 134–57.

Donkin, R.A. (1979) *Agricultural Terracing in the New World*, University of Arizona Press, Tucson.

Doorenbos, J. and Pruitt, W. O. (1977) *Guidelines for Predicting Crop Water Requirements*, FAO, Rome.

Downing, T. E. and Gibson, M. (1974) *Irrigation's Impact on Society*, University of Arizona Press, Tucson. (Anthropological Papers of the University of Arizona No. 25).

Easter, K. W. (1980) *Issues in Irrigation Planning and Development*, University of Minnesota, Institute of Agriculture, Forestry and Home Economics, St Paul Minn. 55108. (University of Minnesota Department of Agriculture and Applied Economics Staff Papers Series P80–5).

Eckholm, E. P. (1976) *Losing Ground: Environmental Stress and World Food Prospects*, Pergamon Press, Oxford and New York.

Eckholm E. P. (1977) 'Of snails and men', in Eckholm, E. P. (ed.), *The Picture of Health: Environmental Sources of Disease*, W. W. Norton, New York and London, pp. 174–88.

Eckholm, E. P. (1982) *Down to Earth: Environmental Stress and Human Needs*, Pluto Press, London.

Ehrlich, P. R., Ehrlich, A. H. and Holdren, J. P. (1977) *Ecoscience: Population, Resources, Environment*, W. H. Freeman, San Francisco.

Eicher, C. K. and Baker, D. C. (1982) *Research on Agricultural Development in Sub-Saharan Africa: a Critical Survey*, Department of Agricultural Economics, Michigan State University, East Lansing, Mich. 48824, (MSU International Development Paper No. 1).

El Gabaly, M. M. (1979) 'Secondary salinization and sodication in Egypt: a case study', *Water Supply and Management*, **3** (3), 179–99.

El Hinnawi, E. and Manzur-Ul-Haque Hashmi (eds) (1982) *Global Environmental Issues*, Tycooly International Publishing, Dublin.

El Kassas, M. (1979) Barren answers, *Development Forum*, VII (3), 6.

Ellen, R. (1982) *Environment, Subsistence and System*, Cambridge University Press, Cambridge and New York.

El Moghraby, A. I. (1982) 'The Jonglei Canal – needed development or potential ecodisaster?', *Environmental Conservation*, **9** (2), 141–8.

El Moghraby, A. I. and El Sammani, M. O. (1985) 'On the environmental and socio-economic impact of the Jonglei Canal Project, Southern Sudan', *Environmental Conservation*, **12** (1), 41–8.

English, P. W. (1958) 'The origin and spread of quanats in the Old World', *Proceedings of the American Philosophical Society*, **12** (3), 175.
Epstein, E. and Norlyn, J. D. (1977) 'Seawater-based crop production: a feasibility study', *Science*, **197** (4300), 249–51.
Epstein, E., Kingsbury, R. W., Norlyn, J. D. and Rush, D. W. (1979) 'Production of food crops and other biomass by seawater culture', in Hollaender, A. (ed.) *The Biosaline Concept: an Approach to the Utilization of Underexploited Resources*, Plenum Press, New York, pp. 77–99 (Environmental Science Research, vol. 14).
Eshman, R. (1983) 'The Jonglei Canal: a ditch too big?', *Environment*, **25** (5), 15–20, 32–3.
Evenari, M. (1981) 'Twenty-five years of research on runoff desert agriculture in the Middle East', in Berkofsky, L., Faiman, D. and Gale, J. (eds), *Settling the Desert*, Gordon and Breach, New York, pp. 5–27.
Evenari, M., Nessler, U., Rogel, A. and Schenk, O. (eds) (1975) *The Management of the Experimental Farm – Wadi Mashash (Fields and Pastures in Deserts: a Low Cost Method for Agriculture in Semi Arid Lands)*, Eduard Roether Buchdruckerei und Verlag, Darmstadt.
Evenari, M., Shanan, L. and Tadmore, N. (1968) 'Runoff farming in the desert, I. Experimental layout', *Agronomy Journal*, **60** (1), 29–32.
Evenari, M., Shanan, L. and Tadmore, N. (1971) *The Negev: the Challenge of a Desert*, Harvard University Press, Cambridge, Mass.
Fahim, H. M. (1981) *Dams, People and Development: the Aswan High Dam Case*, Pergamon Press, Oxford and New York.
Falkenmark, M. (1984) 'New ecological approach to the water cycle: ticket to the future', *Ambio*, XIII (3), 152–60.
Falkenmark, M. and Lindh, G. (1976) *Water for a Starving World*, Westview Press, Boulder, Colo.
Faniran, A. (1980) 'On the definition of planning regions: the case for river basins in developing countries', *Singapore Journal of Tropical Geography*, **1**, 9–15.
FAO (1973) *Trickle Irrigation*, FAO, Rome. (Irrigation and Drainage Paper 14).
FAO (1977) *Conservation Guide 1: Guidelines for Watershed Management*, FAO, Rome.
Farid, M. A. (1977) 'Irrigation and malaria in arid lands', in Worthington, E. B. (ed.), *Arid Land Irrigation in Developing Countries: Environmental Problems and Effects*, Pergamon, Oxford, pp. 413–19.
Farmer, B. H. (1977) *Green Revolution? Technology and Change in Rice-growing Areas of Tamil Nadu and Sri Lanka*, Macmillan Press, London and New York.
Farmer, B. H. (1979) 'The green revolution in South Asian ricefields: environment and production', *Journal of Development Studies*, **15** (4), 304–19.
Farrington, T. S. (1985) *Prehistoric Intensive Agriculture in the Tropics*, (British Archaeological Reports, International Series, No. 232).
Farvar, M. T. and Milton, J. P. (eds) (1972) *The Careless Technology*, Natural History Press, New York.
Fauchon, J. (1980) 'Oasis agriculture: how pumps divide the peasantry', *Ceres*, **13** (4), 41–5.

Fernando, C. H. (1980a) 'Ricefield ecosystems: a synthesis', in Furtado, J. I. (ed.), *Tropical Ecology and Development*, International Society of Tropical Ecology, Kuala Lumpur, Malaysia, pp. 939–42 (Proceedings of the Vth International Symposium of Tropical Ecology, 16–21 April, 1979, Kuala Lumpur).

Fernando, C. H. (1980b) 'Tropical reservoir fisheries: a preliminary synthesis', in Furtado, J. I. (ed.), *Tropical Ecology and Development*, pp. 883–92. (Proceedings of the Vth International Symposium of Tropical Ecology, 16–21 April, 1979, Kuala Lumpur).

Finkel, H. J. (ed.), (1977) *Handbook of Irrigation Technology*, vol. 1, CRC Press, Boca Raton, Fl.

Finkel, H. J. (ed.) (1983) *Handbook of Irrigation Technology*, vol. 2, CRC Press, Boca Raton, Fl.

Franke, R. W. and Chasin, B. H. (1980) *Seeds of Famine: Ecological Destruction and the Development Dilemma in the West African Sahel*, Allanheld, Osmum, Montclair, NJ.

Frasier, G. (1975) 'Water harvesting for livestock, wildlife, and domestic use', in *Proceedings of the Water Harvesting Symposium* (ARS: W-22, 1975), US Water Conservation Laboratory, Science and Education Administration, US Department of Agriculture, Phoenix, Ariz., pp. 40–9 (mimeo.).

Frederick, K. D. (1975) *Water Management and Agricultural Development: a Case Study of the Cuyo Region of Argentina*, published for Resources for the Future, Johns Hopkins University Press, Baltimore.

Freeman, P. H. (1974) *The Environmental Impact of a Large Tropical Reservoir: Guidelines for Policy and Planning (Based On A Case Study of Lake Volta, Ghana, 1973 and 1974*, Office of International and Environmental Programmes, Smithsonian Institution, Washington, DC.

Furon, R. (1963) *The Problem of Water: a World Study*, Faber and Faber, London.

Furtado, J. I. (ed.) (1980) *Tropical Ecology and Development*, International Society of Tropical Ecology, Kuala Lumpur, Malaysia. (Proceedings of the Vth International Symposium of Tropical Ecology, 16–21 April, 1979, Kuala Lumpur).

Gaitskell, A. (1959) *Gezira: a Story of Development in the Sudan*, Faber and Faber, London.

Gale, J. (1981a) 'Controlled environment agriculture for hot desert regions', in Grace, J., Ford, E. D. and Jarvis, P. G. (eds), *Plants and their Atmospheric Environment*, Blackwell Scientific Publications, Oxford and Boston, pp. 391–402.

Gale, J. (1981b) 'High yields and low water requirements in closed system agriculture in arid regions: potentials and problems', in Berkofsky, L., Faiman, D. and Gale, J. (eds), *Settling the Desert*, Gordon and Breach, New York, pp. 81–91.

Gangstad, E. O. (1978) *Weed Control Methods for River Basin Management*, CRC Press, Palm Beach, Fl.

Garlick, J. P. and Keay, R. W. J. (eds) (1970) *Human Ecology in the Tropics*, Pergamon Press, Oxford and New York. (Symposium of the Society for the Study of Human Biology, vol. IX).

Garzon, C. E. (1984) 'Water quality in hydroelectric projects: considerations for planning in tropical forest regions', *World Bank Technical Paper* No. 20, The World Bank, Washington, DC.
Gaudet, J.J. (1979) 'Aquatic weeds in African man-made lakes', *PANS*, **25** (3), 279–86.
Geertz, C. (1963) *Agricultural Involution: the Process of Ecological Change in Indonesia*, University of California Press, Berkeley, Calif.
George, S. (1977) *How the Other Half Dies: the Real Reasons for World Hunger* (rev. edn), Penguin Books, Harmondsworth.
German Foundation for International Development (1977) *Appropriate Technologies for Semiarid Areas: Wind and Solar Water Supply*, Centre for Economic and Social Development, W. Berlin, (Seminar: Centre for Economic and Social Development, W. Berlin, 15–20 Sept. 1975).
Gindel, I. (1965) 'Irrigation of plants with atmospheric water within the desert', *Nature*, **207** (5002), 1173–5.
Gischler, C. E. (1979) *Water Resources in the Arab Middle East and North Africa*, Middle East And North African Studies Press, Cambridge.
Gischler, C. E. (1982) 'Helpful clouds', *Development Forum*, **10** (2), 7.
Gischler, C. and Jauregui, C. F. (1984) 'Low-cost techniques for water conservation and management in Latin America', *Nature and Resources*, XX (3), 11–18.
Glantz, M. H. (ed.) (1977) *Desertification. Environmental Degradation in and Around Arid Lands*, Westview Press, Boulder, Colo. (Westview Special Studies In Natural Resources and Energy Management).
Glantz, M. H. and Katz, R. W. (1985) 'Drought as a constraint in sub-Saharan Africa', *Ambio*, XIV (6), 334–9.
Glaeser, B. (ed.) (1984) *Ecodevelopment, Concepts, Projects, Strategies*, Pergamon Press, New York.
Godana, B. A. (1985) *Africa's Shared Water Resources: Legal and International Aspects of the Nile, Niger and Senegal River Systems*, Francis Pinter, London, and Lynne Rienner, Boulder, Colo.
Goldman, C. R. (1979) 'Ecological aspects of water impoundment in the tropics', *Unasylva*, **31** (123), 2–11.
Goldsmith, E., and Hildyard, N. (eds.) (1984) *The Social and Environmental Effects of Large Dams, vol. I: Overview*, Wadebridge Ecological Centre, Camelford, PL32 977, UK.
Goldsmith, E., and Hildyard, N. (eds.) (1986) *The Social and Environmental Effects of Large Dams*, vol. III: *Annotated Bibliography*, Wadebridge Ecological Centre, Camelford PL32 977, UK. (vol. II: *Case Studies* was published in 1986).
Golley, F. B. and Medina, E. (eds.) (1975) *Tropical Ecological Systems: Trends in Terrestrial and Aquatic Research*, Springer-Verlag, Berlin and New York. (Ecological Studies and Analysis Series, vol. 11).
Golubev, G. and Biswas, A. K. (1978) *Interregional Water Transfer*, Pergamon Press, Oxford and New York.
Gonzales, N. L. (ed.) (1978) *Social and Technological Management in Dry Lands: Past and Present, Indigenous and Improved*, Westview Press, Boulder, Colo. (American Academy for the Advancement of Science, Selected Symposium No. 10).

Goodland, R. G. and Irwin, H. S. (1975) *Amazon Jungle: Green Hell to Red Desert?*, Elsevier, Amsterdam and New York.
Goudie, A. S. and Wilkinson, J. (1977) *The Warm Desert Environment*, Cambridge University Press, Cambridge.
Goudie, A. S. (1981) *The Human Impact: Man's Role in Environmental Change*, Basil Blackwell, Oxford.
Gourou, P. (1980) *The Tropical World: its Social and Economic Conditions and Future Status (5th edn)*, Longman, London and New York.
Grandstaff, T. B. (1978) 'The development of swidden (shifting agriculture)', *Development and Change*, **9** (4), 547–79.
Grandstaff, T. B. (1981) 'Shifting cultivation: a reassessment of strategies', *Ceres*, **14** (4), 28–30.
Grainger, A. (1982) *Desertification: How People Make Deserts, How People Can Stop them and Why they Don't*, International Institute for Environment and Development, London and 1319F Street, NW Washington DC. (Earthscan paperback).
Gribbin, J. (1982) 'The other face of development, *New Scientist*, **96** (1333), 489–95.
Griffin, K. (1979) *The Political Economy of Agrarian Change: an Essay on the Green Revolution (2nd edn)*, Macmillan Press, London and New York.
Grigg, D. (1970) *The Harsh Lands: a Study in Agricultural Development*, Macmillan Press, London and New York.
Grist, D. H. (1959) *Rice*, Longman Green, London.
Gunnerson, C. G. and Kalbermatten, J. M. (eds.) (1978) *Environmental Impacts of International Civil Engineering Projects and Practices* (Proceedings of a Session sponsored by the Research Council on Environmental Impact Analysis of the ASCE Technical Council on Research of the ASCE Convention, San Francisco, 17–21 October, 1977), Published by the American Society of Civil Engineers, New York.
Gupta, I. C. (1979) *Use of Saline Water in Agriculture in Arid and Semi-arid Zones of India*, Oxford and IBH Publishing, New Delhi.
Hagin, J. and Tucker, B. (1982) *Fertilization of Dryland and Irrigated Soils*, Springer-Verlag, Berlin.
Hall, A. E., Cannel, G. H. and Lawton, H. W. (eds) (1979) *Agriculture in Semi-arid Environments*, Springer-Verlag, Berlin (Ecological Studies: Analysis and Synthesis, vol. 34).
Hall, A. L. (1978) *Drought and Irrigation in North-East Brazil*, Cambridge University Press, Cambridge and New York.
Hall, D. O. (1981) 'Oil-yielding plants: food or fuel', *United Nations University Newsletter*, 5 (3), 3.
Hammerton, D. (1972) The Nile River – a case history, in Oglesby, R. T., Carlson, C. A. and McCann, J. A. (eds), *River Ecology and Man*, Academic Press, New York, pp.171–214.
Harris, D. R. (ed.) (1980) *Human Ecology in Savanna Environments*, Academic Press, London and New York.
Harrison, P. (1979) 'The curse of the tropics', *New Scientist*, **84** (1182), 602–4.
Harrison, P. (1983) 'Earthwatch: land and people, the growing pressure', *People*, No. 13, 1–8.

Harrison, P. (1984) 'Population, climate and future food supply', *Ambio*, XIII (3), 161–7.

Harriss, J. (ed.) (1982) *Rural Development: Theories of Peasant Economy and Agrarian Change*, Hutchinson, London.

Hart, D. (1980) *The Volta River Project. A Case Study in Politics and Technology*, Edinburgh University Press, Edinburgh.

Harwood, R. (1979) *Small Farm Development: Understanding and Improving Farming Systems in the Humid Tropics*, Westview Press, Boulder, Colo.

Hay, J. (1972) Salt cedar and salinity on the upper Rio Grande, in Farvar, M. T. and Milton, J. P. (eds), *The Careless Technology*, Natural History Press, New York, pp. 288–230.

Hayton, R. D. (1983) 'The law of international resource systems', in Zaman, M., Biswas A. K., Khan, A. H. and Nishat, A. (eds), *River Basin Development*, (Proceedings of the National Symposium on River Basin Development, 4–10 Dec. 1981, Dacca, Bangladesh), Tycooly International Publishing, Dublin, pp. 195–211.

Hazlewood, A. and Livingstone, I. (1982) *Irrigation Economics in Poor Countries: Illustrated by the Usangu Plains of Tanzania*, Pergamon Press, Oxford and New York.

Heathcote, R. L. (1983) *The Arid Lands: their Use and Abuse*, Longman, London and New York.

Herren, G. B., Hansen, B. K. and Wandesforde-Smith, G. (1982) *Environmental Impact Assessment in the Tropics: Guidelines for Application to River Basin Development*, Centre for Environmental and Energy Policy Research, University of California, Davis, Calif.

Higgins, G. M., Kassam, A. H. and Shah, M. (1984) 'Land, food and population in the developing world', *Nature and Resources*, XX (3), 2–10.

Higgins, G. M., Kassam, A. H., Naiken, L. and Shah, M. M. (1981) 'Africa's agricultural potential', *Ceres*, **14** (5), 13–21.

Hildrew, P. (1984) 'New way of life blooms in Egypt's fertile desert', *The Guardian*, 30 Jan., p. 7.

Hill, R. D. (ed.) (1979) *South-East Asia: a Systematic Geography*, Oxford University Press (Asia), Kuala Lumpur, Malaysia.

HMSO (1956) *Volta River Project, I: Report of the Preparatory Commission, Government of the UK and Gold Coast*, HMSO, London.

Hoddy, E. (1983) 'Weed of great worth', *Development Forum*, XI (2), 4.

Hodges, F. O. C. (1977) 'Water legislation and institutions – their role in water resources development and management', *Water International*, II (3), 3–9, 31.

Hoffman, G. J., Ayres, R. S., Doering, E. J. and McNeal, B. L. (1980) 'Salinity in irrigated agriculture', in Jensen, M. E. (ed.), *Design and Operation of Farm Irrigation Systems*, St Joseph, Mich., pp. 145–85, (ASAE Monograph, No. 3, American Society Of Agricultural Engineers).

Hoffman, G. J. (1981) 'Irrigation management and salinity control', in ASAE, *Irrigation: Challenges of the 80's*, American Society of Agricultural Engineers, PO 410, St Joseph, Mich., pp. 166–74, (Proceedings of the Second National Symposium, 20–23 Oct. 1980, Lincoln, Nebr., ASAE Publication No. 6–81).

Hollaender, A. (ed.) (1979) *The Biosaline Concept: an Approach to the Utilization of Underexploited Resources*, Plenum Press, New York, (Environmental Science Research, vol. 14).
Holmes, J. W. and Talsma, T. (eds) (1981) *Land and Stream Salinity*, Elsevier, Amsterdam and New York. (Developments In Agricultural Engineering, vol. 2).
Howe, C. W. (1982) 'Socially efficient development and allocation of water in developing countries: rates for the public and private sectors', in Howe, C. W. (ed.), *Managing Renewable Natural Resources in Developing Countries*, Westview Press, Boulder, Colo., pp. 95–129.
Howe, C. W. and Easter, K. W. (1971) *Interbasin Transfers of Water: Economic Issues and Impacts*, Johns Hopkins Press, Baltimore and London.
Howell, T. A., Bucks, D. A. and Chesness, J. L. (1981) 'Advances in trickle irrigation', in ASAE, *Irrigation Challenges of the 80s*, American Society of Agricultural Engineers, PO 410, St Joseph, Mich. pp. 69–94. (Proceedings of the Second National Symposium, 20–23 Oct., 1980, Lincoln, Nebr., ASAE Publication No. 6–81).
Hubbard, A. J. and Hubbard, G. (1907) *Dew Ponds and Cattle Ways* (2nd edn), Longman, London.
Hunt, R. C. and Hunt, E. (1976) 'Canal irrigation and local social organization', *Current Anthropology*, **17** (3), 389–411.
Hunter, J. M. and Kwakuntiri, G. (1978) 'Speculations on the future of shifting agriculture in Africa', *Journal of Developing Areas*, **12** (2), 183–201.
Hunter, O. and Munroe, P. A. (eds) (1978) *Hydropower and the Environment*, National Science Research Council, 44 Pere St., Kitty, Greater Georgetown, Guyana. (Proceedings of the International Seminar on Hydropower and the Environment, 4–8 Oct., 1976, Georgetown, Guyana).
Huntington, E. (1915) *Civilization and Climate*, Yale University Press, New Haven, Conn.
Hurst, C. (1984) 'A model of an Indian village: a study of alternative sources of energy for irrigation', *World Development*, **12** (2), 141–56.
Husseiny, A. A. (ed.) (1978) *Iceberg Utilization*, Pergamon Press, New York, Oxford.
Hutchinson, C. F., Dutt, G. R. and Anaya Gardund, M. (eds) (1981) *Rainfall Collection for Agriculture in Arid and Semi-arid Regions*, Commonwealth Agricultural Bureaux, Slough.
Independent Commission on International Development Issues (1980) *North – South: a Programme for Survival (the Brandt Report)*, Pan Books, London.
Interim Mekong Committee. (1982) *Environmental Impact Assessment: Guidelines for Application to Tropical River Basin Development*, Mekong Secretariat, ESCAP-UN Building, Bangkok, 10200, Thailand.
IRRI (1984) *Workshop on Research Priorities in Tidal Swamp Rice*, International Rice Research Institute, Los Banōs, Philippines.
Israelsen, O. W. and Hansen, V. E. (1962) *Irrigation Principles And Practices* (3rd edn), Wiley, New York.

IUCN, UNEP and WWF (1980) *World Conservation Strategy*, International Union for the Conservation of Nature and Natural Resources, 1196 Gland, Switzerland.

IUCN (1981) 'Dams: the African experience', *IUCN Bulletin*, **12** (7, 8, 9), 51–4.

Jackson, I. J. (1977) *Climate, Water and Agriculture in the Tropics*, Longman, London and New York.

Janzen, D. H. (1973) Tropical agro-ecosystems, *Science*, **182** (4118), 1212–19.

Jayaraman, T. K. (1982) 'Malaria impacts of surface irrigation projects: a case study from Gujarat, India', *Agriculture and Environment*, **7** (1), 23–34.

Jennes, J. (1969) Reservoir Resettlement in Africa, FAO, Rome.

Jensen, M. E. (ed.) (1980) *Design and Operation of Farm Irrigation Systems*, American Society of Agricultural Engineers, St Joseph, Mich. (ASAE Monograph No. 3).

Joblin, W. R. (1978) 'Tropical disease bilharzia and irrigation systems in Puerto Rico', *Journal of Irrigation and Drainage Division ASCE*, 104 (IR3), 307–22.

Johl, S. S. (ed.) (1980) *Irrigation and Agricultural Development*, Pergamon Press, Oxford and New York.

Johnson, B. (1979) 'Dammed alternatives', *Development Forum*, VII (7), 7.

Johnson, D. L. (1969) 'The nature of nomadism: comparative study of pastoral migrations in South Western Asia and Northern Africa', *University of Chicago Department of Geography Research Papers*, No. 118, Chicago.

Junk, W. J. (1980) 'Aquatic macrophytes: ecology and use in Amazonian agriculture', in Furtado, J. I. (ed.), *Tropical Ecology and Development*, The International Society of Tropical Ecology, Kuala Lumpur, Malaysia, pp. 763–70 (Proceedings of the Vth International Symposium of Tropical Ecology, 16–21 April, 1979, Kuala Lumpur).

Jurriens, M. and Bos, M. G. (1980) 'Developments in planning of irrigation projects', in International Institute for Land Reclamation and Improvement, *Land Reclamation and Water Management: Developments, Problems And Challenges*, IILRI (PO Box 45, 6700 AA), Wageningen, the Netherlands, pp. 99–112.

Kaduma, J. D. (1982) 'Water as a constraint on agricultural development in the semi-arid areas of Tanzania', *Water Supply and Management*, **6** (6), 417–30.

Kamark, A. M. (1976) *The Tropics and Economic Development*, Johns Hopkins University Press, Baltimore and London.

Kamiar, M. (1983) 'The quanat system: a case study from Iran', *Habitat News*, 5 (1), 40–1.

Karala, K. (1982) 'The pump that changed millions of lives', *The Crown Agents Quarterly Review*, Summer, pp. 15–18.

Kates, R. W. (1981) 'Drought in the Sahel: competing views as to what really happened in 1910–14 and 1968–74, *Mazingira*, 5 (2), 72–83.

Kayastha, S. L. (1981) 'An appraisal of water resources of India and need for National Water Policy', *Geojournal*, 5. **6**, 563–72.

Kazarian, R. (1981) 'Plant scientists get closer to developing buffalo gourd

as a commercial food source', *Environmental Conservation*, **8** (1), 66.
Kellogg, W. W. (1978) 'Global influences of mankind on the climate', in Gribbin, J. (ed.), *Climatic Change*, Cambridge University Press, London and New York, pp. 205–27.
Kershaw, K. A. (1973) *Quantitative and Dynamic Plant Ecology* (2nd edn), Edward Arnold, London.
Khan, I. (1982) 'Managing salinity in irrigated agriculture', *Journal of the Irrigation and Drainage Division ASCE*, **108(IRI)**, 43–56.
Klee, G. A. (ed.) (1980) *World Systems of Traditional Resource Management*, Edward Arnold, London.
Korten, F. F. (1982) *Building National Capacity to Develop Water Users' Associations – Experience from the Philippines*, World Bank, Washington DC. (World Bank Staff Working Papers, No. 528).
Kovda, V. A. (1980) *Land Aridization and Drought Control*, Westview Press, Boulder, Colo. (Westview Special Studies in Natural Resources and Energy Management).
Kovda, V. A. (1983) 'Loss of productive land due to salinization', *Ambio*, XII (2), 91–3.
Kovda, V. A., Van Den Berg, C. and Hagan, R. M. (eds) (1973) *Irrigation, Drainage and Salinity: an International Sourcebook*, published for FAO/UNESCO by Hutchinson, London.
Kowal, J. M. and Kassam, A. H. (1978) *Agricultural Ecology of Savanna*, Clarendon Press, Oxford.
Kramer, P. J. (1969) *Plant and Soil Water Relations: a Modern Synthesis*, McGraw-Hill, New York.
Laconte, P. and Haimes, Y. Y. (eds) (1982) *Water Resources and Land-use Planning: a Systems Approach*, Martinus Nijhoff, The Hague, and Kluwer Boston, Hingham, Massachusetts.
Lagler, K. F. (ed.) (1969) *Man-made Lakes: Planning and Development*, UNDP/FAO, Rome.
Lal, D. (1972) *Wells and Welfare: Exploratory Cost–Benefit Study of the Economics of Small-scale Irrigation for Maharashtra*, OECD, Paris (Development Centre Studies Series on Cost-Benefit Analysis, Case Study No. 1).
Lal, R. and Russell, E. W. (eds) (1981) *Tropical Agricultural Hydrology: Watershed Management and Landuse*, Wiley, Chichester and New York.
Lamb, H. H. (1982) *Climate, History and the Modern World*, Methuen, London and New York.
Lambert, J. D. H. *et al.* (1984) 'Ancient Maya drained field agriculture and its possible application today in the New River Floodplain, Belize', *Agricultural Ecosystems and Environment*, **11,** 67–84.
Land, T. (1980) 'Profits from a beautiful pest', *Development Forum*, VIII (1), 10.
Lanly, J. P. (1985) 'Defining and measuring shifting cultivation', *Unasylva*, **37** (147), 17–21.
Lawson, R. (1968) 'The Volta Resettlement Scheme', *African Affairs*, **67** (267), 124–9.
Lawton, H. W. and Wilkie, P. J. (1979) 'Ancient agricultural systems in dry regions', in Hall, A. E., Cannel, G. H. and Lawton, H. W. (eds),

Agriculture in Semi-arid Environments, Springer-Verlag, Berlin, (Ecological Studies: Analysis and Synthesis, vol. 34).

Lees, S. H. (1978) 'Farmers and technical experts: information flow in irrigated agriculture', in Gonzalez, N. L. (ed.), *Social and Technological Management in Dry Lands: Past and Present Indigenous and Imposed*, Westview Press, Boulder, Colo., pp. 45–57. (American Association for the Advancement of Science Selected Symposium No. 10).

Leng, G. (1982) *Desertification: a Bibliography with Emphasis on Africa*, Universitat Bremen, Presse-und Informationsamt, Druckschriftenlager, Postfach 330 440, Bremen.

Levine, G., Chin, L. T. and Miranda, S. M. (1976) 'Requirements for the successful introduction and management of rotational irrigation', *Agricultural Water Management*, **1** (1) 41–56.

Lewis, J. (1984) '*Baringo Pilot Semi-Arid Area Project*' (summary of Interim Report), Republic of Kenya, BPSAAP, PO Marigat, via Nakuru, Kenya (mimeo.), 45 pp.

Lightfoot, R. P. (1978) 'The cost of resettling reservoir evacuees in Northeast Thailand', *Journal of Tropical Geography*, **47**, 63–74.

Lima, R. R. (1956) 'A agricultura nos várzeas do etuario do Amazonas', *Bol. Tech. do Instituto Agronomico do Norte, Belém*, No. 33, EMBRAPA, Belém (Pará, Brazil).

Lipton, M. (1977) *Why Poor People Stay Poor: a Study of Urban Bias in World Development*, Temple Smith, London.

Liu Changming and Dakang, K. (1983) 'Shifting China's waters north: an environmental challenge', *United Nations University Newsletter*, **7** (2), 2.

Logenbough, R. A. (1980) 'Farm pumps', in Jensen, M. E. (ed.), *Design and Operation of Farm Irrigation Systems*, American Society of Agricultural Engineers, St Joseph, Mich., pp. 347–91. (ASAE Monograph No. 3).

Lowdermilk, M. K. (1985) 'Improved irrigation management: why involve farmers?', *ODI Irrigation Management Network Paper 11c*, 12 pp.

Lowe-McConnel, R. H. (ed.) (1966) *Man-made Lakes*, 1965, London), Academic Press, London, (Proceedings of the Royal Geographical Society Symposium, 30th Sept.–1st Oct., 1965).

Lugo, A. E. and Brown, S. (1981) 'Tropical lands: popular misconceptions', *Mazingira*, 5 (2), 10–19.

Lumsden, D. P. (1973) 'The Volta River Project: village resettlement and attempted rural animation', *Canadian Journal of African Studies*, VII (1), 115–32.

Mabogunje, A. L. (1980) *The Development Process: a Spatial Perspective*, Hutchinson, London.

McAndrews, C. and Chia Lin Sien (eds) (1979) *Developing Economies and the Environmental: the Southeast Asian Experience*, McGraw-Hill International, Singapore.

McCown, R. L., Haaland, G. and DeHaan, C. (1979) 'The interaction between cultivation and livestock production in semi-arid Africa', in Hall, A. E., Cannel, G. H. and Lawton, H. W. (eds), *Agriculture in Semi-arid Environments*, Springer-Verlag, Berlin, pp. 297–332 (Ecological Studies; Analysis and Synthesis vol. 34).

Madeley, J. (1982) 'Fighting the desert profitably: Jojoba's potential', *Mazingira*, **6** (2), 50–7.

Madeley, J. (1983) 'Big dam schemes – value for money or non-sustainable development?', *Mazingira*, 7 (4), 16–25.

Maltby, E. and Turner, R. E. (1983) 'Wetlands of the world', *Geographical Magazine*, 55 (1), 12–17.

Manassah, J. T. and Briskey, E. J. (eds) (1981) *Advances in Food-producing Systems for Arid and Semiarid Lands*, Part A, Academic Press, London and New York, (Proceedings of the Symposium on Advances in Food Producing Systems for Arid and Semiarid Lands, Kuwait, Feb., 1980).

Manshard, W. (1974) *Tropical Agriculture: Geographical Introduction and Appraisal*, (2nd edn), Longman, London (trans. from German).

Mather, T. H. (1984) Water in the service of man, in Institution of Civil Engineers, *World Water '83: the World Problem*, Thomas Telford, London, pp. 181–8. (Proceedings of the Institution of Civil Engineers, Conference, 12–15 July, 1983, London).

Matlock, W. G. (1981) *Realistic Planning for Arid Lands: Natural Resource Limitations to Agricultural Development*, Harwood Academic, London and New York, (Advances in Arid Land Technology and Development, vol. 2).

Medina, J. (1976) 'Harvesting surface runoff and ephemeral streamflow in arid zones', in FAO, *Conservation Guide 3*, FAO, Rome, pp. 61–72.

Merrey, D. J. (1982) 'Reorganizing irrigation: local level management in the Punjab (Pakistan)', in Spooner, B. and Mann, H. S. (eds), *Desertification and Development: Dryland Ecology in Social Perspective*, Academic Press, London and New York, pp. 83–109.

Millikan, M. F. and Hapgood, D. (1967) *No Easy Harvest: the Dilemma of Agriculture in Underdeveloped Countries*, Little, Brown, Boston, Mass.

Mondal, R. C. (1974) 'Pitcher farming', *Appropriate Technology*, **1** (3), 7.

Morales, C. (1977) 'Rainfall variation a natural phenomenon', *Ambio*, VI (1), 30–3.

Monteith, J. L. (1963) 'Dew facts and fallacies', in Rutter, A. J. and Whitehead, F. H. (eds.) *The Water Relations of Plants*, Blackwell Scientific Publications, London, pp. 36–56. (British Ecological Society Symposium No. 3, 5–8 April, 1961).

Montgomery, E., Bennett, J. W. and Scudder, T. (1977) 'The impact of human activities on the physical and social environment: new directives in anthropological ecology', in Deutsch, K. W. (ed.), *Ecosocial Systems and Ecopolitics: a Reader in Human and Social Implications of Environmental Management in Developing Countries*, UNESCO, Paris, pp. 77–144.

Moyes, A. (1979) *The Poor Man's Wisdom*, Oxfam, Oxford.

Munn, R. E. (ed.) (1979) *Environmental Impact Assessment: Principles and Practices, (SCOPE 5)* (2nd edn), Wiley, Chichester and New York.

Munn, R. E. (1982) 'Environmental impact assessment: a useful development tool?', *Mazingira*, **6** (2), 66–73.

Myers, L. E., Frasier, G. W. and Griggs, J. R. (1967) 'Sprayed asphalt pavements for water harvesting', *Journal of the Irrigation and Drainage Division ASCE*, **93** (IR3), 79–97.

Myers, N. (1983) 'Where the next meal is coming from', *The Guardian*, 3 Jan., p. 19.
Myrdal, G. (1968) *Asian Drama: an Enquiry into the Poverty of Nations*, Penguin Books, Harmondsworth. (A Twentieth Century Fund Study).
Myrdal, G. (1970) *An Approach to the Asian Drama*, Pantheon, New York.
Nabhan, G. P. (1986) 'Papago Indian desert agriculture and water control in the Sonaran Desert 1976–1934', *Applied Geography*, **6** (1), 43–59.
National Academy of Sciences (1974) *More Water for Arid Lands: Promising Technologies and Research Opportunities*, National Academy of Sciences, Washington, DC. (Report of a Panel of the Advisory Committee on Technological Innovation, Board of Science and Technology for International Development Commission on International Relations).
National Academy of Sciences (1975) *Underexploited Tropical Plants with Promising Economic Value*, National Academy of Sciences, Washington, DC. (Report of the *ad hoc* Panel of the Advisory Committee on Technology Innovation).
Nir, D. (1974) *The Semi-arid World: Man on the Fringe of the Desert*, (trans. from Hebrew), Longman, London.
Nye, P. H. and Greenland, D. J. (1960) *The Soil under Shifting Cultivation*, Commonwealth Agricultural Bureaux, Harpenden.
OAS (1978) *Environmental Quality and River Basin Development: a Model for Integrated Analysis and Planning*, Organization of American States, Washington, DC. (Govt. Argentina/OAS/UNEP).
Obeng, L. E. (1977) 'Should dams be built?: the Volta Lake experience', *Ambio*, VI (1), 46–50.
Obeng, L. E. (1978) 'Starvation or bilharzia? – a rural development dilemma, *Water Supply and Management*, 2 (4), 343–50.
ODI (1980) *New Irrigation Schemes: Planning, Design and Social Impact*, Overseas Development Institute, London. (Irrigation Management Network Paper 1/80/2).
OECD (1982) *Development Co-operation Efforts and Policies of the Members of the Development Assistance Committee*, Organization for Economic Co-operation and Development, Paris, (A Report by the Committee Chairman: Poates, R. M.).
Oliver, H. (1972) *Irrigation and Water Resources Engineering*, Edward Arnold, London.
Ooi Jin Bee (ed.) (1983) *Natural Resources in Tropical Countries*, Singapore University Press.
Oregon State University (1979) *Dryland Agriculture in Winter Precipitation Regions of the World*, Dryland Agriculture Technical Committee, Oregon State University, Oreg., USA.
Oxby, C. and Bottrall, A. (undated) 'The role of farmers in decision making on irrigation systems', Overseas Development Institute, London (mimeo.), 28 pp.
Pacey, A., *et al*. (1977) 'Technology is not enough: the provision and maintenance of appropriate water supplies', *Aqua*, **1** (1), 1–58. (Paper jointly produced by an 'editorial group' of the Intermediate Technology Development Group, comprised of many individuals).

Pacey, A. and Cullis, A. (1986) *Rainwater Harvesting: the Collection of Rainfall and Run-off in Rural Areas*, Intermediate Technology Publications, 9 King Street, London WC2E 8HW.

Park, C. C. (1981) 'Man, river systems and environmental impacts', *Progress in Physical Geography*, 5 (1), 1–31.

Pasternak, B. (1968) 'Social consequences of equalizing irrigation excess', *Human Organization*, **27** (4), 332–42.

Pearce, F. (1984a) 'Africa's drought revisited', *New Scientist*, **103** (1419), 10–11.

Pearce, F. (1984b) 'Why Gaddafi's wells may run dry', *New Scientist*, **103** (1420), 40.

Pearse, A. (1980) *Seeds of Plenty, Seeds of Want, Social and Economic Implications of the Green Revolution*, Clarendon Press, Oxford and New York.

Pereira, H. C. (1977) *Land Use and Water Resources; in Temperate and Tropical Climates*, Cambridge University Press, London and New York.

Petr, T. (1978) Tropical man-made lakes – the ecological impact, *Archive für Hydrobiologie*, **81** (3), 368–85.

Petr, T. (1980) 'Purari River hydroelectric development and its ecological impact – an attempt at prognosis', in Furtado, J. I. (ed.), *Tropical Ecology and Development*, The International Society of Tropical Ecology, Kuala Lumpur, Malaysia, pp. 871–81. (Proceedings of the Vth International Symposium of Tropical Ecology, 16–21 April, 1979, Kuala Lumpur).

Petrick, C. (1978) 'The complimentary function of floodlands for agriculture utilization: the várzeas of the Brazilian Amazon region', *Applied Geography and Development*, **12**, 24–46.

Petts, G. E. (1984) *Impounded Rivers*, Wiley, Chichester and New York.

Platenkamp, J. D. M. (1978) *The Jonglei Canal, its Impact on an Integrated System in the Southern Sudan*, Institute for Cultural Anthropology and Sociology of non-Western Peoples, University of Leiden, (Publication No. 26).

Porter, E. (1978) *Water Management in England and Wales* (2nd edn), Cambridge University Press.

Postel, S. (1985a) *Conserving Water: the Untapped Alternative*, Worldwatch Institute, NW Washington DC, 20036. (Worldwatch Paper No. 67).

Postel, S. (1985b) When rain only will do, *Development Forum*, **XII** (9), 1, 4.

Postel, S. (1985c) 'The potential for saving water: good management yields significant returns, *Ceres*, No. 106 (**18**, No. 4), 21-23.

Rahman, M. (1981) 'Ecology of karez irrigation: a case of Pakistan', *Geojournal*, **5.1**, 7–15.

Rambo, A. T. (1982) 'Human ecology research in tropical agroecosystems in Southeast Asia', *Singapore Journal of Tropical Geography*, **3**, 86–99.

Rao, V. M. (1978) 'Linking irrigation with development', *Economic and Political Weekly*, XIII (24), 993–7.

Redclift, M. (1984) *Development and the Environmental Crisis: Red or Green Alternatives?*, Methuen, London and New York.

Replogle, J. A. and Merriam, J. L. (1981) 'Scheduling and management of irrigation water delivery systems', in ASAE, *Irrigation Challenges of the 80's*, American Society of Agricultural Engineers P.O. 410, St Joseph, Mich., pp. 112–140, (Proceedings of the Second National Symposium, 20–23 Oct., 1980, Lincoln, Nebr., ASAE Publication No. 6–81).

Reuss, J. O., Skogerboe, G. V. and Merry, D. J. (1979) 'Watercourse improvement strategies for Pakistan', *Water Supply and Management*, **4** (5/6), 409–22.

Richards, P. W. (1964) *The Tropical Rain Forest: an Ecological Study* (2nd edn), Cambridge University Press.

Richards, P. (1985) *Indigenous Agricultural Revolution: Ecology and Food Production in West Africa*, Hutchinson, London.

Riddell, R. (1981) *Ecodevelopment: Economics, Ecology and Development: an Alternative to Growth Imperative Models*, Gower Publishing, Farnborough.

Ritchie, G. A. (ed.) (1979) *New Agricultural Crops*, Westview Press, Boulder, Colo. (American Association for the Advancement of Science – Selected Symposium No. 38).

Rivnay, E. (1972) 'On irrigation-induced changes in insect populations in Israel', in Farvar, M. T. and Milton, J. P. (eds), *The Careless Technology*, Natural History Press, New York, pp. 349–64.

Roberts, N. (1986) 'Dambos in development: management of a fragile ecological resource', Annual Congress of the Institute of British Geographers, University of Reading, 3–6 Jan. (mimeo.).

Rodda, J. C. (ed.) (1976) *Facets of Hydrology*, Wiley, London and New York.

Rothe, J. P. (1968) 'Fill a lake, start an earthquake', *New Scientist*, **39** (605), 75–8.

Rubeli, K. (1976) 'The Tembling Hydroelectric Project from the Taman Negara viewpoint', *Malayan Nature Journal*, **29** (4), 307–13.

Ruddle, K. and Manshard, W. (1981) *Renewable Natural Resources and the Environment: Pressing Problems in the Developing World*, Tycooly International Publishing, Dublin.

Ruthenberg, H. (1980) *Farming Systems in the Tropics* (3rd edn), Clarendon Press, Oxford.

Sachs, I. (1976) 'Environment and styles of development', in Mattews, W. H. (ed.), *Outer Limits and Human Needs: Resource and Environmental Issues of Development Strategies*, Dag Hammerskjold Foundation, Uppsala, pp. 41–61.

Saha, S. K. and Barrow, C. J. (eds) (1981) *River Basin Planning: Theory and Practice*, Wiley, Chichester and New York.

Sandford, S. (1983) *Management of Pastoral Development in the Third World*, Wiley, Chichester and New York.

San Pietro, A. (ed.) (1982) *Biosaline Research: a Look to the Future*, Plenum Press, New York. (Environmental Science Research, vol. 23, Proceedings of the Second International Workshop on Biosaline Research, La Paz, Mexico, 16–20 Nov., 1980).

Saunders, N. (1986) 'Empires ruled by a ring of fire', *New Scientist*, **109** (1498), 55–7.

Schecter, J. (1977) 'Negev: something new under the sun', *Ceres*, **10** (3), 37–40.

Schultz, T. W. (1964) *Transforming Traditional Agriculture*, Yale University Press, Newhaven, Conn.
Schulze, F. E and Van Staveren, J. M. (1980) 'Land and water development in the Third World', in International Institute for Land Reclamation and Improvement, *Land Reclamation and Water Management: Developments, Problems and Challenges*, IILRI (PO Box 45, 6700 AA), Wageningen, the Netherlands, pp. 13–28.
Schumacher, E. F. (1974) *Small is Beautiful: a Study of Economics as if People Mattered*, Sphere Books, London.
SCOPE (1972) *Report 2: Working Group on Man-made Lakes as Modified Ecosystems*, International Council for Scientific Unions, Paris.
Scudder, T. (1973) 'The human ecology of big projects: river basin development and resettlement', *Annual Review of Anthropology*, **2**, 45–61.
Scudder, T. (1980) 'River basin development and local initiative in African savanna environments', in Harris, D. R. (ed.), *Human Ecology in Savanna Environments*, Academic Press, London and New York, pp. 383–405.
Seneviratne, G. (1979) 'Salvation for salty soils', *Development Forum*, VII (2), 5.
Shaffer, G. (1980) 'Ensuring man's food supplies by developing new land and preserving cultivated land', *Applied Geography and Development*, **16**, 7–27.
Shane, J. N. (1979) 'Environmental law in developing nations of Southeast Asia', in McAndrews, C. and Chia Lin Sien (eds), *Developing Economies and the Environment: the Southeast Asian Experience*, McGraw-Hill International, Singapore, pp. 15–45.
Shainberg, I. and Oster, J. D. (1978) *Quality of Irrigation Water*, Pergamon Press, Oxford and New York, (International Irrigation Information Centre Publication No. 2).
Shanan L., Evenari, M. and Tadmore, H. (1969) 'Ancient technology and modern science applied to desert agriculture', *Endeavour*, XXVIII, 68–72.
Shanan, L. and Tadmore, H. (1979) *Microcatchment Systems for Arid Zone Development*, (2nd edn), Centre for International Agricultural Cooperation, Hebrew National University of Jerusalem, Rehovot.
Sharkov, A. A. (1956) *Salt-resistant Plants*, Acad. Nauk. SSSR, Moscow (Russian with English summary).
Sheng, T. C. (1977) 'Terracing steep slopes in humid regions', in FAO, *Conservation Guide 1: Guidelines for Watershed Management*, FAO, Rome, pp. 147–79.
Sheng, T. C. (1981) 'The need for soil conservation structures for steep cultivated slopes in the humid tropics', in Lal, R. and Russell, E. W. (eds), *Tropical Agricultural Hydrology: Watershed Management and Landuse*, Wiley, Chichester and New York, pp. 357–72.
Sigurdson, J. (1977) 'Water policies in India and China', *Ambio*, VI (1), 70–6.
Sirnanda, K. U. (1979) 'The role of applied climatological research in alleviating the hazardous impact of the monsoons in South and Southeast Asia', *Geojournal* **3.2**, 137–46.

Siy, R. Y. jun. (1982) *Community Resource Management: Lessons from the Zanjera*, University of the Philippines Press, Quezon City (distributed outside the Philippines by University of Hawaii Press).
Skogerboe, G. V., Lowdermilk, M. K., Sparling, E. W. and Hautaluoma, J. E. (1982) 'Development process for improving irrigation water management', *Water Supply and Management*, **6** (4), 329–42.
Slater, L. E. and Levin, S. K. (eds) (1981) *Climate's Impact on Food Supplies: Strategies and Technologies for Climate-defensive Food Production*, Westview Press, Boulder, Colo. (American Association for the Advancement of Science, Selected Symposium No. 62).
Smith, N. (1982) 'Triticale: the birth of a new cereal', *New Scientist*, **97** (1340), 98–9.
Spencer, J. E. (1966) *Shifting Cultivation in South East Asia*, University of California Press, Berkeley, Calif. (University of California Publication in Geography, vol. 19).
Spencer, J. E. (1974) 'Water control in terraced rice-field agriculture in Southeast Asia', in Downing, T. E. and Gibson, M. (eds), *Irrigation's Impact on Society*, University of Arizona Press, Tucson, Ariz, pp. 59–65. (Anthropological Paper, the University of Arizona No. 25).
Spooner, B. (1974) 'Irrigation and society: the Iranian Plateau', in Downing, T. E. and Gibson, M. (eds), *Irrigation's Impact on Society*, University of Arizona Press, Tucson, Ariz., pp. 43–57. (Anthropological Paper, the University of Arizona No. 25).
Spooner, B. and Mann, H. S. (eds) (1982) *Desertification and Development: Dryland Ecology in Social Perspective*, Academic Press, London and New York.
Stanley, N. F. and Alpers, P. (eds) (1975) *Man-made Lakes and Human Health*, Academic Press, London and New York.
Steeds, D. (1985) *Desertification in the Sahelian and Sudanic Zones of West Africa*, The World Bank, Washington, DC.
Stern, P. H. (1979) *Small Scale Irrigation: a Manual of Low-cost Water Technology* (Intermediate Technology Publications Ltd, and the International Irrigation Information Centre, London), Russel Press, Nottingham.
Stevens, J. (1972) 'Salt water for the hungry man', *New Scientist*, **54** (797), 447.
Stone, E. L. (1957) 'Dew as an ecological factor. A review of the literature, *Ecology*, **38** (3), 407–17.
Stone, I. (1984) *Canal Irrigation in British India: Perspectives on Technological Change in a Peasant Economy*, Cambridge University Press, Cambridge and New York.
Sudeglin, J. M. (1978) 'Economic impacts of increased water supply on small farms in Iran, *Indian Journal of Economics*, XXXIII (2), 62–9.
Sutcliffe, J. (1979) *Plants and Water* (2nd edn), Edward Arnold, London.
Swaminathan, M. S. (1984) 'Rice', *Scientific American*, **250** (1), 62–72.
Swift, J. (1977a) 'Pastoral development in Somalia: herding co-operatives as a strategy against desertification and famine', in Glantz, M. H. (ed.), *Desertification: Environmental Degradation in and Around Arid Lands*, Westview Press, Boulder, Colo., pp. 275–305.
Swift, J. (1977b) 'Sahelian pastoralists: underdevelopment, desertification

and famine', *Annual Review of Anthropology*, **6**, 457–78.
Syed Hashim Ali (1980) 'Practical experience of irrigation reform, Andhra Pradesh, India', *IDS Discussion Paper No. 153*, Institute of Development Studies, University of Sussex, Brighton, UK.
Syed Hashim Ali (1983) 'One season of integrated water management in Andhra Pradesh', *ODI Irrigation Management Network Paper 7d*, Overseas Development Institute, Percy Street, London.
Tahir, A. A. and El Sammani, M. O. (1980) 'Environmental and socio-economic impact of Jonglei Canal Project', *Water Supply and Management*, **4** (1/2), 45–51.
Taylor, G. C. jun. (1979) 'The United Nations ground water exploration and development programme – a fifteen year perspective', *Natural Resources Forum*, **3** (2), 147–66.
Tenari, S. K. (1978) 'Economics of wind energy use for irrigation in India', *Science*, **202** (3), 481–6.
Thomas, K. J. (1979) 'The extent of *Salvinia* infestation in Kerala (South India): its impact and suggested methods of control', *Environmental Conservation*, **6** (1), 63–9.
Thomi, W. (1984) 'Man-made lakes as human environments; the formation of new socio-economic structures in the region of the Volta Lake', *Applied Geography and Development*, **23**, 109–27.
Tosi, J. A. and Voertman, R. F. (1964) 'Some environmental factors in the economic development of the tropics', *Economic Geography*, **40** (3), 189–205.
Tiffen, M. (ed.) (1985) 'Costs recovery and water tariffs: a discussion', *ODI Irrigation Management Network Paper 11e*, 11 pp.
Tillman, G. (1982) *Environmentally Sound Small-scale Water Projects, Guidelines for Planning*, Coordination in Development Volunteers in Technical Assistance, 79 Madison Ave, New York.
Timberlake, L. (1985) *Africa in Crisis: the Causes, The Cures of Environmental Bankruptcy*, Earthscan Paperback, International Institute for Environment and Development, London and Washington DC.
Tinker, J. (1977) 'Ancient enemies unite against deserts', *New Scientist*, **75** (1068), 580–4.
Todaro, M. P. (1981) *Economic Development and the Third World* (2nd edn), Longman, New York and London.
Tolman, C. F. (1937) *Ground Water*, McGraw-Hill, New York.
Trelease, F. J. (1964) 'Water law' in Chow, Ven Te. (ed.), *Handbook of Applied Hydrology: a Compendium of Water-resources Technology*, McGraw-Hill, New York, pp. 27/1–27/37.
Troll, C. (1963) 'Landscape, ecology and land development with special reference to the tropics', *Journal of Tropical Geography*, **17**, 1–11.
UN (1970) *Integrated River Basin Development: Report of a Panel of Experts* (rev. edn), United Nations, New York.
UN (1975) *Management of International Water Resources: Institutional and Legal Aspects*, United Nations Department of Economic and Social Affairs, Natural Resources/Water Series No. 1, New York. (Report of the Panel of Experts on the Legal and Institutional Aspects of International Water Resources Development).
UNAPDI (1979) *The Environmental Dimensions of Development*, United

Nations Asian and Pacific Development Institute, Bangkok.
UNEP (1983) *Rain and Stormwater Harvesting in Rural Areas: A Report by the United Nations Environment Programme*, Tycooly International Publishing, Dublin, (Water Resources Series, vol. 5).
UNAPDC (1983) *Environmental Assessment of Development Projects*, UN Asian and Pacific Development Centre, Kuala Lumpur, Malaysia.
UNESCO (1962) *The Problems of the Arid Zone, Arid Zone Research*, vol. XVIII, UNESCO, Paris, (Proceedings of the Paris Symposium).
UNESCO/FAO (1973) *Irrigation, Drainage and Salinity: an International Source Book*, published for the United Nations by Hutchinson, New York.
UNESCO (1974) *Natural Resources of Humid Tropical Asia*, UNESCO, Paris, (Natural Resources Research, vol. XII).
Urbina, A. H. (1975) 'The end of the water barons: the nationalization of water as the indispensable complement of agrarian reform', *Ceres*, **8** (4), 42–4.
Utton, A. E. and Teclaff, L. (eds.) (1978) *Water in a Developing World: the Management of a Crucial Resource*, Westview Press, Boulder, Colo.
Van Apeldoorn, J. G. (1981) *Perspectives on Drought and Famine in Nigeria*, George Allen and Unwin, London and Boston.
Van der Meer, C. (1968) 'Changing water control in a Taiwanese ricefield irrigation system', *Annals of the Association of American Geographers*, 58 (4), 720–47.
Van Veldhuizen, L. R. (1982) 'Windmills for small-scale irrigation', *Annual Report of the International Institute for Land Reclamation and Improvement for Year 1981*, pp. 7–23 (PO Box 45, Wageningen, the Netherlands).
Van Raay, H. G. T. (1975) *Rural Planning in a Savanna Region*, Rotterdam University Press, Rotterdam.
Varallyay, G. (1977) 'Soil water problems related to salinity and alkalinity in irrigated lands', in Worthington, E. B. (ed.), *Arid Land Irrigation in Developing Countries: Environmental Problems and Effects*, Pergamon Press, Oxford and New York, pp. 251–64.
Vietmeyer, N. (1982) 'The revival of the Amaranth', *Ceres*, **15** (5), 43–6.
Vohra, B. B. (1975) 'No more gigantism: but a more equitable and efficient utilization of groundwater resources', *Ceres*, **8** (4), 33–5.
Von Maydel, H. J. and Spatz, G. (1981) 'Effects of goats in the development of landscape', *Applied Geography and Development*, **18**, 30–44.
Wade, R. (1981) 'The information problem of South Indian irrigation canals', *Water Supply and Management*, **5** (2), 31–51.
Wade, R. (1982a) *Irrigation and Agricultural Politics in South Korea*, Westview Press, Boulder, Colo.
Wade, R. (1982b) 'Employment, water control, and water supply institutions: South India and South Korea', *IDS Discussion Paper No. 182*, Institute of Development Studies, University of Sussex, Brighton, UK.
Wagner, E. G. and Lanoix, J. N. (1959) *Water Supply for Rural Areas And Small Communities*, World Health Organization, Geneva.
Walker, B. H. (ed.) (1979) *Management of Semi-arid Ecosystems*, Elsevier, Amsterdam. (Developments in Agricultural and Managed Forest Ecology, vol. 7).

Wallace, T. (1979) *Rural Development through Irrigation: Studies in a Town on the Kano River*, Centre for Social and Economic Research, Ahmadu Bello University, Zaria, Nigeria. (Research Report No. 3).

Wallace, T. (1981) 'The Kano River Project, Nigeria: the impact of an irrigation scheme on productivity and welfare', in Heyer, J., Roberts, P. and Williams, G. (eds), *Rural Development in Tropical Africa*, Macmillan Press, London, pp. 281–305.

Walter, H. (1971) *Ecology of Tropical and Sub-tropical Vegetation* (2nd edn), Oliver and Boyd, Edinburgh.

Ward, R. C. (1967) *Principles of Hydrology*, McGraw-Hill, London and New York.

Webster, C. C. and Wilson, P. N. (1966) *Agriculture in the Tropics*, Longman, London.

Welchert, W. T. and Freeman, B. N. (1973) 'Horizontal wells', *Journal of Range Management*, **26** (4), 253–5.

Welcomme, R. L. (1979) *The Fisheries Ecology of Floodplain Rivers*, Longman, London.

White, G. F. (1962) 'Alternative uses of limited water supplies, in UNESCO, *The Problems of the Arid Zone, Arid Zone Research*, vol. XVIII, UNESCO, Paris, pp. 411–421, (Proceedings of the Paris Symposium).

White, G. F. (ed.) (1978) *Environmental Effects of Arid Land Irrigation In Developing Countries*, UNESCO, Paris, (MAB Technical Notes No. 8).

Wiener, A. (1972) *The Role of Water in Development: an analysis of Principles of Comprehensive Planning*, McGraw-Hill, New York.

Wiesner, C. J. (1970) *Climate, Irrigation and Agriculture: a Guide to the Practice of Irrigation*, Angus and Robertson, Sydney and London.

Wijkman, A. and Timberlake, L. (1984) *Natural Disasters: Acts of God or Acts of Man?*, Earthscan, International Institute for Environment and Development, London and Washington DC.

Wilkinson, R. G. (1973) *Poverty and Progress: in Ecological Perspective in Economic Development*, Praeger, New York.

Williams, C. N. and Joseph, K. T. (1973) *Climate, Soil and Crop Production in the Humid Tropics* (rev. edn), Oxford University Press, London and Kuala Lumpur.

Windram, A., Faulkner R., Bell, M., Roberts, N., Hotchkiss, P., and Lambert, R. (1985) *The Use of Dambos in Small Scale Rural Development*, 11th Conference: Water and Sanitation in Africa, Dar es Salaam, 1985 (mimeo.), 4 pp.

Withers, B. and Vipond, S. (1980) *Irrigation: Design and Practice*, (2nd edn), Cornell University Press, Ithaca, NY.

Wittfogel, K. A. (1957) *Oriental Despotism: a Comparative Study of Total Power*, Yale University Press, New Haven, Conn.

Wolff, P. (1977) 'Progress in the use of drip irrigation, *Applied Sciences and Development*, **10**, 114–21.

Wolverton, B. C. and McDonald, R. C. (1979) 'The water hyacinth: from prolific pest to potential provider', *Ambio*, VIII (1), 2–9.

World Bank (1982) *World Development Report 1982*, published for the World Bank by Oxford University Press, New York and Oxford.

Worthington, E. B. (ed.) (1977) *Arid Land Irrigation in Developing Countries: Environmental Problems and Effects*, Pergamon Press, Oxford and New York.

Wrigley, G. (1981) *Tropical Agriculture: the Development of Production*, (4th edn), Longman, London and New York.

Wulff, H. E. (1968) 'The quanats of Iran', *Scientific American*, **218** (4), 94–105.

Yaron, B., Danfors, E. and Vaadia, Y. (eds) (1973) *Arid Zone Irrigation*, Springer-Verlag and Chapman and Hall, Berlin. (Ecological Studies vol. 5).

Yaron, D. (ed.) (1981) *Salinity in Irrigation and Water Resources*, Marcel Dekker, New York and Basle.

Yiqui, C. (1981) 'Environmental impact assessment of China's Water Transfer Project', *Water Supply and Management*, **5** (3), 253–60.

Zaman, M. (1982) 'The Ganges basin and the water dispute', *Water Supply and Management*, **6** (4), 321–8.

Zaman, M., Biswas, A. K., Khan, A. H. and Nishat, A. (eds) (1983) *River Basin Development*, Tycooly International Publishing, Dublin, (Proceedings of the National Symposium on River Basin Development, 4–10, Dec., 1981, Dacca, Bangladesh).

Zonn, I. S. (1979) 'Ecological aspects of irrigated agriculture', *Water Supply and Management*, **3** (4), 285–95.

Index